Key Topics in Fluid Mechanics

Key Topics in Fluid Mechanics

Edited by Dayana Foster

CLANRYE
INTERNATIONAL
www.clanryeinternational.com

Clanrye International,
750 Third Avenue, 9th Floor,
New York, NY 10017, USA

ISBN: 978-1-64726-680-6

Cataloging-in-Publication Data

Key topics in fluid mechanics / edited by Dayana Foster.
    p. cm.
Includes bibliographical references and index.
ISBN 978-1-64726-680-6
1. Fluid mechanics. 2. Fluid dynamics. I. Foster, Dayana.
TA357 .F58 2023
620.106--dc23

For information on all Clanrye International publications
visit our website at www.clanryeinternational.com

Contents

Preface.. VII

Chapter 1   **Prediction of Crosswind Separation Velocity for Fan and Nacelle Systems using Body Force Models (Part 1) Fan Body Force Model Generation without Detailed Stage Geometry** 1
Quentin J. Minaker and Jeffrey J. Defoe

Chapter 2   **Prediction of Crosswind Separation Velocity for Fan and Nacelle Systems using Body Force Models (Part 2) Comparison of Crosswind Separation Velocity with and without Detailed Fan Stage Geometry** 24
Quentin J. Minaker and Jeffrey J. Defoe

Chapter 3   **Cattaneo–Christov Heat Flux Model for Three-Dimensional Rotating Flow of SWCNT and MWCNT Nanofluid with Darcy–Forchheimer Porous Medium Induced by a Linearly Stretchable Surface** 42
Zahir Shah, Asifa Tassaddiq, Saeed Islam, A.M. Alklaibi and Ilyas Khan

Chapter 4   **Effects MHD and Heat Generation on Mixed Convection Flow of Jeffrey Fluid in Microgravity Environment over an Inclined Stretching Sheet** 54
Iskander Tlili

Chapter 5   **MHD Stagnation Point Flow of Nanofluid on a Plate with Anisotropic Slip** 65
Muhammad Adil Sadiq

Chapter 6   **Analytical Study of the Head-On Collision Process between Hydroelastic Solitary Waves in the Presence of a Uniform Current** 79
Muhammad Mubashir Bhatti and Dong Qiang Lu

Chapter 7   **Unsteady Flow of Fractional Fluid between Two Parallel Walls with Arbitrary Wall Shear Stress using Caputo–Fabrizio Derivative** 108
Muhammad Asif, Sami Ul Haq, Saeed Islam, Tawfeeq Abdullah Alkanhal,
Zar Ali Khan, Ilyas Khan and Kottakkaran Sooppy Nisar

Chapter 8   **Integer and Non-Integer Order Study of the GO-W/GO-EG Nanofluids Flow by Means of Marangoni Convection** 119
Taza Gul, Haris Anwar, Muhammad Altaf Khan, Ilyas Khan and Poom Kumam

Chapter 9   **Modified MHD Radiative Mixed Convective Nanofluid Flow Model with Consideration of the Impact of Freezing Temperature and Molecular Diameter** 133
Umar Khan, Adnan Abbasi, Naveed Ahmed, Sayer Obaid Alharbi, Saima Noor,
Ilyas Khan, Syed Tauseef Mohyud-Din and Waqar A. Khan

Chapter 10 **Peristaltic Blood Flow of Couple Stress Fluid Suspended with Nanoparticles under the Influence of Chemical Reaction and Activation Energy** 148
Rahmat Ellahi, Ahmed Zeeshan, Farooq Hussain and A. Asadollahi

Chapter 11   **MHD Slip Flow of Casson Fluid along a Nonlinear Permeable Stretching Cylinder Saturated in a Porous Medium with Chemical Reaction, Viscous Dissipation and Heat Generation/Absorption**..165
Imran Ullah, Tawfeeq Abdullah Alkanhal, Sharidan Shafie, Kottakkaran Sooppy Nisar, Ilyas Khan and Oluwole Daniel Makinde

Chapter 12   **Numerical Solution of Non-Newtonian Fluid Flow Due to Rotatory Rigid Disk**..192
Khalil Ur Rehman, M. Y. Malik, Waqar A Khan, Ilyas Khan and S. O. Alharbi

Chapter 13   **MHD Flow and Heat Transfer over Vertical Stretching Sheet with Heat Sink or Source Effect**..204
Ibrahim M. Alarifi, Ahmed G. Abokhalil, M. Osman, Liaquat Ali Lund, Mossaad Ben Ayed, Hafedh Belmabrouk and Iskander Tlili

Chapter 14   **MHD Nanofluids in a Permeable Channel with Porosity**..218
Ilyas Khan and Aisha M. Alqahtani

Permissions

List of Contributors

Index

Preface

A fluid refers to a state of matter that yields to shearing or lateral forces. Fluid mechanics is a branch of continuum mechanics which is involved in the study of fluid behavior in motion and at rest. It is categorized into fluid dynamics, which studies the effect of forces on fluid motion; and fluid statics, which studies fluids at rest. The energy equation, continuity equation and momentum principle are the basic fluid mechanics principles. Some of the important areas of study within this field are biofilms, dynamics of bubbles and droplets, aerodynamic shape optimization, fire whirls, drag reduction and fish locomotion. The topic of fluid mechanics is studied under various disciplines such as chemical engineering, mechanical engineering, civil engineering and aerospace engineering. This book aims to shed light on the key topics in fluid mechanics. It consists of contributions made by international experts. Scientists and students actively engaged in the study of fluid mechanics will find this book full of crucial and unexplored concepts.

Various studies have approached the subject by analyzing it with a single perspective, but the present book provides diverse methodologies and techniques to address this field. This book contains theories and applications needed for understanding the subject from different perspectives. The aim is to keep the readers informed about the progresses in the field; therefore, the contributions were carefully examined to compile novel researches by specialists from across the globe.

Indeed, the job of the editor is the most crucial and challenging in compiling all chapters into a single book. In the end, I would extend my sincere thanks to the chapter authors for their profound work. I am also thankful for the support provided by my family and colleagues during the compilation of this book.

Editor

Prediction of Crosswind Separation Velocity for Fan and Nacelle Systems using Body Force Models (Part 1) Fan Body Force Model Generation without Detailed Stage Geometry

Quentin J. Minaker and Jeffrey J. Defoe *

Turbomachinery and Unsteady Flows Research Group, Department of Mechanical, Automotive and Materials Engineering, University of Windsor, 401 Sunset Ave., Windsor, ON N9B 3P4, Canada; minakerq@uwindsor.ca
* Correspondence: jdefoe@uwindsor.ca

Abstract: Modern aircraft engines must accommodate inflow distortions entering the engines as a consequence of modifying the size, shape, and placement of the engines and/or nacelle to increase propulsive efficiency and reduce aircraft weight and drag. It is important to be able to predict the interactions between the external flow and the fan early in the design process. This is challenging due to computational cost and limited access to detailed fan/engine geometry. In this, the first part of a two part paper, we present a design process that produces a fan gas path and body force model with performance representative of modern high bypass ratio turbofan engines. The target users are those with limited experience in turbomachinery design or limited access to fan geometry. We employ quasi-1D analysis and a series of simplifying assumptions to produce a gas path and the body force model inputs. Using a body force model of the fan enables steady computational fluid dynamics simulations to capture fan–distortion interaction. The approach is verified for the NASA Stage 67 transonic fan. An example of the design process is also included; the model generated is shown to meet the desired fan stagnation pressure ratio and thrust to within 1%.

Keywords: fan–nacelle interaction; body force models; turbomachinery; CFD

---

1. Introduction

In the design stage of an airframe, the external flow around all components must be considered. This is certainly important around engine nacelles, where the external flow will be affected by the operation of the fan. This interaction is dependent on both the positioning of the fan stage within the nacelle and its operating condition [1]. The state-of-the-art for engine modelling in full airframe computations is to use a simplified model of the propulsion system. This is done to reduce computational costs compared to traditional bladed Reynolds averaged Navier–Stokes (RANS) methods. These simplified models use steady computational fluid dynamics (CFD) simulations for non-uniform inflow where normally unsteady simulations would be required. They also reduce the number of grid cells needed by approximately two orders of magnitude within the turbomachinery blade rows [2].

These modelling approaches were discussed in detail in Godard et al. (2017) [3]. One of the approaches commonly employed in full airframe simulations involves using actuator disks. Actuator disks work by imposing changes to flow direction and stagnation quantities over a single plane, but are limited in their ability to reproduce the effects of the coupling between external flow and the fan. Godard et al. proposed that the main reason for this limitation comes from the fact that the actuator disk takes the inflow as is and computes the outflow accordingly, but lacks feedback effects [3]. Through-flow or body force methods are another approach that is examined to help capture this coupling effect; however, this is a higher fidelity approach and therefore requires more input information. Body force methods work by applying sources of momentum and energy in the swept volumes where the blades would normally be and were found to capture the external flow–fan coupling more accurately than actuator disks [3].

Many variations of body force methods exist; however, they usually require the user to have detailed blade geometry information available. Gong's model [4] and its later refinements in Peters et al. (2015) [1], Hill's model [2], as well as a lift–drag model [5] are examples of these; they require calibration based on experiments or more detailed computations, which include the blade rows in detail. This calibration therefore relies on detailed blade geometry and entails additional computational cost. Models such as those proposed by Hall et al. (2017) [6] and Pazireh and Defoe (2019) [7] have been shown to work without the need for calibration; however, they still require the blade geometry (no thickness information needed in Hall's approach) and the gas path of the fan stage. The modelling approach of Sato et al. (2019) [8] works without the need for fan blade geometries, but still requires information on the gas path and blade leading and trailing edge meridional profiles.

Considering the usage case of an airframer assessing nacelle design early in the design phase, it is possible that the engines to be used have not yet been selected, and even if they have, it may not be possible to obtain the fan geometry and/or bypass duct gas path from the engine manufacturer. This means that the body force methods mentioned above would not be usable. Tools currently exist that allow for the creation of highly detailed stage and blade geometry; however, they generally require more experience with turbomachinery, as well as detailed information about the stage. MULTALL is an open source turbomachinery design suite that takes basic stage information and will generate 3D blades and gas paths [9]. Although simplified, these inputs still require the user to have information on the blade performance, which may be unknown, such as blade rotation speed or the stage work coefficient. The airframer is not interested in the level of fidelity of the blades or the stage; all that is required is that the body force recreates the external–internal flow interaction. Therefore, the airframer desires to create a body force model based on information that they know to some degree of accuracy, such as the required thrust, limitations on engine size, and an estimate of the fan stagnation pressure ratio (FPR) at the design point.

The objective of this paper is to introduce and assess a fan gas path and body force model generation process that enables the simulation of powered nacelles and/or full airframes without any prior detailed fan geometry information. This process consists of 1D analysis to determine the required change in flow quantities through the stage, as well as generating simplified blade camber surfaces and a gas path. The full MULTALL suite is not used since it requires too much input information for the intended level of fidelity; however, certain tools within the suite are utilized, as will be described later. A number of assumptions and simplifications are used during these steps to determine the required information. The key outcomes are that: (1) the process enables the creation of a body force model of a fan stage without a priori knowledge of detailed fan geometry or gas path; (2) this body force model, once implemented in a CFD framework, matches the design intent performance at the design point; (3) the body force model matches the desired spanwise loading at the rotor trailing edge (in this case, uniform) and yields rotor chordwise loading similar to that found in modern machines. With these outcomes met, the resulting model can also be used to assess off-design conditions.

In the first section of this paper, the body force formulation is described in detail, and its validation is demonstrated. Next, the design process is explained, and the selected camber shape is presented and shown to yield the desired loading distribution. The implementation of this process into a commercial CFD framework is discussed, and finally, an example application of the process is demonstrated.

2. Body Force Formulation

The concept of body force modelling involves replacing the physical rotor and stator blades with momentum and energy sources. These sources are added around an annulus covering the radial and axial extent of the physical blades. The sources generate the flow turning, as well as the pressure and temperature changes that occur in the real machine. These sources can be thought as a local blade force smeared across the blade pitch. The concept is visualized in Figure 1.

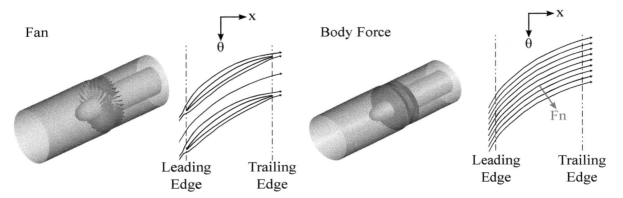

Figure 1. Comparison of the physical blades (**left**) and the source term model (**right**).

The momentum and energy equations are,

$$\frac{\partial \rho \vec{V}}{\partial t} + \nabla \left(\rho \vec{V} \vec{V}^T \right) + \nabla p - \nabla \cdot \tau = \rho \vec{f} \tag{1}$$

$$\frac{\partial \rho e_t}{\partial t} + \nabla \left(\rho h_t \vec{V} \right) - \nabla \cdot (\nabla \cdot \tau) = \rho \left(\vec{r} \times \vec{\omega} \right) \cdot \vec{f} \tag{2}$$

where ρ is the fluid density, $\vec{V}$ is the flow velocity, p is the static pressure, e_t is stagnation energy (where $\rho h_t = \rho e_t + p$), h_t is stagnation enthalpy, r is radius, ω is angular rotation speed, and τ is the viscous shear stress. The equations are modified to account for the momentum and energy source terms, which incorporate the effects of a body force per unit mass f.

The body force method chosen is Hall's model [6], because it requires no calibration and therefore reduces the amount of stage information needed. The required inputs for this approach are the number of blades per row, B, the blade camber surface normal vectors, $\hat{n}$, and the relative velocity vector, $\vec{W}$. The normal force acts to reduce the local deviation δ, which is the angle between the relative velocity vector and a vector tangent to the blade camber surface in the plane shared by $\hat{n}$ and $\vec{W}$. The source term per unit mass, here the incompressible normal force, $f_{n,i}$, is defined as:

$$f_{n,i} = \frac{2\pi \delta \left(\frac{1}{2} W^2 / |n_\theta| \right)}{2\pi r / B} \tag{3}$$

The blade leading and trailing edge meridional profiles and the full machine gas path are also needed. The original Hall model was only intended to be used for low speed machines (incompressible flow), so a correction factor is added to account for compressibility since modern commercial engine fans operate at transonic relative Mach numbers. This takes the form of an added compressibility correction, K, where:

$$f_{n,c} = K f_{n,i} \tag{4}$$

as used in Benichou et al. (2019) [10]. This correction uses the Prandtl–Glauert rule in subsonic relative flow and the Ackeret formula in supersonic relative flow:

$$K' = \begin{cases} \frac{1}{\sqrt{1 - M_{rel}^2}} & M_{rel} < 1 \\ \frac{2}{\pi \sqrt{M_{rel}^2 - 1}} & M_{rel} > 1 \end{cases} \tag{5}$$

and has an upper bound to avoid instabilities as the relative Mach number approaches one, giving,

$$K = \begin{cases} K' & K' \leq 3 \\ 3 & K' > 3 \end{cases} \tag{6}$$

Body force models exist that have added terms to account for the blockage effects caused by the blades; this can be seen in the model used in Benichou et al. (2019) [10]. Including blockage adds complexity as it requires information on blade thickness. This was neglected in the current approach as the aim was not to generate full blade shapes and, as will be shown later, is not required to accurately predict the loading in the body force model. Simple loss models exist that require little or no calibration; however, the upstream influence of a fan on incoming flow is not significantly impacted by the viscous losses in blade rows [6] and therefore neglected in the body force model used in this paper.

3. Assessment of the Body Force Approach

A single passage bladed RANS simulation was compared against a body force model to assess the approach both with and without the compressibility correction at the design flow coefficient,

$$\phi = \frac{V_x}{U_{mid}} \tag{7}$$

of 0.48, where U_{mid} is the rotor blade speed at midspan. The machine used was NASA Stage 67 [11]. The important features of this machine are shown in Table 1. The overall, spanwise, and chordwise loading were examined for both the 70% and 90% speed lines.

Table 1. Important characteristics of the NASA Stage 67 rotor. Operating-point dependent values are at the design flow coefficient at 90% corrected rotational speed [11].

Parameter	Value	Parameter	Value
ω_{corr} (rad/s)	1512	B	22
$M_{rel,tip}$	1.2	$\frac{\overline{V_x}^{-M}}{U_{mid}}$	0.5
FPR	1.48	$\left(\frac{r_{hub}}{r_{tip}}\right)_{inlet}$	0.375
$\dot{m}_{corr}$ (kg/s)	31.1	$\left(\frac{r_{hub}}{r_{tip}}\right)_{outlet}$	0.478
True Chord Aspect Ratio	1.56		

The simulations were run using Ansys CFX 16 [12], using second order spatial discretization. The grids and computational approach were the same as those used in Hill and Defoe (2018) [2]. When the body force regions were relatively small compared to the overall size of the computational domain, it was found that decreasing the pseudo-time step from the default setting in CFX was required to obtain convergence. The bladed Stage 67 simulation used a single passage containing 3.58×10^6 cells. Two grids were used to check grid independence; Table 2 shows the results of the independence study. Less than 1% change was seen between the stagnation pressure ratio and isentropic efficiency, and it was therefore determined that the medium grid was sufficient. The simulations were steady state and used the shear-stress-transport turbulence model. The stagnation quantities were set at the inlet, and a mass flow rate boundary condition was used at the outlet. In this paper, blades with zero shear stress surfaces were used so that a direct comparison could be made against the body force model, which included only the turning (normal) force in the blade rows. The body force model consisted of a 1/16 annulus slice containing 279,760 cells. In Table 3, a summary of the body force grid independence study is shown. The boundary conditions were the same as in the bladed simulations. The computational domains are shown in Figure 2. Further information on the computational setup can be found in Hill and Defoe (2018) [2].

Table 2. Summary of the bladed simulations grid independence study performed by Hill and Defoe (2018) [2].

	Medium Grid	Fine Grid	Percent Change
Rotor Cell Count	1.78×10^6	2.45×10^6	37.6
FPR-1	0.493	0.496	0.71
Rotor η_{is}	92.3%	92.3%	0

Table 3. Summary of the body force grid independence study preformed by Hill and Defoe (2018) [2].

	Medium Grid	Fine Grid	Percent Change
Cell Count	279,760	609,500	117%
$\overline{T}_{t,2}^{M}/T_{t,1} - 1$	0.130	0.130	0%

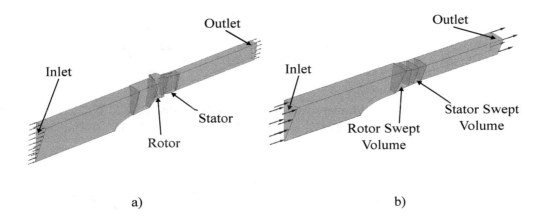

a)   b)

Figure 2. Computational domains of (**a**) the single passage bladed RANS simulations and (**b**) the body force simulations.

The distributions of the work coefficient, defined as:

$$\psi = \frac{h_{\text{t}} - h_{\text{t,inlet}}}{U_{\text{mid}}^2} \tag{8}$$

as a function of chord and along the span at the rotor trailing edge are shown in Figures 3 and 4 for 70% and 90% corrected speed, respectively. The corrected speed is defined as:

$$\omega_{\text{corr}} = \frac{\omega}{\sqrt{\dfrac{T_{\text{t,inlet}}}{T_{\text{t,ref}}}}} \tag{9}$$

where $T_{\text{t,ref}} = 288$ K. The fraction corrected speed is ω_{corr} normalized by the design value.

The figures include results from the Hall body force model with and without the compressibility correction, as well as the single passage results, which were circumferentially averaged. In the hub and tip regions, one would not normally expect a body force model to be particularly accurate since (at least for the simple normal force model used here) there was no consideration of the effects of secondary flows. We therefore restricted our chordwise plots to 20–80% span.

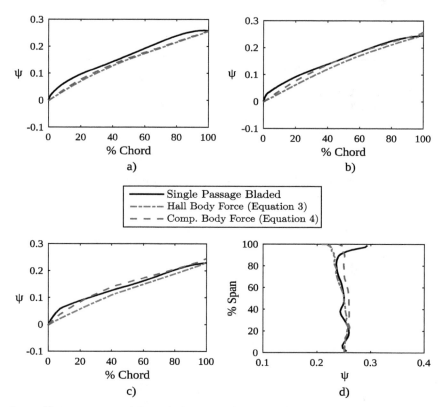

Figure 3. Work coefficient vs. meridional distance through the rotor at: (**a**) 20% span, (**b**) 50% span, (**c**) 80% span, and (**d**) rotor trailing edge at 70% corrected speed and $\phi = 0.48$.

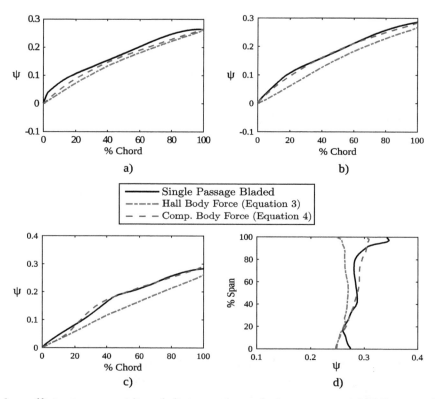

Figure 4. Work coefficient vs. meridional distance through the rotor at: (**a**) 20% span, (**b**) 50% span, (**c**) 80% span, and (**d**) rotor trailing edge at 90% corrected speed and $\phi = 0.48$.

The root mean squared (RMS) difference in the local chordwise and spanwise work coefficient as a percentage of total work coefficient for the bladed simulation (mass averaged in the spanwise case) between the body force (BF) models and the bladed simulations (BS), defined as:

$$\%\text{RMS} = 100 \times \sqrt{\frac{\int_0^c (\psi_{\text{BS}} - \psi_{\text{BF}})^2 \, dx}{c}} \bigg/ \psi_{\text{BS,TE}} \tag{10}$$

is shown in Tables 4 and 5 for the two rotational speeds for the incompressible and compressibility corrected body force approaches.

Table 4. Work coefficient RMS errors for NASA Stage 67 at 70% corrected speed and $\phi = 0.48$.

	$f_{n,i}$ %RMS	$f_{n,c}$ %RMS	Improvement
20% Span	7.78%	6.62%	1.16%
50% Span	7.27%	2.74%	4.53%
80% Span	7.98%	3.75%	4.23%
Spanwise	4.50%	4.30%	0.20%

Table 5. Work coefficient RMS errors for NASA Stage 67 at 90% corrected speed and $\phi = 0.48$.

	$f_{n,i}$ %RMS Difference	$f_{n,c}$ %RMS Difference	Improvement
20% Span	9.25%	5.37%	3.88%
50% Span	10.9%	1.81%	9.09%
80% Span	12.7%	2.15%	10.5%
Spanwise	8.34%	3.42%	4.92%

The correction factor improved the accuracy of the body force model in both the overall and chordwise loadings. The correction factor had a greater influence the larger the relative Mach number became; this was seen as the span fraction increased and as the corrected rotational speed increased. In the 70% corrected speed case, the improvement increased by approximately 3% as the span fraction increases, and in the 90% corrected speed case, this increased even further to approximately 6%. The reason for this larger improvement in the 90% corrected speed case was due to the fact that the relative Mach numbers were increased throughout the rotor. This was also seen by the fact that the improvement more than doubled throughout in this higher corrected speed case. The results showed that the body force model with a compressibility correction was capable of matching the chordwise loading to within 7%, which was deemed as an acceptable level of accuracy for this design process. The spanwise loading showed even better agreement.

4. Fan Stage Design Approach

We took the fan design point to be cruise, as this was the typical design condition for low fan pressure ratio commercial aircraft engines [13], which are of increasing interest in modern design. The benefit of selecting this typical design condition was that this is usually where the designer will have the most information about the required performance. This condition requires the specification of a cruise altitude and flight Mach number. These are quantities an airframer would normally know and provide the information needed to find inlet stagnation quantities. We employed 1D analysis to determine the flow properties through the stage to meet the desired performance at cruise. Based on the resulting flow properties, as well as a series of assumptions and geometric constraints, the gas path was defined.

The fan pressure ratio and net thrust were required inputs. The input geometric parameters were the fan blade tip leading edge radius, rotor hub-to-tip ratio ($r_{\text{hub}}/r_{\text{tip}}$) at the leading edge, blade aspect ratios (b_R/c_R, b_S/c_S) based on the axial chords, axial distances upstream of, between, and downstream of the blade rows (L_1/c_R, L_2/c_R, L_3/c_R), nozzle contraction length (L_N/c_R), hub curvature length (L_A/c_R), and fractional tip radius change through the rotor ($\Delta r_\text{tip}/r_\text{tip}$). A diffusion factor was specified to determine the number of rotor and stator blades, or these can be directly specified. If an elliptical spinner nose was desired, the axial length of the spinner nose, L_spin, was also needed and was specified as a fraction of a linear spinner nose length. The body force model was created to generate a set fan stagnation-to-stagnation pressure ratio at a corrected mass flow, which combined with the gas path geometry, achieved the desired thrust. Figure 5 shows the generic meridional profile of the gas path and illustrates the definitions of the geometric parameters.

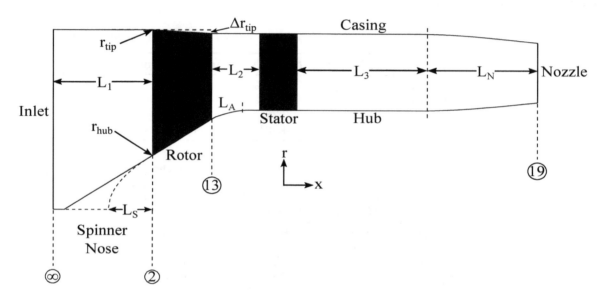

Figure 5. Meridional profile displaying the geometric parameters required for gas path generation and station numbering.

The assumptions made were:

1. the axial velocities at the leading and trailing edge of the blade rows were equal and constant along the span,
2. the bypass ratio was high enough that the core flow contribution to thrust generation and the core suction effect on flow in the fan rotor were negligible,
3. the turbomachinery and duct flow were isentropic, and
4. the flow was in the meridional direction at the fan inlet.

Assumption (1) is in practice not accurate at the rapidly contracting rotor hub and would also be expected to become less accurate near the tip as the fan pressure ratio and, thus, relative Mach numbers rise. It is later shown that for the fan design space of interest, over most of the span, the variation of axial velocity was less than 10% of the assumed value. From Assumption (2), we did not include a bifurcation into a core duct in the gas path.

A simple linear scaling was used to set the blade tip relative Mach number. From the literature, it was found that modern day fans with pressure ratios of 1.6 would be expected to have a tip relative Mach number of approximately 1.4 [13,14]. We applied this scaling to set our tip relative Mach number based on the design fan pressure ratio (FPR):

$$M_\text{rel,tip} = \frac{1.4}{1.6} FPR \qquad (11)$$

At design, hub and casing boundary layers were thin and fully attached due to the high Reynolds numbers in practical engine fans, and thus, we assumed no changes in stagnation quantities up to the fan face; these were then set by the flight condition.

4.1. Stage Performance and Gas Path

Application of control volume analysis to the flow going through the engine yielded the standard expression for thrust:

$$F = \dot{m}(V_{19} - V_\infty) + A_{19}(p_{19} - p_\infty) \tag{12}$$

where F is thrust, $\dot{m}$ is the mass flow rate, and A is the passage area. The thrust, flight velocity, and freestream static pressure were known at the outset, with the other quantities to be determined; this was done using a quasi-1D approach. Two cases could exist, depending on whether the exhaust nozzle was choked or not. The nozzle is choked if:

$$FPR\frac{p_{t,\infty}}{p_\infty} \geq 1.893 \tag{13}$$

for air with specific heat ratio $\gamma = 1.4$. If the nozzle is choked, the nozzle exit static pressure is:

$$p_{19} = p_\infty \frac{FPR}{1.893}\left(1 + \left(\frac{\gamma - 1}{2}\right)M_\infty^2\right)^{\frac{\gamma}{\gamma-1}} \tag{14}$$

If the nozzle is not choked, the nozzle exit static pressure is equal to the atmospheric static pressure:

$$p_{19} = p_\infty \tag{15}$$

The nozzle velocity, assuming isentropic flow, is:

$$V_{19} = M_{19}\left[\gamma R\left(\frac{FPR^{\frac{\gamma-1}{\gamma}} T_\infty \left(1 + \left(\frac{\gamma-1}{2}\right)M_\infty^2\right)}{1 + \left(\gamma - \frac{1}{2}\right)M_{19}^2}\right)\right]^{\frac{1}{2}} \tag{16}$$

If the nozzle is choked, then $M_{19} = 1$, and if it is unchoked, it is determined by:

$$M_{19} = \left[\left(\frac{p_{t,\infty}FPR}{p_\infty}\right)^{\frac{\gamma-1}{\gamma}} - \frac{1}{\gamma - \frac{1}{2}}\right]^{\frac{1}{2}} \tag{17}$$

If the flow is choked, the mass flow and nozzle area are then given by the simultaneous solution of Equation (12) and the corrected flow per unit area equation applied at Station 19,

$$\dot{m} = \frac{A_{19}p_{t,19}}{\sqrt{T_{t,19}}}\sqrt{\frac{\gamma}{R}}M_{19}\left(1 + \frac{\gamma-1}{2}M_{19}^2\right)^{-\frac{\gamma+1}{2(\gamma-1)}}. \tag{18}$$

In Equation (18), the stagnation quantities are the mass weighted averaged values. To keep the body force model as simple as possible, we designed for uniform spanwise work input so that the local values were everywhere equal to the mass weighted averages.

If the flow was unchoked, the mass flow rate was directly calculated from Equation (12) since $p_\infty = p_{19}$. This along with the nozzle exit Mach number were used in Equation (18) to determine the nozzle exit area.

The axial Mach number at the fan face (Station 2) was found from Equation (18) given the fan inlet area (computed from the tip radius and hub-to-tip ratio) and the now known mass flow rate. This Mach number was then used to determine the static temperature at the fan face.

The assumption of equal leading and trailing edge axial velocities along with the selection of FPR allowed the rotor trailing edge area to be calculated. In doing so, we neglected the effect of swirl on the required rotor exit area; however, within the design space typically of interest, swirl angles will normally be well under $30°$, and there is only a minor effect on the required passage area [15].

The gas path shape through the rotor was generated using straight line hub and casing curves. This means that the axial velocity would vary within the blade row, but greatly simplified the generation of the gas path. A parameter, Y, which is the fraction of the fan leading edge span, set the amount of tip radius change through the rotor,

$$\Delta r_{\text{tip}} = Y(r_{\text{tip,LE}} - r_{\text{hub,LE}}) \tag{19}$$

Downstream of the rotor, the casing radius was constant.

The slope of the hub through the rotor was set to meet the required decrease in passage area while keeping the leading and trailing edge axial velocities equal.

Downstream of the rotor trailing edge, the hub radius curved back towards axial over some desired fraction of the distance between rotor and stator (L_A; the default value was $L_2/2$). The stator span was set to be constant along the chord. In reality, the removal of swirl would require a decrease in passage area, but by the same logic applied to the determination of the rotor trailing edge area, this effect is normally small.

The spinner length determined its shape. If the axial length was less than that of a straight line with the rotor hub slope extended to zero radius, then the spinner nose was assumed to be elliptical in shape. It matched the rotor hub slope downstream and extended to zero radius upstream with the tangent to the ellipse at the nose purely radial. Otherwise, a conical spinner was used with the rotor hub line extended directly down to zero radius.

4.2. Blade Performance and Camber

The rotor inlet velocity triangle at the tip, which was determined by the axial Mach number found using Equation (18) and the relative Mach number found using Equation (11), determined the rotation speed of the rotor blades.

A camber surface was needed for the body force model. Camber lines were determined at set span fractions; in the current approach, the hub, midspan, and tip were used. The camber surface was generated by fitting a 3D surface, which passed through these lines, as described in detail later in this section.

The chordwise loading distribution was shown to have an effect on inlet distortion interaction [6]; therefore, one of the aims was to generate a body force model with a camber surface that produced realistic chordwise loading distributions, while remaining relatively simple. The solution employed was to use camber shapes defined by a combination of a circular arc and a straight line. An example of this camber shape is shown in Figure 6. In physical blades, the highest loading tends to be in the leading edge region; however, in the Hall body force model (Equation (3)), the loading scales with deviation, which tends to increase towards the trailing edge at design. The intent of pushing all camber curvature forward was to combat this effect. The straight line in the rear section of the chord worked to ensure that the required overall flow turning was met as the Hall model acted to reduce the deviation. A range of circular-straight line dividing locations was tested, and it was found that a 50:50 split between circular arc and straight line provided the best combination of guaranteeing the correct flow turning and chordwise loading distribution accuracy, as shown later. It should be noted that the model design approach did not produce realistic blade shapes, but increased the accuracy of the loading distribution of the body force model. This was a significant difference compared to the no blade information process used by Sato, Spotts, and Gao (2019) [8], as no real attempt was made to capture realistic chordwise loading in that paper.

In the design velocity triangles, the meridional velocity was used as opposed to the axial velocity. This was important because of the significant radial velocities in the rotor, especially near the hub. The consequence was that the velocity triangles and hence camber angles were dependent on the stream surface inclination since the leading and trailing edge axial velocities were assumed constant.

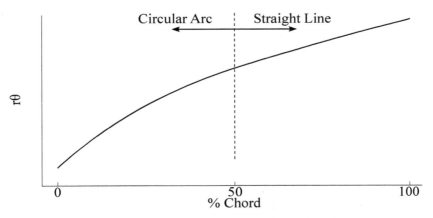

Figure 6. Example of the blade camber line.

At the rotor leading edge, a small positive incidence of 2° was used; this along with the design velocity triangles set the rotor inlet camber angle. The incidence was added to provide a more realistic chordwise loading. It increased the blade loading and flow deflection in the rotor leading edge region. This also helped ensure that the chordwise loading distributions matched predicted trends when the assumption of constant axial velocity was not realized when employing the model within a CFD simulation; if the axial velocity exceeded the assumed value, it would cause negative incidence at the leading edge, which could result in local work removal. The positive incidence acted to counteract this trend. In Figure 7, the chordwise loading is shown at 80% rotor span with and without the added incidence from the example design described later to demonstrate the difference in work addition. In the blade with 0° incidence, the stagnation enthalpy in the first 20% chord dropped below the freestream stagnation enthalpy; this could alter the expected distortion interaction behaviour. Adding the incidence eliminated this decrease in the leading edge region. The stator leading edge camber angle was set by assuming zero incidence. Zero incidence was used for the stator leading edge because there was no change in stagnation quantities across the stator, which eliminated the need to add incidence to improve the chordwise loading distribution.

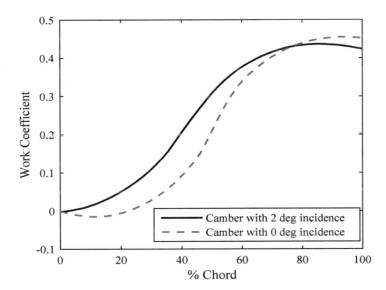

Figure 7. Rotor chordwise work coefficient at 80% span comparison between a blade with and without added incidence from the CFD simulations of the example design shown later.

The rotor trailing edge flow angles were set based on the required work input using the Euler turbine equation,

$$c_{\mathrm{p}}(T_{t,\mathrm{TE}} - T_{t,\mathrm{LE}}) = \omega(r_{\mathrm{TE}}V_{\theta,\mathrm{TE}} - r_{\mathrm{LE}}V_{\theta,\mathrm{LE}}), \tag{20}$$

the design choice to have a constant spanwise work input/pressure rise and flow deviation. The stator trailing edge was set to remove the swirl from the flow. The trailing edge angles in both blade rows accounted for the deviation. Trailing edge deviation was estimated using a modified form of Carter's rule [13]:

$$\delta_{TE} = \left(0.23 \left(\frac{2a}{c}\right)^2 + \left(\frac{\alpha}{500}\right)\right) \xi \left(\frac{s}{c}\right)^{0.5} \quad \text{(degrees)}. \quad (21)$$

where the maximum camber of the blade was at an axial distance a from the leading edge, c is the axial chord length, α is the exit flow angle (β in the rotor), ξ is the overall change in blade angle, and s is the pitch (spacing between blades). Carter's rule is intended for fully circular blade camber shapes; a modification was made to a to account for the adjustment of the location of maximum camber from the mid point to the new value of 37.5% chord ($a/c = 0.375$). The relationship between the flow angles (α and β), blade angle (κ), and deviation (δ) at a rotor trailing edge are illustrated using the generic velocity triangle shown in Figure 8.

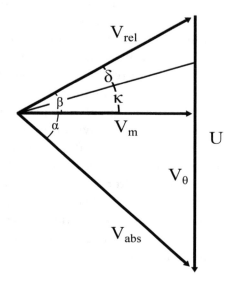

Figure 8. Velocity triangle at the rotor trailing edge.

The 2D camber line sections were stacked using the open-source turbomachinery design suite MULTALL's geometry and the grid generator Stagen. The information required for Stagen was the camber distribution and the corresponding axial and radial coordinates through the gas path at the set span fractions (0% and 100% were required; however, additional span fractions can be supplied to further constrain the design), as well as the leading and trailing edge meridional coordinates. Stagen creates a single passage grid based on the information given; the number of spanwise sections generated was equal to the number of radial grid points requested. The current maximum of 64 points was used here, which was shown to be an adequate number of radial grid points in body force models [2]. The thickness distribution was set within Stagen to produce blades of negligible thickness such that the maximum thickness was less than 1% of the blade chord. Blade thickness information was not required for the body force model, and therefore, this was done so the camber surface extraction could be simplified. The blade sections were stacked with their centroids lying along a radial line through the centroid of the hub blade section. Shown in Figure 9 are the camber lines that were produced by Stagen. The grid points on the blade surfaces were then extracted, and this was used to generate the 3D blade shapes. The camber surface was found by extracting the average of the $r\theta$ coordinates of the pressure and suction sides of the 3D blade shapes at each axial location. Finally, the camber surface normal vectors used in the compressibility corrected Hall body force model were calculated using the MATLAB [16] built-in function "surfnorm". For more information on how Stagen works, refer to Denton (2017) [9].

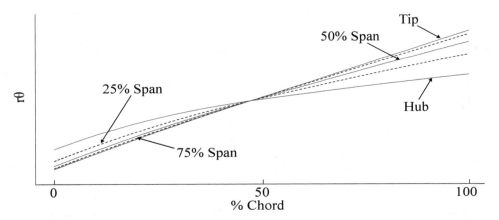

Figure 9. Camber lines for the example design shown later produced by running Stagen. The Hub, 50% span, and tip camber lines are supplied to Stagen.

5. Assessment of Camber Distribution

The ability of the half circular arc, half straight line camber shape (used in the procedure described in Section 4) to produce realistic chordwise loading distributions was assessed by using NASA Stage 67's gas path and overall performance specifications at $\phi = 0.48$ and 90% corrected speed. Leading and trailing edge meridional profiles, as well as the gas path were kept the same as the existing NASA Stage 67 to isolate loading changes caused by camber shape. The leading edge blade angles were the same as those in the actual Stage 67 blade rows, and the rotor trailing edge blade angles for the simplified camber distribution were iteratively adjusted to minimize the %RMS error in the spanwise work coefficient distribution at the trailing edge between the simplified and original camber shapes. This ensured that the overall performance of the blade rows was similar and enabled isolation of the impact of camber shape on the loading distribution. The loading distributions are shown in Figure 10. The spanwise loading (shown in Figure 10d) %RMS difference was 3.19%; with the outer region having a larger contribution to this difference. The work coefficient in the outer region had a higher sensitivity to adjustments in blade angle, which led to increased computational costs to reduce the %RMS difference in this region; therefore, the accuracy of the trailing edge blade angle was iterated to $\pm 0.05°$. The chordwise loadings displayed similar overall trends, especially at lower span fractions with the %RMS difference being 6.48% and 7.43% along the 20% and 50% span lines, respectively. This slight difference was due to the modified distribution having larger work addition within the first 50% chord, but this was expected as this was where blade turning occurred. As span fraction increased, this difference became more evident as the %RMS difference increased to 13.2%; again, this stemmed from the increased sensitivity due to blade angle changes in the outer span region. The %RMS difference provided a way to compare quantitatively the loading distributions to those of a real machine; however, it was expected that the loading distributions would not be exactly the same, as the camber distributions were different. The general trends in the rate of work addition showed similarities, which was a good indication that although the camber distribution was relatively simple, it produced loadings similar to those in real machines.

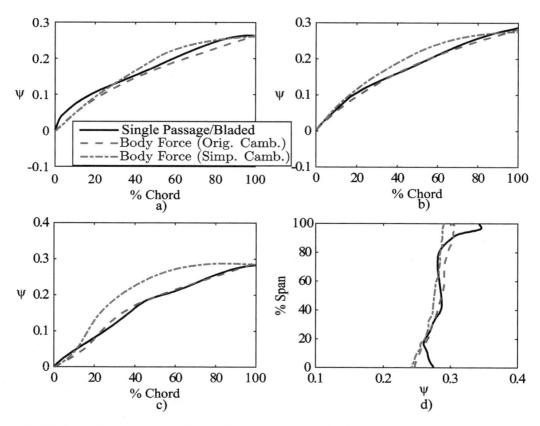

Figure 10. Work coefficient vs. meridional distance through the Stage 67 rotor at: (**a**) 20% span, (**b**) 50% span, (**c**) 80% span, and (**d**) rotor trailing edge for comparison of chordwise loading between a real machine and the simplified stage for 90% corrected speed and $\phi = 0.48$.

6. Estimating Operating Conditions for Off-Design Thrust and Mass Flow

One of the intended uses of the models produced by the design process was to allow airframers to investigate external–fan interactions at a variety of different conditions. To investigate off-design conditions, the user must know the fraction of design inlet corrected mass flow,

$$\dot{m}_{\text{corr}} = \dot{m}\sqrt{\frac{T_t}{T_{t,\text{ref}}}}\left(\frac{p_{t,\text{ref}}}{p_t}\right) \tag{22}$$

as well as the flight and atmospheric conditions. Inlet corrected mass flow is a function of the fan pressure ratio and thus rotational speed. When investigating the off-design conditions, it was assumed that the fan was operating along the working line; in CFD, this required the outlet boundary condition to be set at a constant pressure. The fact that, in general, the flow coefficient is nearly constant along a fan's working line was used to assume a linear relationship between the fraction of design inlet corrected mass flow and the fraction of design corrected speed:

$$\frac{\dot{m}_{\text{corr}}}{\dot{m}_{\text{corr,des}}} = \frac{\omega_{\text{corr}}}{\omega_{\text{corr,des}}} \tag{23}$$

This yielded an initial guess for the rotational speed required to drive a certain mass flow through the machine.

With the mass flow supplied, the fan inlet axial Mach number can be found using Equation (18) at the fan face, and subsequently, the static temperature can be found. With the axial Mach number and static temperature, the axial velocity was known. It was again assumed that the axial velocity was equal at the rotor leading and trailing edges. While this assumption was acceptable for the design condition, it would be far less accurate off-design; however, this was only used as an initial estimate, which was then later corrected through an iterative process. Using the new velocity triangles and the blade angles set at design, the Euler turbine equation, Equation (20), was used to determine the rotor outlet stagnation temperature, and this was then used to determine the rotor outlet stagnation pressure. The stagnation quantities were found at the hub, 50% span, and the casing; a parabolic curve was then fit to these points, and that was used to mass average analytically the rotor outlet stagnation conditions. Using the stagnation quantities at the rotor outlet, the same steps as before were used to determine the axial velocity, static temperature, and static pressure at the nozzle exit. With the static pressure and temperature known, the density at the nozzle exit was found, which allowed for the mass flow rate to be computed. This mass flow rate was compared to the desired mass flow rate, and the process was repeated with the rotational speed altered until the desired mass flow rate was achieved. This provided an initial guess for the rotational speed; however, CFD simulations must be run and the rotational speed adjusted to verify that the off-design operating point has been correctly found. An example of this process is shown later in this paper, and the off-design predictions were compared to those found using CFD; the overall 1D performance prediction matched to within 2%.

7. Implementation of the Body Force Model Generation Approach

The design process was implemented as a MATLAB [16] code, but could be implemented in any scientific computing system. It generated the hub and casing curves, as well as the 2D blade camber lines at the hub, 50%, and tip span fractions. The blade camber surface extraction process was also done within MATLAB. The process ran on a personal workstation and was computationally inexpensive. Computational run times for all steps (including Stagen) were typically under two minutes.

8. Implementation in 3D CFD

Hub and casing curves, as well as the the blade leading and trailing edge profiles were imported into grid generation software (here, Pointwise v18 [17]) to generate the gas path and demarcate blade swept volumes, which must be designated as separate cell zones.

Shown in Figure 11 is a computational domain created using this process. The upstream and downstream boundaries were placed 1.2 and one fan diameters from the rotor leading edge, respectively. These were set far enough away to provide clean inflow and avoid possible interactions with the blade rows. The process created a constant radius (equal to blade tip radius) casing curve upstream of the rotor blades. It should also be noted that in the CFD grid, the outlet nozzle was manually cut slightly before the throat area (A^*); in the example case, this was at $A/A^* = 1.08$ (nozzle length cut by 10% before the throat area). This was done to avoid having a Mach number equal to one occurring at a boundary condition, which was found to lead to stability issues in some solvers.

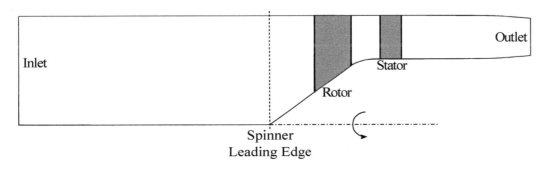

Figure 11. Computational domain created for internal flow simulations.

Four grid levels were used to assess grid independence. A five degree slice of the full machine was used with uniform inflow. This saved computational cost as the body force model produced circumferentially uniform flow when the inflow was uniform. Simulations were run at the design operating point for the design detailed in the next section. All grids were fully structured using hexahedral cells with higher mesh density in the bladed areas. A summary of the grids tested is in Table 6. Only a 0.9% change in pressure rise coefficient was seen from the second finest grid to the finest grid, and therefore, the second finest grid was chosen as the final grid level.

Table 6. Summary of the grid independence study for the example design in the next section.

Overall Cell Count	(x, r, θ) Grid Count	Percent Change in Pressure Rise Coefficient
30,024	(147, 25, 10)	N/A
123,480	(288, 50, 10)	8.1%
464,310	(474, 100, 11)	1.5%
914,860	(621, 150, 11)	0.9%

The computations were carried out using Ansys CFX v18.2 [12], using the same boundary condition types described in the NASA Stage 67 body force simulations at the design operating point; however, at off-design operating points, the outlet static pressure was fixed based on the atmospheric conditions being tested, and the rotational speed was adjusted until the desired mass flow was reached. This was done to remain on the fan working line. The hub and casing were set as zero shear stress walls as the model assumed no losses.

9. Example Application of the Process

In this section, we present an example application of the model generation process and its implementation into CFD. The main purpose of this example was to show the level of expected accuracy of the desired performance at design and therefore assumed that there was no inlet distortion or separation. In Part 2 of this paper [18], the ability to predict these flow phenomena is examined. The example design discussed was based a on high bypass ratio turbofan engine for a medium range jet airliner. Where possible, the parameter values were based on publicly available information for the Pratt & Whitney 1524G engine and the Airbus A220-300 (formerly Bombardier CS300) aircraft. The cruise FPR for the 1524G was estimated to be 1.4 [14]. This and other key design parameters that were used in this example are shown in Table 7. Data sources for most specific values are cited within the table. Static thrust at sea level is 103 kN [19]. As described in Cumpsty and Heyes [14], for a high bypass ratio, low fan pressure ratio engines, typically, the ratio between sea level static and cruise thrust is between six and 6.5; in this paper, we used a ratio of 6.15. The aspect ratios were based on midspan chord values. L_1 was chosen to ensure the imposition of flow direction at the computational inlet boundary did not directly affect the flow within the rotor, as discussed in the previous section. L_2 was based on the distance between the rotor at the stator at the hub for the 1524G engine. $L_3 + L_N$ was set as discussed in the previous section and was evenly divided between the two parameters.

Table 7. Key design parameters.

Parameter	Value	Parameter	Value
FPR	1.4	b_R/c_R	2.33 [20]
Thrust at cruise	16.75 kN	b_S/c_S	2.25 [20]
Fan tip diameter	1.85 m [21]	L_1/c_S	8.00
Fan hub-to-tip ratio	0.3 [20]	L_2/c_S	0.813 [20]
Cruise Mach number	0.78 [22]	L_3/c_S	2.00
Cruise altitude	10,668 m [14]	L_N/c_S	2.00

In this example application, the number of rotor and stator blades was supplied based on best estimates for the Pratt & Whitney 1524G engine and were 18 and 36, respectively [20]. The spinner nose was set such that the rotor hub slope line was extended to zero radius.

9.1. Results at the Design Point

These inputs produced a stage with the gas path shown in Figure 11. The rotational speed of the rotor (camber lines shown in Figure 9) was 374 rad/s, and the required mass flow was 182 kg/s. The computed increase in the mass averaged fan stagnation pressure ratio across the rotor (FPR-1) was found to be 0.395, which was 1.25% below the design intent, and the mass averaged stage work coefficient was found to be 1.9% lower than the desired value. Using the mass averaged Mach number at the outlet boundary, it was found that the area that would create choked flow was 0.8% lower than that generated by the design code. This resulting smaller area required was a result of the stage slightly under-predicting the FPR and work coefficient. If the flow was to be isentropically brought to the nozzle area for choke, the thrust generated would be 0.14% lower than the desired thrust. In Figure 12, the rotor chordwise and trailing edge spanwise work coefficients are shown. The spanwise trailing edge work distribution had an RMS difference of 0.6% from the mass averaged overall work coefficient, so the goal of uniform outlet stagnation temperature was largely achieved.

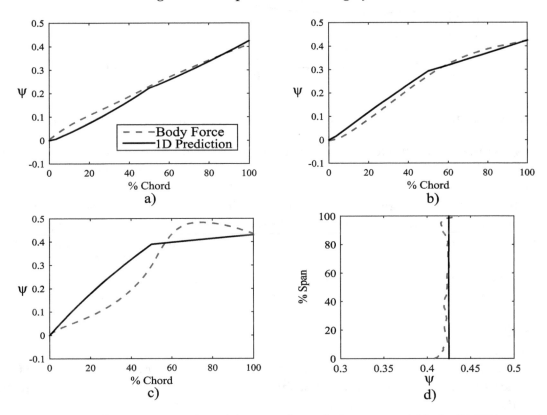

Figure 12. Work coefficient vs. meridional distance through the rotor at: (**a**) hub, (**b**) 50% span, (**c**) tip, and (**d**) rotor trailing edge at design corrected speed and $\phi = 0.64$.

The loading was compared against a 1D prediction generated using the Euler turbine equation, which assumed constant axial velocity through the blade row, as well as a linear build-up of deviation along the chord. The 1D prediction had a discontinuity of slope at the transition from circular arc to straight line camber. The chordwise loading was well predicted at lower span fractions with %RMS differences between the 1D prediction and the CFD being 4.12% and 4.09% at the 20% and 50% span fractions, respectively; however, as the span fraction increased to 80% span, the accuracy decreased, and the %RMS difference increased to 14.5%. The error in the prediction stemmed from the difference in axial velocity between the 1D prediction and the CFD simulations. The assumption of equal axial velocity at the blade leading and trailing edge and along the span was not borne out upon model implementation, in part due to the redistribution of the transonic relative flow. In Figure 13, the chordwise flow coefficients at set spans within the rotor are compared; equal axial velocities at the leading and trailing edge would result in all the lines starting and ending at a value of 0.64.

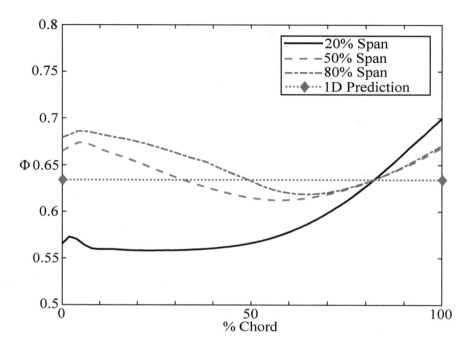

Figure 13. Chordwise flow coefficient through the rotor at design corrected speed and $\phi = 0.64$.

The flow coefficient and the local work input were inversely related to one another for a given rotational speed. This trend matched what was displayed at the leading and trailing edge regions in the chordwise loading distributions shown in Figure 12. This further confirmed the need for positive incidence at the rotor leading edge, as this variation in velocities caused a change in inlet relative flow angle. This was the result of the lack of consideration of 2D/3D effects in the design process.

The largest difference in flow coefficient was found to be a deficit of 0.22 in the rotor hub leading edge region, which corresponded to an increase in the relative flow angle of 11°. This rather large difference in relative flow angle had minimal impact on the local work coefficient when compared to the predicted value, as shown in Figure 12a. This change had a minor impact on the overall work input because it occurred at a low span fraction, and also, the hub leading edge blade angle was relatively small.

The goal was to create a chordwise loading distribution similar to those seen in modern machines. This is a qualitative goal, as there is no one correct solution. The lower 80% span fractions seemed to be producing acceptable chordwise loading distributions as work was increasing almost linearly along the chord, similar to what was seen in Stage 67. However, in the upper 20% span, the work coefficient increased to a maximum and then slightly decreased through the last 10–15% chord. This behaviour was caused by a combination of two effects. The first was the variation in spanwise axial velocity mentioned prior, and the second was the deviation buildup along the chord. Shown in Figure 14 is the deviation through the rotor at the tip. The local flow coefficient trend at the tip was similar to that shown in Figure 13 at 80% span, so we used the 80% span curve from Figure 13 and the deviation behaviour shown in Figure 14 to explain the work addition trend from Figure 12c. In the prediction code, the deviation linearly changed from the 2° caused by the incidence to the final predicted value. In the body force simulation, the deviation was initially higher than assumed, but the body force always acted to reduce deviation; this occurred over the first 10% chord. The deviation then rose through a combination of two effects: the flow coefficient decreased (see Figure 13), and the camber surface turned away from the relative flow direction over the first 50% chord. Thus, the slower flow was unable to turn fast enough, and the deviation rose. At 50% chord, the camber shape became straight, and the body force caused the relative flow to move towards the camber surface direction, decreasing the deviation. From 50% to 80% chord, the flow coefficient was less than the design intent value, so the deviation dropped rapidly. At 80% chord, the flow coefficient rose again to be higher than the design intent value. Since when deviation was zero (near 80% chord), the body force was also zero, there was little change in tangential velocity. The increase in axial velocity associated with a rising flow coefficient means that the relative flow angle decreased, in this case to less than the camber angle, and the deviation became negative. Between 80% chord and the trailing edge, the body force acted to turn the flow back towards the camber surface, leading to work removal. The buildup of deviation was not known a priori and therefore made it difficult to predict the severity of this work removal. The overall work addition was very well predicted and showed that keeping the rear 50% camber angle constant worked as intended to ensure the correct work addition by the rotor trailing edge.

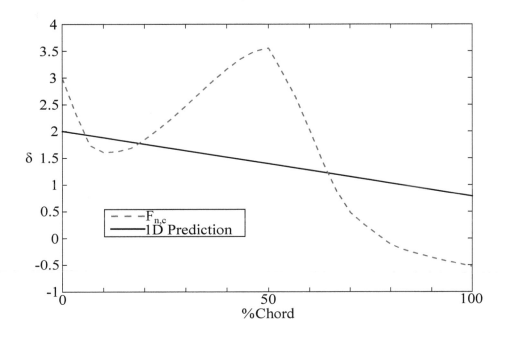

Figure 14. Deviation (degrees) through the rotor tip.

9.2. Results Off-Design

The off-design point of interest in this example was the start of take-off roll condition, with no forward movement of the aircraft. The desired corrected mass flow at take-off was chosen to be 114% of the design; the initial prediction for corrected rotation speed was found to be 119%. After running detailed CFD, it was found to be 116%. The thrust was lower than the thrust found by the 1D mass averaged take-off prediction code by 1.79%, which stemmed from the CFD producing an FPR of 0.016 less than the 1D prediction.

In Figure 15, the trailing edge work coefficient is shown, and as can be seen, the prediction code did not accurately capture the spanwise loading correctly. This inaccuracy was not unexpected as running at off-design conditions modified the velocity triangles to an extent that the simplifying assumptions within the prediction code were no longer valid. A large decrease in the meridional velocity at the tip caused by a large increase in density, which was not captured within the prediction code due to the simplifying assumptions, was the cause of the increased work coefficient.

The prediction code struggled at reproducing the chordwise and spanwise distributions off-design due to the simplifying assumptions; however, it was still beneficial since its estimated overall off-design performance and angular velocity were accurate to within 2% and 3%, respectively. This saved computational time by reducing the amount of iterations required in the CFD.

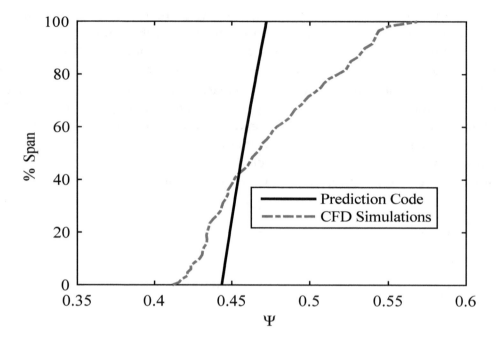

Figure 15. Work coefficient at the rotor trailing edge during off-design (take-off) comparison.

10. Summary and Conclusions

In this paper, a process was described that generated a fan gas path and body force model that met the performance requirements specified. This process differed from existing approaches in that it did not require information about blade or gas path geometry. The body force model used Hall's formulation with an added compressibility correction. This approach was assessed by comparing the results to those found with traditional single passage simulations using the NASA Stage 67 fan geometry. Good agreement was seen with the max %RMS difference in the chordwise and spanwise work coefficient being 6.62% and 4.40%, respectively.

CFD simulations using the models produced matched the desired performance well at the design point, where the desired fan stagnation pressure ratio and thrust agreed to within 1.25%. The design objective of constant spanwise work addition was met as the RMS variation was 0.6% from the mass averaged overall work coefficient. The simplified camber distribution was used with NASA Stage 67 information to assess if the chordwise loading produced was realistic. The chordwise loading within the inner 80% span matched closely to those produced by the NASA Stage 67 with the %RMS being below 8%, with the remaining outer span showing reduced agreement with a %RMS of 13.2%. The agreement in the outer span could be improved by adjusting the ratio between the circular arc and straight line camber distribution; in this case, increasing the fraction of chord, which was a circular arc section, would create a loading distribution more similar to that of the NASA Stage 67. This suggests that the camber distribution could be a function of span, but there is no a priori way to know how one should do this; this is the reason we retained the constant 50:50 split in this paper. The rates of work addition through the rotor showed similarities, which indicates that the simplified camber distributions were producing realistic loading distributions when considering the intended use and level of fidelity. The intended use of this process was not for fan design, but for assessing external flow–fan interactions when limited fan information is available. This process created the fan stage gas path and body force model, which would be integrated into a nacelle and run in full wheel simulations. This is useful when airframers wish to investigate coupling between non-uniform inflows caused by off-design operation and the engine fan, such as a take-off with crosswind. That is precisely the focus of Part 2 of this paper [18], where we show that a fan model developed according to the procedure described in the current paper predicts crosswind separation velocity to within 5% of the value found when a detailed fan body force model is employed.

Author Contributions: Conceptualization, Q.J.M. and J.J.D.; data curation, Q.J.M.; formal analysis, Q.J.M.; funding acquisition, J.J.D.; investigation, Q.J.M.; methodology, Q.J.M. and J.J.D.; project administration, J.J.D.; resources, J.J.D.; software, Q.J.M.; supervision, J.J.D.; validation, Q.J.M.; visualization, Q.J.M.; writing, original draft, Q.J.M.; writing, review and editing, J.J.D.

Acknowledgments: Computational resources were provided by the facilities of the Shared Hierarchical Academic Research Computing Network (SHARCNET) (www.sharcnet.ca) and Compute/Calcul Canada (www.computecanada.ca). The authors would like to thank Dr. Ewan Gunn for the Stage 67 stator model used in this work.

Abbreviations

The following abbreviations are used in this manuscript:

Symbols

A	Area
a	Location of maximum camber
B	Number of blades in a row
b	Blade span
c	Blade chord
c_p	Pressure coefficient
e	Energy
F	Thrust
f	Force
h	Enthalpy
K,K'	Compressibility correction
L	Length
M	Mach number
$\dot{m}$	Mass flow rate
$\hat{n}$	Blade camber surface normal unit vector
p	Pressure
R	Gas constant
r	Radius
s	Spacing between blades
T	Temperature
U	Blade speed
$\vec{V}$	Velocity
$\vec{W}$	Relative velocity
α	Absolute flow angle
γ	Specific heat ratio
δ	Deviation
η	Efficiency
ξ	Change in blade angle
ρ	Density
τ	Shear stress
Y	Tip radius change parameter
ϕ	Flow coefficient
ψ	Work coefficient
ω	Angular velocity

Subscripts

abs	Absolute quantity
c	Compressible
corr	Corrected value
hub	Hub span location
i	Incompressible
LE	Blade leading edge

m	Meridional component
mid	Mid span location
is	Isentropic
inlet	Inlet boundary value
R	Rotor
ref	Reference quantity
rel	Relative quantity
S	Stator
spin	Spinner nose
TE	Blade trailing edge
t	Stagnation quantity
tip	Tip span location
x	Axial component
2	Fan face quantity
13	Fan trailing edge quantity
19	Nozzle exit quantity
θ	Tangential component
∞	Freestream quantity

Abbreviations

RANS	Reynold averaged Navier–Stokes
CFD	Computational fluid dynamics
RMS	Root mean squared
FPR	Fan stagnation pressure ratio
BMA	Blade metal angle

References

1. Peters, A.; Spakovszky, Z.S.; Lord, W.K.; Rose, B. Ultrashort Nacelles for Low Fan Pressure Ratio Propulsors. *J. Turbomach.* **2015**, *137*, 021001. [CrossRef]
2. Hill, D.; Defoe, J. Innovations in Body Force Modeling of Transonic Compressor Blade Rows. *Int. J. Rotat. Mach.* **2018**, *2018*, 6398501. [CrossRef]
3. Godard, B.; Jaeghere, E.; Nasr, N.; Marty, J.; Barrier, R.; Gourdain, N. Methodologies for Turbofan Inlet Aerodynamics Prediction. In Proceedings of the 35th AIAA Applied Aerodynamics Conference, Denver, CO, USA, 5–9 June 2017; AIAA: Reston, VA, USA, 2017. [CrossRef]
4. Gong, Y.; Tan, C.S.; Gordon, K.A.; Greitzer, E.M. A Computational Model for Short-Wavelength Stall Inception and Development in Multistage Compressors. *J. Turbomach.* **1999**, *121*, 726–734. [CrossRef]
5. Thollet, W.; Dufour, G.; Carbonneau, X. Assessment of Body Force Methodologies for the Analysis of Intake-Fan Aerodynamic Interactions. In Proceedings of the ASME Turbo Expo 2016, Seoul, Korea, 13–1 June 2016; ASME: New York, NY, USA, 2016. [CrossRef]
6. Hall, D.; Greitzer, E.; Tan, C. Analysis of Fan Stage Conceptual Design Attributes for Boundary Layer Ingestion. *J. Turbomach.* **2017**, *139*, 071012. [CrossRef]
7. Pazireh, S.; Defoe, J. A No-Calibration Approach to Modeling Compressor Blade Rows with Body Forces Employing Artificial Neural Networks. In Proceedings of the ASME Turbo Expo 2019, Phoenix, AZ, USA, 17–21 June 2019; ASME: New York, NY, USA, 2019. [CrossRef]
8. Sato, S.; Spotts, N.; Gao, X. Validation of Fan Source Term Model Constructed without Blade Geometry. In Proceedings of the AIAA Scitech 2019 Forum, San Diego, CA, USA, 7–11 January 2019; AIAA: Reston, VA, USA, 2019. [CrossRef]
9. Denton, J. Multall—An Open Source, Computational Fluid Dynamics Based, Turbomachinery Design System. *J. Turbomach.* **2017**, *139*, 121001. [CrossRef]
10. Benichou, E.; Dufour, G.; Bousquet, Y.; Binder, N.; Ortolan, A.; Carbonneau, X. Body Force Modelling of the Aerodynamics of a Low-Speed Fan Under Distorted Inflow. *Int. J. Turbomach. Propuls. Power* **2019**, *4*, 29. [CrossRef]

11. Strazisar, A.; Wood, J.; Hathaway, M.D.; Suder, K. *Laser Anemometer Measurements in a Transonic Axial-Flow Fan Rotor*; Techreport 2879; NASA: Washington, DC, USA, 1989.

12. ANSYS Inc. *ANSYS v18.2 User's Guide*; ANSYS Inc.: Canonsburg, PA, USA, 2017.

13. Dixon, S.; Hall, C. *Fluid Mechanics and Thermodynamics of Turbomachinery*, 7th ed.; Butterworth-Heinemann: Oxford, UK, 2014. [CrossRef]

14. Cumpsty, N.; Heyes, A. *Jet Propulsion*, 3rd ed.; Cambridge University Press: Cambridge, UK, 2015. [CrossRef]

15. Greitzer, E.M.; Tan, C.; Graf, M.B. *Internal Flow*, 1st ed.; Cambridge University Press: Cambridge, UK, 2004. [CrossRef]

16. Mathworks Inc. *MathWorks MATLAB 2017 Documentation*; Mathworks Inc.: Natick, MA, USA, 2017.

17. Pointwise. *Pointwise v18.0R4 User's Manual*; Pointwise: Forth Worth, TX, USA, 2017.

18. Minaker, Q.J.; Defoe, J.J. Prediction of Crosswind Separation Velocity for Fan and Nacelle Systems Using Body Force Models: Part 2: Comparison of Crosswind Separation Velocity with and without Detailed Fan Stage Geometry. *Int. J. Turbomach. Propuls. Power* **2019**, 4, 41. [CrossRef]

19. EASA. *TYPE-CERTIFICATE DATA SHEET No. IM.E.090 for PW1500G Series Engines*; Techreport; European Union Aviation Safety Agency: Cologne, Germany, 2019.

20. Transportation Safety Board of Canada. *Aviation Investigatation Report A14Q0068*; Techreport; Government of Canada: Gatineau, QC, Canada, 2016.

21. Pratt and Whitney. PW1500G, AIRBUS A220 PRODUCT CARD. Available online: http://newsroom.pw.utc.com/download/PW1500G.pdf (accessed on 13 December 2019).

22. Bombardier. C Series. Available online: http://commercialaircraft.bombardier.com:80/content/dam/Websites/bca/literature/cseries/Bombardier-Commercial-Aircraft-CSeries-Brochure-en.pdf.pdf (accessed on 8 September 2015).

Prediction of Crosswind Separation Velocity for Fan and Nacelle Systems using Body Force Models (Part 2) Comparison of Crosswind Separation Velocity with and without Detailed Fan Stage Geometry

Quentin J. Minaker and Jeffrey J. Defoe *[ID]

Turbomachinery and Unsteady Flows Research Group, Department of Mechanical, Automotive and Materials Engineering, University of Windsor, 401 Sunset Ave., Windsor, ON N9B 3P4, Canada
* Correspondence: jdefoe@uwindsor.ca

Abstract: Modern aircraft engines must accommodate inflow distortions entering the engines as a consequence of modifying the size, shape, and placement of the engines and/or nacelle to increase propulsive efficiency and reduce aircraft weight and drag. It is important to be able to predict the interactions between the external flow and the fan early in the design process. This is challenging due to computational cost and limited access to detailed fan/engine geometry. In this, the second part of a two part paper, we apply the fan gas path and body force model design process from Part 1 to the problem of predicting flow separation over an engine nacelle lip caused by crosswind. The inputs to the design process are based on NASA Stage 67. A body force model using the detailed Stage 67 geometry is also used to enable assessment of the accuracy of the design process based approach. In uniform flow, the model produced by the design process recreates the spanwise loading distribution of Rotor 67 with a 7% RMS error. Both models are then employed to predict crosswind separation velocity. The two approaches are found to agree in their prediction of the crosswind separation velocity to within 5%.

Keywords: fan–nacelle interaction; body force models; turbomachinery; CFD; crosswind; nacelle

1. Introduction

Modern aircraft engine design is moving towards using larger bypass ratios with lower fan stagnation pressure ratios. This improves propulsive efficiency; however, it comes with the negative trade off of larger and heavier nacelles. To combat this negative side effect, manufacturers are using shorter inlets with thinner nacelle lips [1]; however, these changes increase the chance of flow separation occurring in the inlet. The airframers' ability to determine the impact of these changes is important during design. To model such situations, the airframer must be able to also model the engine fan stage as its operation will greatly impact the intake performance [1]; however, two major concerns arise while modelling these flows. The first is that it requires full wheel simulations to capture the non-uniform flow caused by inlet separation; using traditional bladed full wheel simulations to model non-uniform flow is very computationally expensive. These bladed full wheel simulations can contain over 100 million cells for the internal flow alone, usually require 20–30 rotor revolutions to obtain statistically stationary results [2,3], and can take over two months to complete even with modern computing power. The second issue is that access to detailed fan stage geometry may not be possible and that most airframers lack the expertise or time required to reproduce this geometry.

The solution to the first issue is to use a simplified model of the propulsion system. This reduces the computational cost, as it allows for steady simulations and reduces the number of cells required.

Godard et al. (2017) [4] discussed multiple simplified modelling approaches and concluded that a body force approach was needed to capture the coupling effect between the external flow and the fan's operation. Body force methods work by applying sources of momentum and energy in the swept volumes where the blades would normally be. Another study conducted by Burlot et al. (2018) [5] compared simplifying methods to high fidelity unsteady Reynolds averaged Navier–Stokes (RANS) simulations of a nacelle casing with non-uniform inflows; it agreed with the findings of Godard et al. (2017) and showed that body force models worked better than other simplifying methods at reproducing the results of the unsteady RANS simulations. The body force method captures radial distributions of stagnation pressure and downstream distortion maps as smeared out averages of that seen in the bladed simulations.

Many variations of the body force method exist; however, they commonly require calibration based on experimental results or detailed bladed simulations. As described in Part 1 of this paper [6], the model produced by Hall et al. (2017) [7] requires no calibration; however, it still requires blade geometry and gas path information. This relates to the second issue mentioned earlier as this information would typically not be available. Part 1 described a process that creates a simplified fan stage gas path and body force model using Hall's method with limited stage information, and it was shown to produce the desired results at design conditions.

In this process, the desired thrust, fan stagnation pressure ratio (FPR), and geometric parameters, all of which would commonly be known by an airframer, are inputs used to generate a body force model. The process consists of using 1D analysis through the fan stage, simplified blade camber shapes, and simplifying assumptions to find the required information needed for the Hall body force approach. How the parameters are found, what assumptions are made, and the steps used to create the body force model were described in Part 1 [6].

An important non-uniform flow studied is the crosswind around a nacelle because it is the most likely scenario to result in inlet separation. In Yeung et al. (2019) [8], different nacelles were tested with crosswind flows, and the effects on the separation velocity and the stagnation pressure distributions were described. In Lee et al. (2018) [9], the effect that crosswind had on the operation of the fan stage was analysed. It was found that this inlet distortion caused a loss in stall margin, even in cases where separation had not occurred. The authors also discussed the suppression effect that the fan applied to the inlet separation. These findings further emphasized the importance of designers being able to incorporate this external flow–fan stage interaction into the nacelle design process.

The objective of this paper is to assess the ability of a model produced using the simplified process detailed in Part 1 to predict the crosswind separation velocities seen in a real machine operating with crosswind. The model design process is used to create a gas path and body force model based on a real machine, NASA Stage 67. Both the created stage and the original NASA Stage 67 are run using the body force approach (Hall's model) with varying crosswind speeds, and the results are compared. In Hall et al. (2017) [7], the model's ability to capture upstream flow redistribution and distortion transfer correctly showed good agreement with higher fidelity models. Comparing the results from NASA Stage 67 to those of the simplified stage will allow for a quantification of the accuracy lost by simplifying the stage design. The key outcomes are that the design simplification has minor effects on the rotor and the nacelle performance prediction.

The first section of this paper will discuss how the simplified stage is created. This simplified stage is run at the design condition and compared to the original NASA Stage 67. Next, the numerical setup of the full wheel crosswind simulation is described. In the Results Section, the full wheel crosswind body force simulations using the original NASA Stage 67 and the simplified stage are compared. This includes the difference in separation velocities and the effect on the fan stage performance.

2. Simplified Stage Creation

The simplified generation process was used to create a stage based on the performance and geometry of NASA Stage 67. The FPR supplied to the simplified design process was found using the results of NASA Stage 67 running at 70% corrected speed at the design flow coefficient using Hall's model [7] with an added compressibility correction, as described in Part 1 [6]. Normally, the process would not require any prior simulations, but as the intent was to create a stage as similar to NASA Stage 67 as possible, this was done here. The corrected speed of 70% was used as this provided similar tip relative Mach numbers to those seen in modern engine fan stages. The details of the Stage 67 simulation, including the setup, validation, and results can be found in Part 1 [6]. This was found to produce an FPR of 1.31; this value was then supplied for the simplified design process.

The process described in Part 1 used the desired thrust to determine the required mass flow for the machine; however, in this case, the mass flow from the NASA Stage 67 simulation was used so an objective comparison could be made between the two models. This means that the process was slightly altered such that mass flow was supplied, instead of a desired thrust. This was done by substituting the mass flow rate as opposed to the thrust in the thrust equation,

$$F = \dot{m}(V_{19} - V_{\infty}) + A_{19}(p_{19} - p_{\infty}) \qquad (1)$$

where F is thrust, $\dot{m}$ is the mass flow rate through the fan, V is the velocity in the aircraft reference frame, A is flow area normal to the engine rotational axis, and p is static pressure. The subscripts represent the quantities at the stations shown in Figure 1. The thrust then became an output of the process, though it was not important in this instance.

The design process required several geometric parameters to be specified to enable the generation of the gas path. In this case, all of the parameters required were found using the Stage 67 geometry and applying simplifications as required. Similar to the FPR, these parameters could be determined without detailed geometry information; however, Stage 67 data [10] were used here so that a comparison could be made and the effects of simplifying the stage design could be seen. The parameters were generated as follows:

- The inlet rotor tip radius of $r_{\text{tip}} = 0.255$ m and hub-to-tip ratio of 0.375 were set equal to NASA Rotor 67.

- The simplified process generated constant axial coordinate leading and trailing edge blade profiles. The blade aspect ratios were set such that these axial coordinates were equal to the average axial coordinates of the leading and trailing edge profiles from NASA Stage 67, which gave rotor and stator blade aspect ratios of $b_R/c_R = 2.53$ and $b_S/c_S = 2.22$, respectively.

- The parameter that determined the amount of casing radius change through the rotor was set so that there was an equal decrease in radius between the two machines and was $Y = 0.04$.

- The spacing between the blades was set so that the stator leading edge was set at an equal axial coordinate to that in NASA Stage 67 and was $L_2/c_R = 0.706$.

- The axial length of the spinner nose was set so that the upstream distance from the rotor leading edge (averaged for NASA Stage 67 case) was equal in both cases and was 94% of the length if the rotor hub slope was continued to zero radius.

A comparison between the original NASA Stage 67 gas path and the gas path generated using the simplified design process is shown in Figure 1. It is clear from the figure that the most significant difference was in the shapes of the spinners. This was due to a constraint of continuity of the hub slope between the rotor and spinner in the stage design process employed. This was one factor that contributed to the differences in the flow field upstream of the rotor, which will be discussed later in the paper.

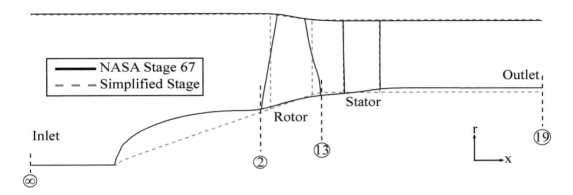

Figure 1. Comparison between NASA Stage 67 and simplified stage gas paths and meridional blade profiles.

There was a major difference between the blade shapes and their operation in the NASA Stage 67 compared to the simplified stage. The simplified process generated blade angles at the hub, midspan, and tip. These were used to create a camber surface as described in Part 1 [6]. The change in rotor blade angle from leading to trailing edge found at these spans was 33.2°, 7.5°, and 2.3°, respectively. The change in blade angles at the same span locations in NASA Stage 67 was 63.2°, 17.8°, and 12.2°. It was seen that the simplified process generated blades with significantly less turning; however, it created the required overall flow turning by increasing the rotational speed of the blades. The corrected rotational speed of the rotor blades in the simplified stage was 1435 rad/s compared to 1176 rad/s used by NASA Stage 67. This difference in blade shapes was not unexpected or considered an issue though, as the aim was to reproduce stage performance rather than blade shapes. The different rotational speeds implied different velocity triangles and differing local flow characteristics between the two fan models. As will be shown, however, the good agreement in the results obtained using the two models suggested that these effects had only a minor impact on fan upstream influence.

Computational fluid dynamics (CFD) simulations were run with the simplified stage at the design point. A 1/8 slice of the simplified stage was used with uniform inflow, with the same boundary conditions as the NASA Stage 67 simulation used to determine the supplied FPR [6]. The grid used contained 559,520 cells and had a distribution similar to what was used by the NASA Stage 67 simulations, and therefore, the grid independence study conducted for NASA Stage 67 in Part 1 was considered sufficient to treat the new results as grid independent. The solver and grid generation tools were the same as those used in Part 1.

Comparison of Stages at the Design Condition

The overall performance of the simplified stage was compared against NASA Stage 67 at 70% corrected speed. The simplified stage had a mass averaged fan stagnation pressure ratio (FPR-1) of 0.325, which was an 3.83% increase from the desired value of 0.313. In Figure 2, the chordwise and trailing edge spanwise work coefficient,

$$\psi = \frac{h_{\mathrm{t}} - h_{\mathrm{t},\infty}}{U^2_{\mathrm{mid},67}} \tag{2}$$

are compared between the simplified stage and NASA Stage 67, using the blade velocity in NASA Stage 67 for normalisation. h_{t} is the stagnation enthalpy, and $U_{\mathrm{mid},67}$ is the midspan blade speed for Stage 67. The rates of work input in the first 20% chord were generally in good agreement between the two stages across the span, but the overall work input was high for the simplified stage: the spanwise mass averaged work coefficient was 5.15% higher than the target value. The root mean squared (RMS) local difference in work coefficient normalised by the total work addition along the chord at 20%, 50%, and 80% span, as well as at the rotor trailing edge are shown in Table 1. In general, agreement worsened with increasing span fraction (outside of the endwall regions); however, the trailing edge spanwise profile of work input had only a 7% RMS error, indicating reasonable agreement in overall loading prediction.

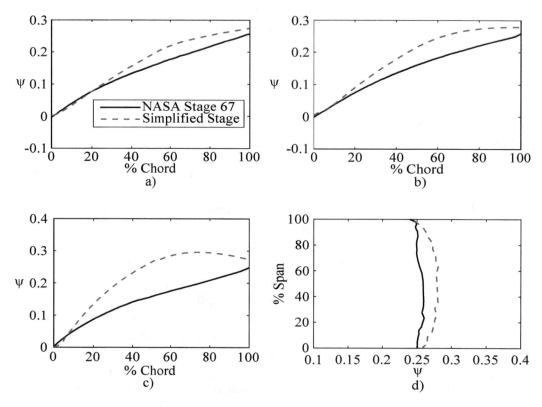

Figure 2. Work coefficient vs. meridional distance through the rotor at: (**a**) 20% span, (**b**) 50% span, (**c**) 80% span, and (**d**) rotor trailing edge.

Table 1. Comparison of NASA Stage 67 and the simplified stage.

	RMS Local ψ Error/ψ_{TE}
20% Span	0.091
50% Span	0.147
80% Span	0.284
Spanwise	0.068

The spanwise trailing edge work distribution in the simplified stage had an RMS difference of 2.6% from the mass averaged overall work coefficient for this same case. One of the aims of the design process was to create a spanwise uniform trailing edge work coefficient, and this level of agreement indicated that this had largely been achieved.

A decrease in the accuracy of the actual vs. desired performance was observed for the case shown here compared to the example fan stage studied in Part 1 [6]. The accuracy of FPR-1 decreased by 2.58%, and the non-uniformity of the spanwise trailing edge work distribution increased by 2%. This decrease in accuracy was primarily caused by the decrease in tip radius through the rotor for the case presented in the current paper. This change caused the axial velocity in the tip region to increase, which led to a radially outward shift in mass flux, as shown in Figure 3; this shift lowered the axial velocity in the midspan region, which modified the velocity triangles all along the span and had the main effect of increasing the work coefficient near midspan.

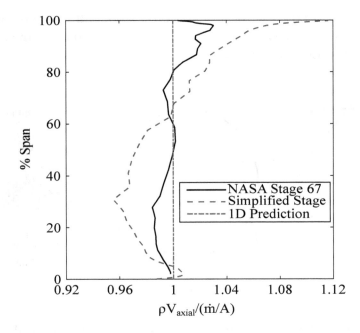

Figure 3. Mass flux along the rotor trailing edge.

This simplified stage showed comparable performance to NASA Stage 67 at the selected operating point and therefore provided an adequate foundation to continue with the performance comparison of the two stages with non-uniform inflow.

3. Numerical Setup of Full Wheel Crosswind Simulations

The simulations modelled pure crosswind at sea level with no forward movement of the engine/nacelle, similar to the conditions seen at the start of a take-off roll; Ansys CFX v18.2 [11] was used as the solver. The simulations were steady state and used the shear-stress-transport turbulence model [11]. For crosswind velocities, which result in flow separation on the inner part of the nacelle, steady computations provided only an approximate solution to the flow in and downstream of the separated region. If the details of the fan–nacelle interaction were of interest in such cases, then time resolved computations would be more suitable. However, in this paper, we were interested in comparing the ability of the fan models to capture fan–nacelle interaction, and so, the steady treatment was sufficient. In our computations, at even the highest crosswind velocities investigated, the solutions obtained were truly steady, showing good convergence including all fluxes (mass, momentum, and energy) having less than 0.1% imbalance across both the full computational domain and a smaller control volume encompassing just the flow near the nacelle and the internal flow. The computational domain had outer boundaries at 25 engine diameters away from the fan axis, as schematically illustrated in Figure 4. The cylindrical computational domain comprised four separate boundary conditions on its outer surface, since CFX lacks a true far-field boundary condition. The curved surface was divided into two equally sized boundaries, one inlet and one outlet. At the inlet, the static pressure and crosswind velocity were imposed, and the other was an outlet with the same imposed static pressure. The boundaries perpendicular to the fan axis were openings, which in CFX means the boundary can act as either an inlet or an outlet. When the flow left the domain across these boundaries, the static pressure was set to the same value as on the inlet and outlet boundaries. Flow entering across this boundary was treated as total pressure (set as the same value as the static pressure), from which the static pressure was calculated. In these simulations, the static pressure was taken as the sea level standard. A mass flow outlet boundary condition was set at the outlet within the fan stage. The hub and casing curves were set as no slip walls everywhere except downstream of the rotor leading edge, where they switched to zero shear stress walls. This was done as the design process described in Part 1 made the assumption of zero losses within the turbomachinery and its associated ducting. The spinner nose was a no slip rotating wall with the same rotation speed as the rotor blades. The crosswind velocity was varied.

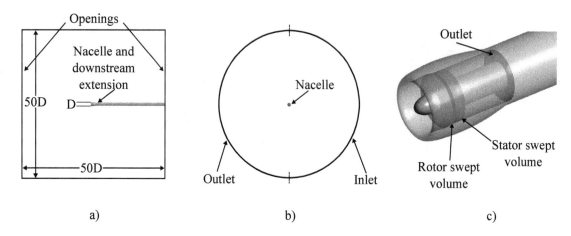

Figure 4. Computational domain for crosswind simulations. (a) Side view, (b) front view, and (c) zoomed in view of the nacelle and fan stage.

The nacelle geometry was generated by Bombardier Aerospace for this analysis and was representative of those seen on modern wide body aircraft. Figure 5 shows the meridional profile of the nacelle used. To avoid sharp corners that have been known to cause issues within numerical methods [11], the outside of the nacelle was extended axially outwards to the downstream boundary.

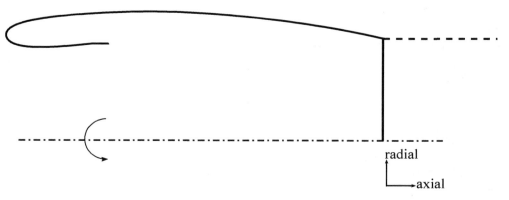

Figure 5. Nacelle casing used within full wheel simulations, where the dashed line represents the casing extended axially downstream and the dashed-dotted line is the rotation axis.

An initial structured mesh containing 13.5×10^6 nodes was used for the full domain and was generated using Pointwise v18.0 [12]. A cell growth rate of 1.1 was used at all wall boundaries. The grid used downstream of the fan leading edge was similar to that used in the uniform inflow slice from the prior NASA Stage 67 simulations described in Part 1 [6]. The automatic wall treatment option within Ansys CFX v18.2 was used; this treatment was y^+ insensitive; however, it is recommended that at least ten cells exist in the boundary layer if this option is used [11]; this condition was met for all the simulations.

These full wheel simulations were relatively computationally expensive due to larger cell counts (compared to internal flow only computations). This caused two main issues, the first being that the grid independence study itself could become expensive, and the second being that to find the separation point, many iterations on the crosswind velocity may be necessary. To maintain a lower computational cost, the method presented by Roache [13], which was based on the use of Richardson extrapolation, could be used to determine the expected error based on the gird used. This method involves performing two or more simulations with successively finer grids and assumes that the change in the results should asymptotically approach zero as the number of grid points tends towards infinity.

Three grid levels were assessed for the simplified stage with $U^* = 0.232$, where U^* is the ratio between the crosswind velocity and mass averaged fan leading edge axial velocity. A formal definition of U^* is provided in the next section. One grid was finer than the one described earlier, and one was coarser. $U^* = 0.232$ was selected as it was large enough to generate flow separation, but low enough that the flow reattached before the rotor leading edge. The variables of interest in the grid study were the separation size (defined by the percentage of area between the nacelle lip and the rotor leading edge where the shear stress was less than zero in the axial direction) and FPR. Figure 6 shows these parameters at the three grid levels tested, along with the value predicted using Richardson extrapolation with an infinite number of grid points. The extrapolations predicted a separation size of 3.96% with a possible error of $\pm 0.06\%$ and an FPR of 1.308 with a possible error of ± 0.0004 when the cell count tended to infinity. To maintain lower computational cost, the medium grid level was selected. Therefore, the expected errors in separation size and FPR between this grid and the values that would be obtained with an infinite grid resolution were $0.91\% \pm 0.06\%$ and 0.0004 ± 0.0004, respectively.

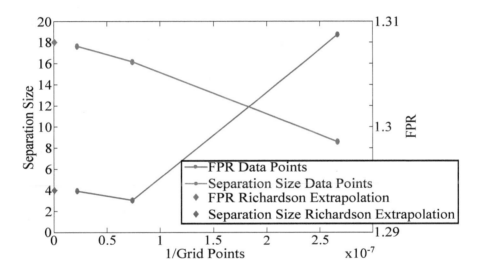

Figure 6. Separation size and FPR as a function of the grid points used to assess error caused by the use of a finite grid size.

4. Results and Discussion of Crosswind Simulations

The crosswind velocity was increased until separation was observed. We define the non-dimensional crosswind velocity to be:

$$U^* = \frac{V_y}{\overline{V}_{x,2}^M} \tag{3}$$

where V_y is the crosswind velocity and $\overline{V}_{x,2}^M$ is the mass averaged axial velocity at the fan face (Station 2). Since discrete crosswind velocities are required in CFD, identifying the precise velocity required for separation was challenging. Instead, we aimed to find the value of U^* to within ± 0.005 of the value at which separation first occurred. Due to possible hysteresis in the behaviour of the computations, we always moved in the direction of increasing crosswind velocity from one computation to the next. With the conditions tested, the mass averaged axial velocity at the fan face was approximately constant, which translated to a precision of roughly 0.5 m/s for the crosswind velocity.

Flow separation was identified by a region where the flow was locally travelling upstream; this could be identified by a negative value of the wall shear stress in the axial direction on the inside of the nacelle. The point where separation first occurred in NASA Stage 67 was at $U^* = 0.22$. The same condition occurred at a value of 0.23 in the simplified stage. This translated to approximately a difference in crosswind velocity of 1.5 m/s between the two stages (5.13% increase). It was possible that the actual difference was smaller since the difference in these two values was not much larger than the possible error due to the discrete values of U^* assessed. The difference stems from the facts that the simplified stage produced a slightly larger FPR, that there were differences in the local characteristics of the two fans, and that the spinner shapes for the two machines were different. At this condition, the separation occurred over a short region before the flow reattached. The crosswind velocity was increased further until the flow remained separated up to the fan face within the NASA Stage 67 case; this occurred at $U^* = 0.30$. In Figure 7, a 180° slice of the inner nacelle casing is unwrapped and shows the regions of separation at the conditions described here, as well as when the flow is fully attached ($U^* = 0.21$) for both stages.

The regions of separated flow were similar between the two stages. The largest difference when comparing separation was seen in the highly separated case. For NASA Stage 67, the flow separated and remained separated until the fan face; however, in the simplified stage, the flow reattached slightly before the fan face. This occurred for the same reasons as the delayed separation: the slightly larger FPR created by the simplified stage provided additional suction, which gave rise to a more favourable pressure gradient (aiding flow reattachment), as well as the different local flow characteristics of the fan models and varying meridional velocity distributions caused by different spinner shapes. These effects were also responsible for other differences between the two stages' performance with crosswind, as will be discussed later in this section.

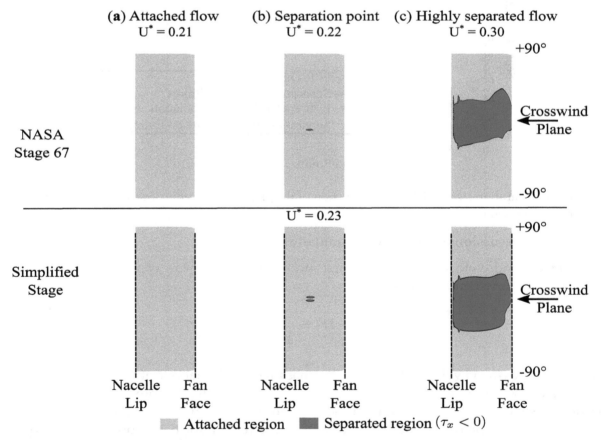

Figure 7. Areas of separated flow within the nacelle at the (a) fully attached condition, (b) separation point, and (c) highly separated condition.

Figure 8 shows the contours of the Mach number on a plane tangent with the crosswind direction (and passing through the fan axis) for increasing crosswind velocities. In the attached flow case, the effects of the crosswind were apparent with larger flow acceleration around the nacelle lip followed by a decrease along the nacelle wall. In the highly separated case, the low Mach number region increased in thickness and represented an area of recirculating flow, as shown by the streamlines. The effect of the simplified stage's larger FPR and altered characteristic slope was seen in the decreased thickness and axial length of the recirculating region along the nacelle wall. In the highly separated flow case, the radial thickness of the recirculating flow region decreased by 15.4%, and the axial length decreased by 12.2% for the simplified stage.

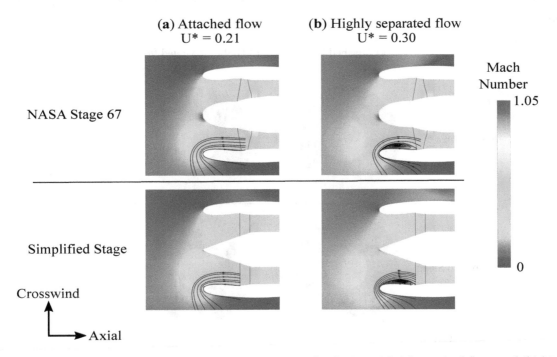

Figure 8. Mach number in the plane tangent to crosswind velocity with (**a**) attached flow and (**b**) highly separated flow.

The effect of this flow separation at the rotor leading edge was investigated next. Separation caused a decrease in stagnation pressure, which was convected downstream to the rotor. In Figure 9, the ratio of local stagnation pressure to freestream stagnation pressure is shown at the fan leading edge for both stages in the attached and highly separated flow conditions. In Frame (a), the stagnation pressure distribution was mostly uniform, except within the region adjacent to the wall surfaces due to boundary layer losses, which was larger on the side of crosswind flow. In both stages, the maximum radial thickness of the distortion region ($P_t / P_{t,atm} < 1$) was increased by approximately 6% on the side of crosswind flow. The increase in thickness of the distortion region was relatively small and indicated that the flow was fully attached by the fan leading edge and that the crosswind did not have a large adverse effect on fan performance, as will be confirmed later. In Frame (b), the stagnation pressure dropped due to the large separation. The difference in the minimum fan leading edge stagnation pressure between the two stages in the highly separated case was 3.4%, with the larger drop in stagnation pressure being in NASA Stage 67. The two stages showed similar stagnation pressure distributions, with the location of the minimum stagnation pressure at 5.8% span closer to the casing for NASA Stage 67.

To determine the impact of the flow separation on the fan rotor, we next accessed mass flux distributions at the fan leading edge. Figure 10 compares the mass flux distributions at the fan face for both stages in the fully attached condition, as well as the highly separated condition. At $U^* = 0.21$, the mass distribution was relatively uniform with a slight decrease along the walls as the velocity went to zero, which caused a small increase in mass flux near midspan. However, in the highly separated flow case, there were larger differences in the mass flux deviations for the two stages. In the regions of separation, the mass flux was decreased as there was flow recirculation/reversal; this had the effect of causing the adjacent regions to have increased mass flux as the flow accelerated around the edges of the recirculating regions. This was apparent in the NASA Stage 67 case, where the flow recirculation reached the fan (seen as a negative value in the contour plot), and a large increase was seen surrounding this region. The reason behind the slight drop in prediction performance in the highly separated case was due to the simplified stage experiencing flow reattachment slightly before the fan. This indicated that the simplified stage fan had a larger corrective effect on the distribution of mass flux, which stemmed from it having a larger FPR, and also possibly steeper loading in the outer span.

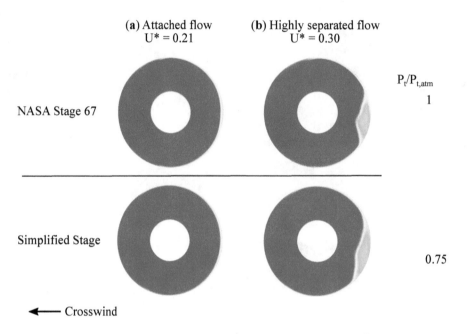

Figure 9. Stagnation pressure ratio at rotor leading edge with (a) attached flow and (b) highly separated flow.

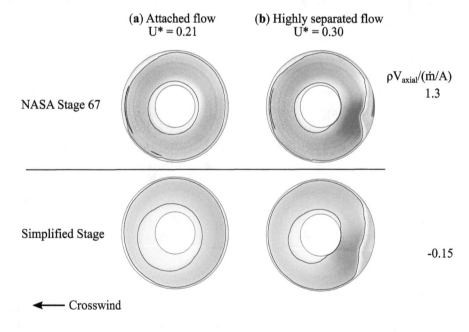

Figure 10. Mass flux distribution at the fan face for the (a) fully attached condition and (b) highly separated condition. Isolines of $\rho V_{\mathrm{axial}} / (\dot{m}/A) = 1$ also shown in black.

To further examine the effects that the separation had on the performance on the fan stage, as well as how accurately the simplified stage captured this, the absolute flow angles and changes in incidence angles were investigated at the rotor leading edge. In Figure 11, the absolute flow angles, α, are shown along the 20%, 50%, and 80% span lines at the rotor leading edge for both stages in the attached ($U^* = 0.21$) and separated conditions ($U^* = 0.30$). In the attached case, the maximum and minimum flow angles were $-3.4°$ and $5.7°$; this confirmed that the flow was largely uniform as there were only small changes in flow angles as seen both around the annulus and along the span. The maximum difference in the absolute flow angle between the two stages was $0.92°$, $1.8°$, and $3.2°$ at the 20%, 50%, and 80% span lines, respectively. In the separated case, the absolute flow angles had larger fluctuations in the separated region. The maximum and minimum flow angles were $20.6°$ and $-14.9°$, and these occurred along the 80% span line, as expected, as this travelled through the region directly affected by the separation. The maximum difference in the absolute flow angles at the 20% and 50% span fractions was $3.4°$ and $3.3°$. A maximum difference of $23.0°$ in the absolute flow angle was seen along the 80% span line at $\theta = 20°$; the reason for this large difference was due to the distortion region being extended over a slightly larger θ range. However, the value of the maximum flow angle was in good agreement between the two stages even at 80% span.

Equally important to consider is the incidence, as high incidence can lead to premature stall when it occurs near the tip of the fan blades [9]. In Figures 12 and 13, the changes in incidence angle, i, from the incidence angle with no crosswind, $i_{U^*=0}$, are shown. The comparison to incidence with no inlet distortion, as opposed to using the circumferential variations in incidence, allowed for the effect of operating at an off-design condition (crosswind) to be more clearly seen. In Figure 12, the change in incidence angle for the simplified stage is calculated using the absolute flow angles and the meridional velocity (plotted in Figure 14 normalised by blade speed) produced by the simplified stage CFD simulations; however, the rotational speed used to transform the flow to the relative frame was that of NASA Stage 67.

The opposite is true for Figure 13: here, the relative flow angles are determined using the simplified stage's rotational speed for both machines. This approach allowed for a direct comparison of how simplifying the stage affected the prediction of changes in incidence due to the crosswind for a real machine. The incidence further confirmed the earlier observations: that the fan was largely unaffected at the attached condition even with significant crosswind ($U^* = 0.21$), as the change in incidence was no more than $+0.3°/-2.3°$ anywhere (for either reference rotational speed). Incidence changes were heavily affected in the regions of separated flow at $U^* = 0.30$: for Stage 67 at 80% span, the peak incidence was $14.3°$ based on Stage 67 rotational speed; $12.1°$ based on the simplified fan rotational speed. The simplified stage captured the locations of the maximum excursions in incidence and was able to predict the largest reduction in incidence to within $0.3°$ regardless of the reference rotational speed. The simplified stage under-predicted the maximum increased incidence at 80% span by $4.8°/3.8°$ (Stage 67/simplified fan rotational speeds, respectively).

While this was a significant difference, the peak incidence for the simplified fan (using its own rotational speed) was still $8.3°$. Such a large increase in incidence would almost certainly initiate stall in an actual fan rotor, though the circumferential region over which this occurred was relatively small and would most likely result in decay of the stall cell before it could fully travel around the annulus. Thus, while the simplified fan stage provided only an estimate of the peak incidence associated with the inlet distortion, it still indicated that a problem may exist for rotor operability. Since the simplified modelling approach was intended for use at the preliminary design stage, this was adequate to flag a potential issue that should be investigated with higher fidelity methods once the detailed fan geometry is known.

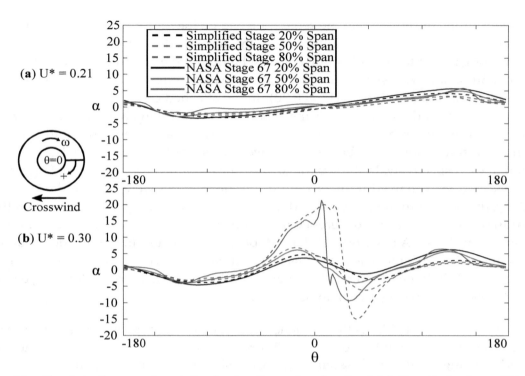

Figure 11. Absolute flow angles at the fan leading edge for the (**a**) fully attached condition and (**b**) highly separated condition.

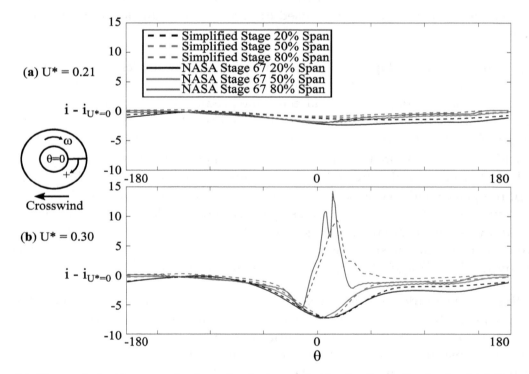

Figure 12. Change in incidence angles from the design at the fan leading edge for the (**a**) fully attached condition and (**b**) highly separated condition. Relative frame moving at the rotational speed of Stage 67.

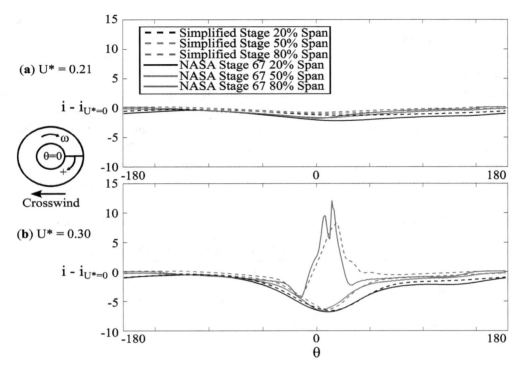

Figure 13. Change in incidence angles from the design at the fan leading edge for the **(a)** fully attached condition and **(b)** highly separated condition. Relative frame moving at the rotational speed of the simplified fan.

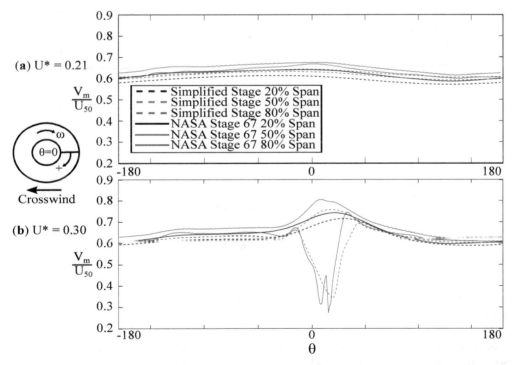

Figure 14. Meridional velocity normalised by the blade rotation speed at 50% span at the fan leading edge for the **(a)** fully attached condition and **(b)** highly separated condition.

For airframers, it is important to be able to measure the performance of a nacelle. To quantify the performance, the metric known as $DC60$ is commonly used [8,9] and is defined as:

$$DC60 = \frac{\overline{p}_{t,60}^A - \overline{p}_t^A}{\overline{p}_t^A - \overline{p}^A} \tag{4}$$

where $\bar{p}_t$ and $\bar{p}^A$ are the area averaged stagnation and static pressures at the fan leading edge and $\bar{p}^A_{t,60}$ is the lowest value of area averaged stagnation pressure over any $60°$ circumferential sector at the fan leading edge. $DC60$ is a measure of inlet distortion and decreases as the fan leading edge separation increases. Though normally mass weighted averages are more meaningful than area weighted ones, use of area weighting is explicit in the common definitions of $DC60$, so we retained that approach here. The $DC60$ values, as well as their corresponding circumferential locations are shown in Table 2 for $U^* = 0.21$ and $U^* = 0.30$. In Figure 15, the absolute values of $DC60$ as U^* varies are shown for both stages. Note that an additional computation was carried out at $U^* = 0.26$, but since the separation did not extend to the fan leading edge for that case, the results were used only to initialize the solutions at $U^* = 0.30$. Thus, this case was not fully run to convergence, so no $DC60$ values can be reported.

Table 2. $DC60$ values for crosswind velocities just before separation and with strong separation.

	NASA Stage 67	Simplified Stage
$DC60_{U^*=0.21}$	−0.0583	−0.0479
$\theta_{U^*=0.21}$ Range (degrees)	−32 to 28	−31 to 29
$DC60_{U^*=0.30}$	−0.3461	−0.3776
$\theta_{U^*=0.30}$ Range (degrees)	−25 to 35	−19 to 41

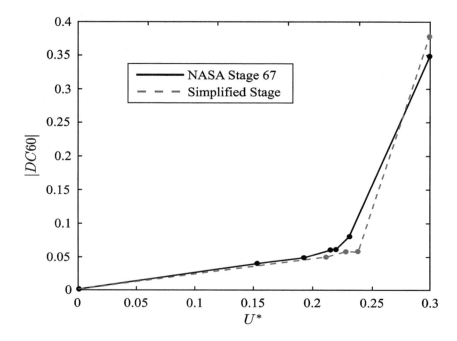

Figure 15. $DC60$ comparison between NASA Stage 67 and the simplified stage as crosswind velocity varies.

The largest difference in the $DC60$ prediction between the two stages occurred at $U^* = 0.30$ where the simplified stage over-predicted $DC60$ by 9%. The simplified stage had a greater rise in $DC60$ magnitude when $U^* = 0.30$ due to the circumferential thickness of the separated region. Although the simplified stage under-predicted the maximum decrease in stagnation pressure at the fan face, as well as the axial length of the separated region, the increase in circumferential extent of the separated region caused the area averaged stagnation pressure of the $60°$ sector to decrease, which led to the higher $DC60$ magnitude. Figure 15 shows that the effect of stage simplification on the prediction of $DC60$ was minor over a wide range of U^* with the only major difference being that the larger crosswind separation velocity caused the sharp increase in $DC60$ to be delayed by $U^* = 0.01$.

5. Summary and Conclusions

In this paper, the simplified fan stage gas path and body force model design approach described in Part 1 [6] were applied to predict crosswind separation velocities, as well as the associated effects on fan and nacelle performance. A stage based on NASA Stage 67 was generated. A comparison between this simplified stage and NASA Stage 67 was done with uniform inflow. The simplified stage produced a 1.14% higher FPR and approximately a 5% difference in the other variables of interest, such as the thrust and overall work coefficient. The stage produced did not match the desired performance as closely as the example fan stage created in Part 1 due to the increase in gas path complexity. It nevertheless allowed for the effects of simplifying the stage to be seen and provided a foundation for investigating non-uniform inflow without detailed stage information.

The intended use for the process described in Part 1 [6] is to allow for assessing external flow–fan interactions when limited fan information is available. This is useful, for example, when airframers wish to investigate coupling between non-uniform inflows caused by off-design operation and the fan. The simplified stage and the original NASA Stage 67 were inserted into a nacelle. Both stages were run in full wheel body force simulations with varying crosswind speeds. The results were compared, and the impacts of simplifying the stage design were investigated. The simplified stage was able to predict the separation velocity to within $5.1 \pm 1.6\%$. The Mach number in the crosswind plane was examined, and although the height and width of the recirculation region decreased in the simplified stage, the overall flow showed the same trends. Flow at the fan leading edge was reproduced well, with the minimum fan stagnation pressure and its location varying by 3.4% and 5.8% respectively between the two stages. The simplified stage can estimate the effect of non-uniformities on the fan, as it captured the maximum change in incidence angle to within $4.8°$. To measure the nacelle performance, the $DC60$ metric was used; the simplified stage showed a maximum difference in $DC60$ when $U^* = 0.30$, with an increase of 9%.

The intended level of fidelity of this process was such that it could be used for early design stages or in the case where no detailed geometry is available; the quantities examined in this paper showed that the simplified stage was capable of reproducing the overall flow found when detailed geometry was used. We showed that the simplified fan model worked well despite changes in blade shape and rotational speed compared to Stage 67. This supports the utility of the approach since the simplified model was intended to be used when the fan rotational speed may not be known. The results also suggested that this process would work under other non-uniform inflow conditions, for example a nacelle with forward movement at the angle of attack. With an angle of attack, U^* would be larger when separation occurred, and the simplified stage reproduced the flow more accurately before the flow separated, meaning that the case modelled in this paper, pure crosswind, was one of the most challenging scenarios possible for assessing the fan–nacelle interaction. The goodness of agreement obtained despite the significant differences in the fan geometries suggested that an even simpler fan model may in some cases be adequate for capturing fan–nacelle interaction. Determining exactly what the limits are for good agreement is beyond the scope of this paper, but may be an interesting topic for future research.

Author Contributions: Conceptualization, Q.J.M. and J.J.D.; data curation, Q.J.M.; formal analysis, Q.J.M.; funding acquisition, J.J.D.; investigation, Q.J.M.; methodology, Q.J.M. and J.J.D.; project administration, J.J.D.; resources, J.J.D.; software, Q.J.M.; supervision, J.J.D.; validation, Q.J.M.; visualization, Q.J.M.; writing, original draft, Q.J.M.; writing, review and editing, J.J.D.

Acknowledgments: Computational resources were provided by the facilities of the Shared Hierarchical Academic Research Computing Network (SHARCNET) (www.sharcnet.ca) and Compute/Calcul Canada (www.computecanada.ca). The authors would like to thank Ewan Gunn for the stage 67 stator model used in this work. The authors would also like to thank Bombardier Aerospace for providing the generic nacelle geometry used in this work.

Abbreviations

The following abbreviations are used in this manuscript:

Symbols

A	Area
a	Location of maximum camber
B	Number of blades in a row
b	Blade span
c	Blade chord
c_p	Pressure coefficient
$DC60$	Distortion coefficient based on $60°$-sector
e	Energy
F	Thrust
f	Force
h	Enthalpy
i	Incidence angle
K,K'	Compressibility correction
L	Length
M	Mach number
$\dot{m}$	Mass flow rate
$\hat{n}$	Blade camber surface normal unit vector
p	Pressure
R	Gas constant
r	Radius
s	Spacing between blades
T	Temperature
U	Blade speed
U^*	Velocity ratio
$\vec{V}$	Velocity
$\vec{W}$	Relative velocity
α	Absolute flow angle
γ	Specific heat ratio
δ	Deviation
η	Efficiency
ξ	Change in blade angle
ρ	Density
τ	Shear stress
Y	Tip radius change parameter
ϕ	Flow coefficient
ψ	Work coefficient
ψ_{pt}	Stagnation pressure rise coefficient
ω	Angular velocity

Subscripts

i	Incompressible
LE	Blade leading edge
m	Meridional component
mid	Midspan location
R	Rotor
S	Stator
TE	Blade trailing edge
t	Stagnation quantity
x	Axial component
y	Crosswind direction quantity
2	Fan face quantity

13	Fan trailing edge quantity
19	Nozzle exit quantity
θ	Tangential component
∞	Freestream quantity

Abbreviations

RANS	Reynold averaged Navier–Stokes
RMS	Root mean squared
FPR	Fan stagnation pressure ratio

References

1. Peters, A.; Spakovszky, Z.S.; Lord, W.K.; Rose, B. Ultrashort Nacelles for Low Fan Pressure Ratio Propulsors. *J. Turbomach.* **2015**, *137*, 021001. [CrossRef]

2. Gunn, E.J.; Hall, C.A. Aerodynamics of Boundary Layer Ingesting Fans. In Proceedings of the ASME Turbo Expo 2014, Dusseldorf, Germany, 16–20 June 2014. [CrossRef]

3. Fidalgo, V.; Hall, C.; Colin, Y. A Study of Fan-Distortion Interaction Within the NASA Rotor 67 Transonic Stage. *J. Turbomach.* **2012**, *134*, 051011. [CrossRef]

4. Godard, B.; Jaeghere, E.; Nasr, N.; Marty, J.; Barrier, R.; Gourdain, N. Methodologies for Turbofan Inlet Aerodynamics Prediction. In Proceedings of the 35th AIAA Applied Aerodynamics Conference, Denver, CO, USA, 5–9 June 2017. [CrossRef]

5. Burlot, A.; Sartor, F.; Vergez, M.; Méheut, M.; Barrier, R. Method Comparison for Fan Performance in Short Intake Nacelle. In Proceedings of the 2018 Applied Aerodynamics Conference, Atlanta, GA, USA, 25–29 June 2018. [CrossRef]

6. Minaker, Q.; Defoe, J. Prediction of Crosswind Separation Velocity for Fan and Nacelle Systems Using Body Force Models: Part 1: Fan Body Force Model Generation Without Detailed Stage Geometry. *Int. J. Turbomach. Propuls. Power* **2019**, *4*, 43. [CrossRef]

7. Hall, D.; Greitzer, E.; Tan, C. Analysis of Fan Stage Conceptual Design Attributes for Boundary Layer Ingestion. *J. Turbomach.* **2017**, *139*, 071012. [CrossRef]

8. Yeung, A.; Vadlamani, N.R.; Hynes, T. Quasi 3D Nacelle Design to Simulate Crosswind Flows: Merits and Challenges. *Int. J. Turbomach. Propuls. Power* **2019**, *3*, 25. [CrossRef]

9. Lee, K.; Wilson, M.; Vahdati, M. Effects of Inlet Disturbances on Fan Stability. In Proceedings of the ASME Turbo Expo 2018, Oslo, Norway, 11–15 June 2018. [CrossRef]

10. Strazisar, A.; Wood, J.; Hathaway, M.D.; Suder, K. *Laser Anemometer Measurements in a Transonic Axial-Flow Fan Rotor*; Techreport 2879; NASA: Washington, DC, USA, 1989.

11. ANSYS Inc. *ANSYS v18.2 User's Guide*; ANSYS Inc.: Canonsburg, PA, USA, 2017.

12. Pointwise. *Pointwise v18.0R4 User's Manual*; Pointwise: Forth Worth, TX, USA, 2017.

13. Roache, P. *Fundamentals of Computational Fluid Dynamics*; Hermosa Publishers: Socorro, NM, USA, 1998.

Cattaneo–Christov Heat Flux Model for Three-Dimensional Rotating Flow of SWCNT and MWCNT Nanofluid with Darcy–Forchheimer Porous Medium Induced by a Linearly Stretchable Surface

Zahir Shah [1], Asifa Tassaddiq [2], Saeed Islam [1], A.M. Alklaibi [3] and Ilyas Khan [4,*]

[1]   Department of Mathematics, Abdul Wali Khan University, Mardan 23200, Pakistan; zahir1987@yahoo.com (Z.S.); saeedislam@awkum.edu.pk (S.I.)

[2]   College of Computer and Information Sciences, Majmaah University, Al-Majmaah 11952, Saudi Arabia; a.tassaddiq@mu.edu.sa

[3]   Department of Mechanical and Industrial Engineering, College of Engineering, Majmaah University, P.O. Box 66 Majmaah 11952, Saudi Arabia; a.alklaibi@mu.edu.sa

[4]   Faculty of Mathematics and Statistics, Ton Duc Thang University, Ho Chi Minh City 72915, Vietnam

*   Correspondence: ilyaskhan@tdt.edu.vn

Abstract: In this paper we investigated the 3-D Magnetohydrodynamic (MHD) rotational nanofluid flow through a stretching surface. Carbon nanotubes (SWCNTs and MWCNTs) were used as nano-sized constituents, and water was used as a base fluid. The Cattaneo–Christov heat flux model was used for heat transport phenomenon. This arrangement had remarkable visual and electronic properties, such as strong elasticity, high updraft stability, and natural durability. The heat interchanging phenomenon was affected by updraft emission. The effects of nanoparticles such as Brownian motion and thermophoresis were also included in the study. By considering the conservation of mass, motion quantity, heat transfer, and nanoparticles concentration the whole phenomenon was modeled. The modeled equations were highly non-linear and were solved using homotopy analysis method (HAM). The effects of different parameters are described in tables and their impact on different state variables are displayed in graphs. Physical quantities like Sherwood number, Nusselt number, and skin friction are presented through tables with the variations of different physical parameters.

Keywords: SWCNTs; MWCNTs; stretched surface; rotating system; nanofluid; MHD; thermal radiation; HAM

1. Introduction

Heat transfer phenomenon is important in manufacturing and life science applications, for example in freezing electronics, atomic power plant refrigeration, tissue heat transfer, energy production, etc. Fluids that flow on a stretched surface are more significant among researchers in fields such as manufacturing and commercial processes, for instance in making and withdrawing polymers and gum pieces, crystal and fiber production, food manufacturing, condensed fluid layers, etc. Considering these applications, heat transfer is an essential subject for further investigation in order to develop solutions to stretched surface fluid film problems. The flow of a liquefied sheet was initially considered to obtain a viscid stream and was further extended to stretched surface for non-Newtonian liquids. Choi [1] examined the enhancement of thermal conductivity in nanoparticles deferrals. For the enhancement of thermal conductivity and heat transfer, Hsiao [2,3] performed a successful survey using the Carreau-Nanofluid and Maxwell models, and obtained some interesting results. Ramasubramaniam et al. [4] treated a homogeneous carbon nanotube composite for electrical purposes. Xue [5] work as presenting a CNT model for grounded compounds. Nasir et al. [6] deliberate the nanofluid tinny liquid flow of SWCNTs using an optimal approach. Ellahi et al. [7] presented the usual transmitting nanofluids based on CNTs. Shah et al. [8,9] investigated nanofluid flow in a rotating frame with microstructural and inertial properties with Hall effects in parallel plates. Hayat and others [10] examined Darcy–Forchheimer flow carbon nanotube flow due to a revolving disk. Recently, scholars have been working on finding a rotational flow close to the flexible or non-expandable geometries due its wide array of uses in rotating-generator systems, food handling, spinning devices, disc cleaners, gas transformer designs, etc. Wang [11] presented a perturbation solution for rotating liquid flow through an elastic sheet. The magnetic flux features of rotating flow above a flexible surface was premeditated by Takhar et al. [12]. Shah et al. [13–15] studied nanofluid and heat transfer with radiative and electrical properties using an optimal approach. Rosali et al. [16] presented a numerical survey for flow with rotation over porous surface with exponential contraction. Hayat et al. [17] used the non-Fourier heat fluctuation hypothesis to get a three-dimensional turning stream of the Jeffrey substances. Mustafa. [18] discussed non-linear aspects of rotating nano-fluid flow through the flexible plane. Sheikholeslami et al. [19] inspected the consistent magnetic and radiated effect on water-based nanofluid in a permeable enclosure. Hsiao [20,21] researched the microploar nanofluid stream with MHD on a stretching surface. Khan et al. [22,23] examined nanofluid of micrpoler fluid with the Darcy–Forchheimer and irregular heat generation/absorption between two plates.

The classical Fourier law of conduction [24] is one best model for explanation of the temperature transmission process under numerous relevant conditions. Cattaneo [25] successfully extended the Fourier model in combining significant properties of the temperature reduction period. Cattaneo's work produced a hyperbolic energy equation for the temperature field which allows heat to be transferred by transmission of heat waves with finite velocity. Heat transfer has many practical applications, from the flow of nanofluids to the simulation of skin burns (see Tibullo and Zampoli [26]). Christov [27] discussed the Cattaneo–Maxwell model for finite heat conduction. Straughan and Ciarletta [28] demonstrated the rareness of the solution of the Cattaneo–Christov equation. Straughan et al. [29] presented the heat transfer analysis for this model with a brief discussion of a solution for the model. Recently, Han et al. [30] deliberated the sliding stream with temperature transmission through the Maxwell fluids for the Christov–Cattaneo model. A numerical comparative survey was also presented for the validation of their described results. The current exploration of nanofluid with entropy analysis can be studied in References [31–36].

This paper is based on the features of the Christov–Cattaneo heat flux in rotating nano liquid. A three-dimensional nanofluid flow is considered over a stretching surface with carbon nanotubes (CNTs). An effective thermal conductivity model was used in the enhancement of heat transfer. The problem was modeled from the schematic diagram with concentration. These modeled equations were transformed into a system of non-linear ordinary differential equations. The modeled equations were coupled and highly non-linear and were tackled by an analytical and numerical approach. Homotopy analysis method (HAM) (a high-precision analytical technique proposed by Liao [37]) was used for the solution of the reduced system. Many researcher [38–46] used HAM due to it to it excellent results. Various parameters are presented via graphs. Different physical parameters (thermal relaxation time, skin friction, etc.) with the variations of other physical constraints are presented via graphs and discussed in detail.

2. Effective Thermal Conductivity Models Available in the Literature

Maxwell's [47] proposed a thermal conductivity model as

$$\frac{k_{nf}}{k_f} = 1 + \frac{3\left(\varsigma - 1\right)\psi}{\left(\varsigma + 2\right) - \left(\varsigma - 1\right)\psi}.$$

(1)

where $\varsigma = \frac{k_{nf}}{k_f}$ and ψ is a volumetric fraction. Also, Jeffery [48] proposed the following model:

$$\frac{k_{nf}}{k_f} = 1 + 3\chi\psi + \left(3\chi^2 + \frac{3\chi^2}{4} + \frac{9\chi^3}{16}\left(\frac{\varsigma+2}{2\varsigma+3}\right) + \ldots\ldots\right)\psi^2. \tag{2}$$

where $\chi = \frac{(\varsigma-1)}{(\varsigma+2)}$. After a little modification, Davis [49] presented a model defined as:

$$\frac{k_{nf}}{k_f} = 1 + \frac{3(\varsigma-1)\,\psi}{(\varsigma+2)-(\varsigma-1)\psi}\left\{\psi + \psi(\varsigma)\psi^2 + O(\psi^3)\right\}. \tag{3}$$

This model gives a good approximation of thermal conductivity even for a very small capacity and is independent of the atom's form.

Hamilton and Crosser [50] presented a particle-form-based model defined as:

$$\frac{k_{nf}}{k_f} = \frac{\varsigma + (\hbar-1) - (\varsigma-1)(\hbar-1)\psi}{\varsigma + (\hbar-1) + (1-\varsigma)\psi}. \tag{4}$$

Here $\hbar$ denotes the particle form used. The main limitation of the models discussed above is that they can only be used for rotating or circular components and cannot be used for CNTs, especially for their spatial distribution. To overcome this deficiency, Xue [5] presented a model of very large axel relation and used it for the spatial distribution of CNTs. This model has a mathematical description given as:

$$\frac{k_{nf}}{k_f} = \frac{1 - \psi + 2\left(\frac{k_{nf}}{k_{nf}-k_f}\ln\frac{k_{nf}+k_f}{2k_f}\right)\psi}{1 - \psi + 2\left(\frac{k_f}{k_{nf}-k_f}\ln\frac{k_{nf}+k_f}{2k_f}\right)\psi}. \tag{5}$$

In the present work we implement the Xue [5] model to calculate thermal conductivity.

3. Formulation of the Problem

A three-dimensional rotational flow of CNTs was carried through a linear flexible surface. The temperature distribution was deliberated by the Xue model [5]. The compact fluid describing the Darcy–Forchheimer relationship saturates the permeable area. The stretching surface was adjusted in the Cartesian plane that plates associated in the xy plane. We assumed only the positive values of liquid for z. Surface is extended in the x-direction with a positive rate c. In addition, the liquid is uniformly rotated at a continuous uniform speed ω around the z-axis. Surface temperature is due to convective heating, which is provided by the high temperature of the fluid T_f. The coefficient of this heat transfer is h_f. The relevant equations after applying assumptions are [13–18]:

$$\vec{u}_x + \vec{v}_y + \vec{w}_z = 0 \tag{6}$$

$$u\vec{u}_x + v\vec{u}_y + w\vec{u}_z - 2\omega v = v_{nf}\vec{u}_{zz} - \frac{v_{nf}}{K^*}u - Fu^2 \tag{7}$$

$$u\vec{v}_x + v\vec{v}_y + w\vec{v}_z - 2\omega u = v_{nf}\vec{v}_{zz} - \frac{v_{nf}}{K^*}v - Fv^2 \tag{8}$$

$$\rho c_p\left(uT_x + vT_y + wT_z\right) = -\nabla \cdot \vartheta \tag{9}$$

$$\vartheta + \lambda_2(\vartheta_t + V.\nabla\vartheta - \vartheta.\nabla V + (\nabla.V)\vartheta) = -k\nabla T \tag{10}$$

$$uT_x + vT_y + wT_z = \frac{k}{\rho c_p}T_{zz} - \lambda_2\left[\begin{array}{c} u^2T_{xx} + v^2T_{yy} + w^2T_{zz} + 2uvT_{xy} + 2vwT_{yz} \\ +2uwT_{xz} + \left(u\vec{u}_x + v\vec{u}_y + w\vec{u}_z\right)T_x \\ +\left(u\vec{v}_x + v\vec{v}_y + w\vec{v}_z\right)T_y + \left(u\vec{w}_x + v\vec{w}_y + w\vec{w}_z\right)T_z \end{array}\right] \tag{11}$$

$$u\,C_x + v\,C_y + w\,C_z = D_B\,C_{zz} + \frac{D_T}{T_0}\,T_{zz}. \tag{12}$$

The related boundary conditions are:

$$\begin{aligned} u = u_w(x) = cx, \quad v = 0, \quad w = 0, \quad -k_{nf}T_z = h_f\left(T_f - T\right), \\ -k_{nf}C_z = h_f\left(C_f - C\right) \quad at \quad z = 0 \\ u \to 0, \quad v \to 0, \quad T \to T_\infty, \quad C \to C_\infty \quad as \quad z \to \infty \end{aligned} \tag{13}$$

K is the permeability, $F = \frac{C_b}{xK^{*1/2}}$ is the irregular inertial coefficient of the permeable medium, C_b represents drag constant and T_∞ represents the ambient fluid temperature. The basic mathematical features of CNTs are [5]:

$$\begin{aligned} \mu_{nf} = \frac{\mu_f}{(1-\phi)^{2.5}}, \quad v_{nf} = \frac{\mu_{nf}}{\rho_{nf}}, \quad \alpha_{nf} = \frac{k_{nf}}{(\rho c_p)_{nf}}, \quad \rho_{nf} = \rho_f(1-\phi) + \rho_{CNT}\phi, \\ (\rho c_p)_{nf} = (\rho c_p)_f(1-\phi) + (\rho c_p)_{CNT}\phi, \quad \frac{k_{nf}}{k_f} = \frac{(1-\phi)+2\phi\frac{k_{CNT}}{k_{CNT}-k_f}\ln\frac{k_{CNT}+k_f}{2k_f}}{(1-\phi)+2\phi\frac{k_f}{k_{CNT}-k_f}\ln\frac{k_{CNT}+k_f}{2k_f}} \end{aligned} \tag{14}$$

Transformations are taken as follows:

$$\begin{aligned} u = cxf'(\eta), \quad v = cxg(\eta), \quad w = -(cv_f)^{1/2}f(\eta), \\ \Theta(\eta) = \frac{T-T_\infty}{T_f-T_\infty}, \quad \Phi(\eta) = \frac{C-C_\infty}{C_f-C_\infty}, \quad \eta = \left(\frac{c}{v_f}\right)^{1/2}z. \end{aligned} \tag{15}$$

Now Equation (6) is identically satisfied and Equations (7), (8), (11)–(13) were reduced to

$$\frac{1}{(1-\phi)^{2.5}\left(1-\phi+\frac{\rho_{CNT}}{\rho_f}\phi\right)}\left(f''' - \lambda f'\right) + ff'' + 2Kg - (1+F_r)f'^2 = 0 \tag{16}$$

$$\frac{1}{(1-\phi)^{2.5}\left(1-\phi+\frac{\rho_{CNT}}{\rho_f}\phi\right)}\left(g'' - \lambda g\right) + fg' - f'g - 2Kf' - F_rg^2 = 0 \tag{17}$$

$$\frac{k_{nf}}{k_f}\Theta'' + Pr\left[(1-\varphi)+\frac{(\rho c_p)_{nf}}{(\rho c_p)_f}\right]\left[\Theta'(\Phi')Nb + (\Theta')^2Nt + f\Theta'\right] = 0 \tag{18}$$

$$\Phi'' + Scf\Phi' + \frac{Nt}{Nb}\Theta'' = 0. \tag{19}$$

$$\begin{aligned} f = 0, \; f' = 1, \; g = 0, \; \Theta' = -\frac{k_f}{k_{nf}}\gamma(1-\Theta), \; \Phi' = \frac{k_f}{k_{nf}}\gamma(1-\Phi) \; at \; \eta = 0 \\ f' \to 0, \; g \to 0, \; \Theta \to 0, \; \Phi \to 0, \; when \; \eta \to \infty \end{aligned} \tag{20}$$

The dimensionless parameters are defined as

$$\begin{aligned} \lambda = \frac{v_f}{cK^*}, \quad F_r = \frac{C_b}{K^{*1/2}}, \quad K = \frac{\omega}{c}, \quad Pr = \frac{v_f}{\alpha_f}, \quad \gamma = \frac{h_f}{k_f}\sqrt{\frac{v_f}{c}}, Sc = \frac{v_f}{D_B}, \\ Pr = \frac{(\rho\,c_p)_f}{k_f}, Nb = \frac{\tau\,D_B\,(C_f-C_\infty)}{v_f}, Nt = \frac{\tau\,D_T\,(T_f-T_\infty)}{T_0v_f}. \end{aligned} \tag{21}$$

where λ represents porosity K rotation parameter, F_r signifies coefficient of is the inertia, Pr, signifies Prandtl number, Nb is the parameter of Brownian motion, Sc signifies Schmidt number, and γ is Biot number and Nt is the thermophoresis parameter which are defined in Equation (21).

4. Results and Discussion

3-D Magnetohydrodynamic (MHD) rotational nanofluid flow through a stretching surface is modeled. The Cattaneo–Christov heat flux model was used for heat transport phenomenon. By considering the conservation of mass, motion quantity, heat transfer, and nanoparticles concentration the whole phenomenon was modeled. The modeled equations were solved using homotopy analysis method (HAM). Figure 1 Show the geometry of the flow pattern.

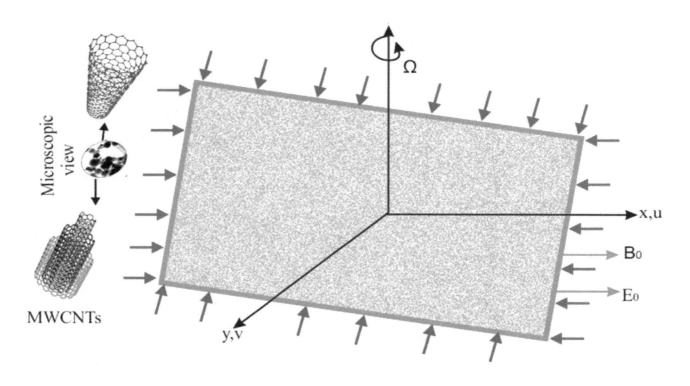

Figure 1. Schematic physical geometry.

Figure 2a–d presents the impact of λ on $f'(\eta)$, $g(\eta)$ & $\Theta(\eta)$ and Biot number γ on $\Theta(\eta)$. Figure 2a displays the deviation of $f'(\eta)$ for different numbers of λ. It was observed that greater porosity parameter λ values indicate a decline in velocity field $f'(\eta)$. Figure 2b reflects the $g(\eta)$ for dissimilar values of the permeability constraint λ. It is detected that for greater permeability constraint λ, the velocity field $g(\eta)$ increased. Figure 2c shows the impact of permeability parameters λ on $\Theta(\eta)$. It is observed that $\Theta(\eta)$ enhanced by increasing the permeability constraint λ for SWCNTs and MWCNTs. Figure 2d illustrates that a greater Biot number γ yields stronger convection, which results in a greater temperature field $\Theta(\eta)$ and hotter sheet wideness.

Figure 3a–d present the impact of K on $f'(\eta)$, $g(\eta)$, $\Theta(\eta)$, and Biot number γ on $\Phi(\eta)$. Figure 3a shows in what way the K affects $f'(\eta)$. A rise in K produced a lesser velocity field $f'(\eta)$ and a smaller momentum sheet wideness of the SWCNTs and MWCNTs. Greater rotational parameter K values resulted in greater rotational rates than tensile rates. Thus, a greater turning effect relates to inferior velocity field $f'(\eta)$ and smaller momentum sheet wideness. Figure 3b describes $g(\eta)$ for K. Larger values of the rotation parameter K, caused a decrease in the velocity field $g(\eta)$. Figure 3c illustrates $\Theta(\eta)$ variations for dissimilar values of K. Greater rotational parameter K decreases the temperature field $\Theta(\eta)$ and supplementary thermal layer width. Figure 3d demonstrates the concentration distribution $\Phi(\eta)$ for varying Biot numbers γ. Higher values of γ indicate enhancement in $\Phi(\eta)$.

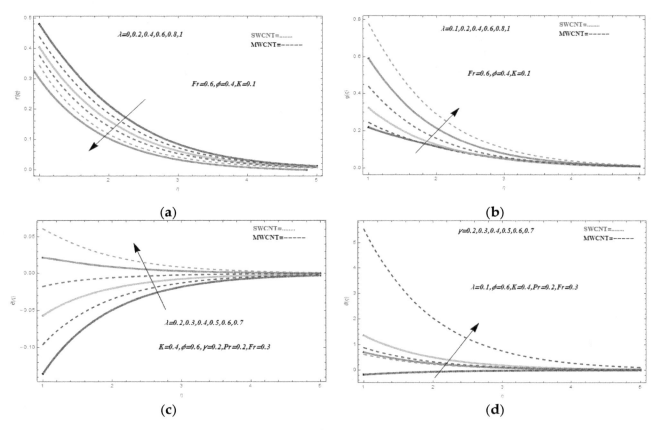

Figure 2. Impression of λ on $f'(\eta)$, $g(\eta)$ & $\Theta(\eta)$, and γ on $\Theta(\eta)$.

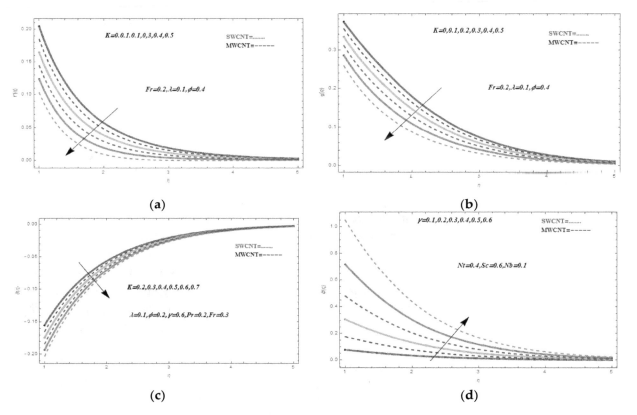

Figure 3. Impression of K on $f'(\eta)$, $g(\eta)$ & $\Theta(\eta)$, and γ on $\Phi(\eta)$.

The influences of inertia coefficient F_r on $f'(\eta)$, $g(\eta)$, $\Theta(\eta)$, and Prandtl number P_r on $\Theta(\eta)$ are shown in Figure 4a–d. Figure 4a shows the inertia coefficient F_r on $f'(\eta)$. It is observed that greater values of inertia coefficients F_r resulted in the decline of $f'(\eta)$. Figure 4b depicts the effect of inertia coefficient F_r over the velocity field $g(\eta)$. For greater inertia coefficients F_r of SWCNTs and MWCNTs, there is an increase in the velocity field $g(\eta)$. The influence of the inertia constant F_r on $\Theta(\eta)$ is shown in Figure 4c. Greater rates of inertia factor F_r resulted in powerful temperature field $\Theta(\eta)$ and additional thermal layer thicknesses for SWCNTs and MWCNTs. Figure 4d shows that greater Prandtl number P_r resulted in the decline of the temperature field $\Theta(\eta)$ of SWCNTs and MWCNTs.

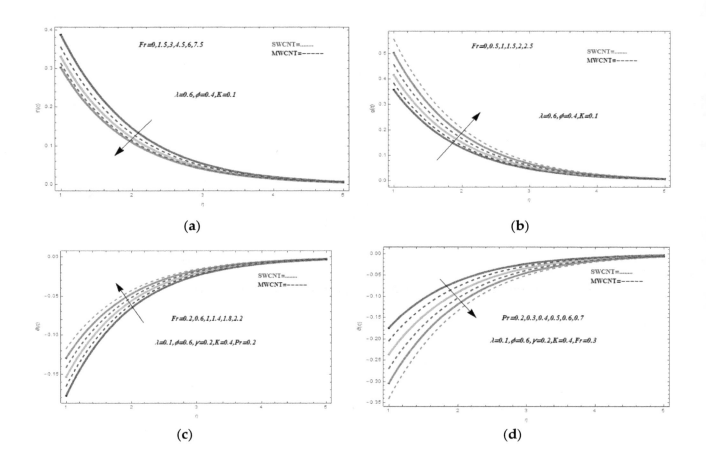

Figure 4. Impression of F_r on $f'(\eta)$, $g(\eta)$, $\Theta(\eta)$, and P_r on $\Theta(\eta)$.

The influences of nanoparticle capacity fraction ϕ on $f'(\eta)$, $g(\eta)$ & $\Theta(\eta)$, Sc on $\Phi(\eta)$ are shown in Figure 5a–d. Figure 5a demonstration the modification in $f'(\eta)$ of the changing nanoparticle capacity fraction ϕ. It was noted that with the rise of the nanoparticle capacity fraction ϕ, an increase $f'(\eta)$ is observed. Results of nanoparticle capacity fraction ϕ on the $g(\eta)$ is shown in Figure 5b. The higher values of the nanoparticle capacity fraction ϕ caused a decreases $g(\eta)$. Figure 5c represents $\Theta(\eta)$ for different nanoparticles volume fraction ϕ. It is observed that greater nanoparticle capacity fraction ϕ resulted in the decline of the temperature field $\Theta(\eta)$. Figure 5d displays the consequence of Sc on $\Phi(\eta)$ of the nanoparticles. It is noticed that an increase in Sc caused a decline in $\Phi(\eta)$.

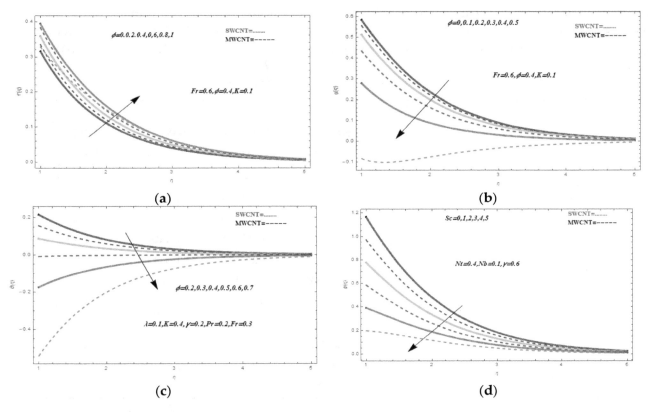

Figure 5. Impression of ϕ on $f'(\eta)$, $g(\eta)$, $\Theta(\eta)$ and Sc on $\Phi(\eta)$.

Figure 6a depicts the concentration distribution $\Phi(\eta)$ for dissimilar values of thermophoretic parameter Nt for SWCNTs and MWCNTs. Higher values of Nt designate the augmentation in $\Phi(\eta)$. Figure 6b depicts the concentration distribution $\Phi(\eta)$ for the varying Brownian motion parameter Nb of SWCNTs and MWCNTs. We noted that greater values of Nb show a reduction in $\Phi(\eta)$ and the connected boundary film thickness.

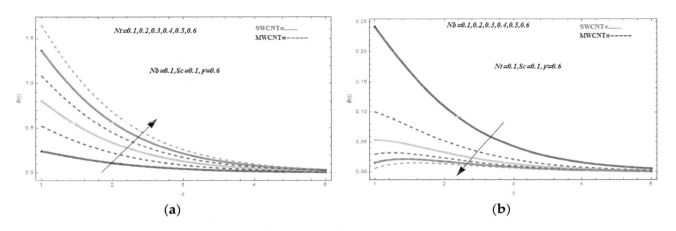

Figure 6. Impression of Nt on $\Theta(\eta)$ and Nb. on $\Phi(\eta)$.

4.1. Table Discussion

Physical values of skin friction for dissimilar values of SWCNTs and MWCNTs in the case of different parameters for C_{fx} and C_{fy} are calculated numerically in Table 1. It was perceived that amassed values of F_r, λ and γ increasing C_{fx} and C_{fy} for SWCNTs nanofluid. Similar results were obtained for MWCNTs. The higher value of K reduces C_{fx} and C_{fy} for SWCNT nanofluid, while for MWCNTs the result was opposite. Physical values for the heat and mass fluxes for dissimilar parameters at $Pr = 7.0$ are calculated in Table 2. Greater values of F_r and K augmented the heat flux as well as the mass flux for both SWCNTs and MWCNTs. The higher value of Nt and Nb reduced the heat flux as well as the mass flux while increasing γ decreased it for both SWCNTs and MWCNTs.

Table 1. Variation in skin friction.

F_r	K	λ	γ	$-C_{fx}$		$-C_{fy}$	
				SWCNT	MWCNT	SWCNT	MWCNT
0.0	0.1	0.1	0.1	1.08621	1.03466	1.24014	1.14536
0.1	–	–	–	1.15376	1.12687	1.27592	1.19880
0.3	–	–	–	1.19736	1.14350	1.34626	1.24368
0.1	0.0	–	–	1.19574	1.04576	1.19574	1.13667
–	0.3	–	–	0.98487	0.23571	1.44073	1.17462
–	0.5	–	–	0.85351	0.02547	1.61115	1.32445
–	0.1	0.0	–	0.96939	0.23419	1.10121	1.03462
–	–	0.3	–	1.54123	1.45483	1.23261	1.100344
–	–	0.5	–	1.68412	1.52445	1.69793	1.37352
–	–	0.1	0.1	1.69222	1.57245	1.72350	1.22359
–	–	–	0.3	1.73432	1.62355	1.88193	1.31912
–	–	–	0.5	1.92365	1.81199	1.90843	1.59331

Table 2. Variation in Nusselt number and Sherwood Number at $Pr = 7.0$.

F_r	K	Nt	Nb	γ	$-Nu_x$		$-Sh_x$	
					SWCNT	MWCNT	SWCNT	MWCNT
0.0	0.1	0.1	0.1	0.1	0.116964	0.231567	0.120208	0.243362
0.1	–	–	–	–	0.116964	0.231390	0.120212	0.243021
0.3	–	–	–	–	0.116965	0.233321	0.120212	0.243204
0.1	0.0	–	–	–	0.116965	0.134136	0.120202	0.243204
–	0.3	–	–	–	0.116966	0.134342	0.120196	0.243192
–	0.5	–	–	–	0.116967	0.134351	0.128786	0.243145
–	0.1	0.3	–	–	0.116953	0.261532	0.13737	0.243145
–	–	0.5	–	–	0.116943	0.261531	0.15026	0.950126
–	–	0.8	–	–	0.116926	0.261530	0.118472	0.950139
–	–	0.1	0.5	–	0.118451	0.156382	0.116457	0.936457
–	–	–	1.0	–	0.120623	0.234521	0.114607	0.934607
–	–	–	1.5	–	0.122502	0.267373	0.11678	0.932678
–	–	–	0.1	0.5	0.116945	0.234536	0.116352	0.566352
–	–	–	–	1.0	0.116928	0.198342	0.116209	0.566209
–	–	–	–	1.5	0.116895	0.162455	0.116198	0.526198

5. Conclusions

Three-dimensional MHD rotational flow of nanofluid over a stretching surface with Cattaneo–Christov heat flux was numerically investigated. Nanofluid is formed as a suspension of SWCNTs and MWCNTs. The modeled equations under different physical parameters were analyzed via graphs for SWNTs. The following main points were concluded from this work.

- It was observed that greater porosity parameter λ values reduced $f'(\eta)$ while increasing $g(\eta)$ and also increasing temperature field $\Theta(\eta)$ for SWCNTs and MWCNTs.

- Greater Biot number γ yields stronger convection, which results in a greater temperature field $\Theta(\eta)$ and hotter sheet wideness.

- Greater rotational parameter K values resulted in greater rotational rates than tensile rates. The higher value of K increased the fluid velocity.

- It was observed that greater values of inertia coefficients F_r resulted in the decline of the velocity field.

- The higher value of the nanoparticle capacity fraction ϕ increased the velocity field and decreased temperature.

- Higher values of Nt indicated enhancement in $\Phi(\eta)$.

- Greater values of Nb showed reduction in $\Phi(\eta)$.

- Larger values of Sc demoted nanoparticle concentration $\Phi(\eta)$.

- It was perceived that increasing values of F_r, λ, and γ increased C_{fx} and C_{fy} for SWCNT nanofluid. Similar results were obtained for MWCNTs.

- The higher values of Nt and Nb reduced the heat flux as well as the mass flux, while increasing γ decreased it for both SWCNTs and MWCNTs.

Author Contributions: Z.S. and S.I. modeled the problem and wrote the manuscript. A.T. and I.K. thoroughly checked the mathematical modeling and English corrections. Z.S. and A.M.A. solved the problem using Mathematica software. I.K., S.A. and A.T. contributed to the results and discussions. All authors finalized the manuscript after its internal evaluation.

References

1. Choi, S.U.S.; Zhang, Z.G.; Yu, W.; Lockwood, F.E.; Grulke, E.A. Anomalous thermal conductivity enhancement in nanotube suspensions. *Appl. Phys. Lett.* **2001**, *79*, 2252–2254. [CrossRef]

2. Hsiao, K. To promote radiation electrical MHD activation energy thermal extrusion manufacturing system efficiency by using Carreau-Nanofluid with parameters control method. *Energy* **2017**, *130*, 486–499. [CrossRef]

3. Hsiao, K. Combined Electrical MHD Heat Transfer Thermal Extrusion System Using Maxwell Fluid with Radiative and Viscous Dissipation Effects. *Appl. Therm. Eng.* **2017**, *112*, 1281–1288. [CrossRef]

4. Shah, Z.; Dawar, A.; Islam, S.; Khan, I.; Ching, D.L.C. Darcy-Forchheimer Flow of Radiative Carbon Nanotubes with Microstructure and Inertial Characteristics in the Rotating Frame. *Case Stud. Therm. Eng.* **2018**, *12*, 823–832. [CrossRef]

5. Xue, Q.Z. Model for thermal conductivity of carbon nanotube-based composites. *Phys. B Condens. Matter* **2005**, *368*, 302–307. [CrossRef]

6. Nasir, N.; Shah, Z.; Islam, S.; Bonyah, E.; Gul, T. Darcy Forchheimer nanofluid thin film flow of SWCNTs and heat transfer analysis over an unsteady stretching sheet. *AIP Adv.* **2019**, *9*, 015223. [CrossRef]

7. Ellahi, R.; Hassan, M.; Zeeshan, A. Study of natural convection MHD nanofluid by means of single and multi-walled carbon nanotubes suspended in a salt-water solution. *IEEE Trans. Nanotechnol.* **2015**, *14*, 726–734. [CrossRef]

8. Shah, Z.; Islam, S.; Ayaz, H.; Khan, S. The electrical MHD and hall current impact on micropolar nanofluid flow between rotating parallel plates. *Results Phys.* **2018**, *9*, 1201–1214. [CrossRef]

9. Shah, Z.; Islam, S. Radiative heat and mass transfer analysis of micropolar nanofluid flow of Casson fluid between two rotating parallel plates with effects of Hall current. *ASME J. Heat Transf.* **2019**, *141*, 022401. [CrossRef]

10. Hayat, T.; Haider, F.; Muhammad, T.; Alsaedi, A. On Darcy-Forchheimer flow of carbon nanotubes due to a rotating disk. *Int. J. Heat Mass Transf.* **2017**, *112*, 248–254. [CrossRef]

11. Wang, C.Y. Stretching a surface in a rotating fluid. *Zeitschrift für angewandte Mathematik und Physik ZAMP* **1988**, *39*, 177–185. [CrossRef]

12. Takhar, H.S.; Chamkha, A.J.; Nath, G. Flow and heat transfer on a stretching surface in a rotating fluid with a magnetic field. *Int. J. Therm. Sci.* **2003**, *42*, 23–31. [CrossRef]

13. Shah, Z.; Bonyah, E.; Islam, S.; Gul, T. Impact of thermal radiation on electrical mhd rotating flow of carbon nanotubes over a stretching sheet. *Aip Adv.* **2019**, *9*, 015115. [CrossRef]

14. Shah, Z.; Dawar, A.; Islam, S.; Khan, I.; Ching, D.L.C.; Khan, Z.A. Cattaneo-Christov model for Electrical MagnetiteMicropoler Casson Ferrofluid over a stretching/shrinking sheet using effective thermal conductivity model. *Case Stud. Therm. Eng.* **2018**. [CrossRef]

15. Shah, Z.; Gul, T.; Khan, A.M.; Ali, I.; Islam, S. Effects of hall current on steady three dimensional non-newtonian nanofluid in a rotating frame with brownian motion and thermophoresis effects. *J. Eng. Technol.* **2017**, *6*, 280–296.

16. Rosali, H.; Ishak, A.; Nazar, R.; Pop, I. Rotating flow over an exponentially shrinking sheet with suction. *J. Mol. Liquids* **2015**, *211*, 965–969. [CrossRef]

17. Hayat, T.; Qayyum, S.; Imtiaz, M.; Alsaedi, A. Three-dimensional rotating flow of Jeffrey fluid for Cattaneo-Christov heat flux model. *AIP Adv.* **2016**, *6*, 025012. [CrossRef]

18. Mustafa, M.; Mushtaq, A.; Hayat, T.; Alsaedi, A. Rotating flow of magnetite-water nanofluid over a stretching surface inspired by non-linear thermal radiation. *PLoS ONE* **2016**, *11*, e0149304. [CrossRef] [PubMed]

19. Sheikholeslami, M.; Shah, Z.; Shafi, A.; Khan, I.; Itili, I. Uniform magnetic force impact on water based nanofluid thermal behavior in a porous enclosure with ellipse shaped obstacle. *Sci. Rep.* **2019**, *9*, 1196. [CrossRef] [PubMed]

20. Hsiao, K. Micropolar Nanofluid Flow with MHD and Viscous Dissipation Effects Towards a Stretching Sheet with Multimedia Feature. *Int. J. Heat Mass Transf.* **2017**, *112*, 983–990. [CrossRef]

21. Hsiao, K. Stagnation Electrical MHD Nanofluid Mixed Convection with Slip Boundary on a Stretching Sheet. *Appl. Therm. Eng.* **2016**, *98*, 850–861. [CrossRef]

22. Khan, A.; Shah, Z.; Islam, S.; Khan, S.; Khan, W.; Khan, Z.A. Darcy–Forchheimer flow of micropolar nanofluid between two plates in the rotating frame with non-uniform heat generation/absorption. *Adv. Mech. Eng.* **2018**, *10*, 1–16. [CrossRef]

23. Khan, A.; Shah, Z.; Islam, S.; Dawar, A.; Bonyah, E.; Ullah, H.; Khan, Z.A. Darcy-Forchheimer flow of MHD CNTs nanofluid radiative thermal behaviour andconvective non uniform heat source/sink in the rotating frame with microstructureand inertial characteristics. *AIP Adv.* **2018**, *8*, 125024. [CrossRef]

24. Fourier, J. Theorie Analytique de la Chaleur, par M. Fourier; Chez Firmin Didot, père et Fils. 1822. Available online: https://gallica.bnf.fr/ark:/12148/bpt6k1045508v.texteImage (accessed on 1 January 2019).

25. Cattaneo, C. Sulla conduzione del calore. *Atti Sem. Mat. Fis. Univ. Modena* **1948**, *3*, 83–101.

26. Tibullo, V.; Zampoli, V. A uniqueness result for the Cattaneo–Christov heat conduction model applied to incompressible fluids. *Mech. Res. Commun.* **2011**, *38*, 77–79. [CrossRef]

27. Christov, C.I. On frame indifferent formulation of the Maxwell–Cattaneo model of finite-speed heat conduction. *Mech. Res. Commun.* **2009**, *36*, 481–486. [CrossRef]

28. Straughan, B. Thermal convection with the Cattaneo–Christov model. *Int. J. Heat Mass Transf.* **2010**, *53*, 95–98. [CrossRef]

29. Ciarletta, M.; Straughan, B. Uniqueness and structural stability for the Cattaneo–Christov equations. *Mech. Res. Commun.* **2010**, *37*, 445–447. [CrossRef]

30. Han, S.; Zheng, L.; Li, C.; Zhang, X. Coupled flow and heat transfer in viscoelastic fluid with Cattaneo–Christov heat flux model. *Appl. Math. Lett.* **2014**, *38*, 87–93. [CrossRef]

31. Feroz, N.; Shah, Z.; Islam, S.; Alzahrani, E.O.; Khan, W. Entropy Generation of Carbon Nanotubes Flow in a Rotating Channel with Hall and Ion-Slip Effect Using Effective Thermal Conductivity Model. *Entropy* **2019**, *21*, 52. [CrossRef]

32. Alharbi, S.O.; Dawar, A.; Shah, Z.; Khan, W.; Idrees, M.; Islam, S.; Khan, I. Entropy Generation in MHD Eyring–Powell Fluid Flow over an Unsteady Oscillatory Porous Stretching Surface under the Impact of Thermal Radiation and Heat Source/Sink. *Appl. Sci.* **2018**, *8*, 2588. [CrossRef]

33. Khan, N.S.; Shah, Z.; Islam, S.; Khan, I.; Alkanhal, T.A.; Tlili, I. Entropy Generation in MHD Mixed Convection Non-Newtonian Second-Grade Nanoliquid Thin Film Flow through a Porous Medium with Chemical Reaction and Stratification. *Entropy* **2019**, *21*, 139. [CrossRef]

34. Ishaq, M.; Ali, G.; Shah, Z.; Islam, S.; Muhammad, S. Entropy Generation on Nanofluid Thin Film Flow of Eyring–Powell Fluid with Thermal Radiation and MHD Effect on an Unsteady Porous Stretching Sheet. *Entropy* **2018**, *20*, 412. [CrossRef]

35. Dawar, A.; Shah, Z.; Khan, W.; Idrees, M.; Islam, S. Unsteady squeezing flow of MHD CNTS nanofluid in rotating channels with Entropy generation and viscous Dissipation. *Adv. Mech. Eng.* **2019**, *10*, 1–18. [CrossRef]

36. Sheikholeslami, M.; Shah, Z.; Tassaddiq, A.; Shafee, A.; Khan, I. Application of Electric Field for Augmentation of Ferrofluid Heat Transfer in an Enclosure including Double Moving Walls. *IEEE Access* **2019**. [CrossRef]

37. Liao, S. Notes on the homotopy analysis method: Some definitions and theorems. *Commun. Nonlinear Sci. Numer. Simul.* **2009**, *14*, 983–997. [CrossRef]

38. Shah, Z.; Bonyah, E.; Islam, S.; Khan, W.; Ishaq, M. Radiative MHD thin film flow of Williamson fluid over an unsteady permeable stretching. *Heliyon* **2018**, *4*, e00825. [CrossRef] [PubMed]

39. Jawad, M.; Shah, Z.; Islam, S.; Islam, S.; Bonyah, E.; Khan, Z.A. Darcy-Forchheimer flow of MHD nanofluid thin film flow with Joule dissipation and Navier's partial slip. *J. Phys. Commun.* **2018**. [CrossRef]

40. Khan, N.; Zuhra, S.; Shah, Z.; Bonyah, E.; Khan, W.; Islam, S. Slip flow of Eyring-Powell nanoliquid film containing graphene nanoparticles. *AIP Adv.* **2018**, *8*, 115302. [CrossRef]

41. Fiza, M.; Islam1, S.; Ullah, H.; Shah, Z.; Chohan, F. An Asymptotic Method with Applications to Nonlinear Coupled Partial Differential Equations. *Punjab Univ. J. Math.* **2018**, *50*, 139–151.

42. Ali, A.; Sulaiman, M.; Islam, S.; Shah, Z.; Bonyah, E. Three-dimensional magnetohydrodynamic (MHD) flow of Maxwell nanofl uid containing gyrotactic micro-organisms with heat source/sink. *AIP Adv.* **2018**, *8*, 085303. [CrossRef]

43. Dawar, A.; Shah, Z.; Idrees, M.; Khan, W.; Islam, S.; Gul, T. Impact of Thermal Radiation and Heat Source/Sink on Eyring–Powell Fluid Flow over an Unsteady Oscillatory Porous Stretching Surface. *Math. Comput. Appl.* **2018**, *23*, 20. [CrossRef]

44. Hammed, K.; Haneef, M.; Shah, Z.; Islam, I.; Khan, W.; Asif, S.M. The Combined Magneto hydrodynamic and electric field effect on an unsteady Maxwell nanofluid Flow over a Stretching Surface under the Influence of Variable Heat and Thermal Radiation. *Appl. Sci.* **2018**, *8*, 160. [CrossRef]

45. Khan, A.S.; Nie, Y.; Shah, Z.; Dawar, A.; Khan, W.; Islam, S. Three-Dimensional Nanofluid Flow with Heat and Mass Transfer Analysis over a Linear Stretching Surface with Convective Boundary Conditions. *Appl. Sci.* **2018**, *8*, 2244. [CrossRef]

46. Maxwell, J.C. *Electricity and Magnetism*, 3rd ed.; Clarendon: Oxford, UK, 1904.

47. Jawad, M.; Shah, Z.; Islam, S.; Majdoubi, J.; Tlili, I.; Khan, W.; Khan, I. Impact of Nonlinear Thermal Radiation and the Viscous Dissipation Effect on the Unsteady Three-Dimensional Rotating Flow of Single-Wall Carbon Nanotubes with Aqueous Suspensions. *Symmetry* **2019**, *11*, 207. [CrossRef]

48. Jeffrey, D.J. Conduction through a random suspension of spheres. *Proc. R. Soc. Lond. A* **1973**, *335*, 355–367. [CrossRef]

49. Davis, R.H. The effective thermal conductivity of a composite material with spherical inclusions. *Int. J. Thermophys.* **1986**, *7*, 609–620. [CrossRef]

50. Hamilton, R.L.; Crosser, O.K. Thermal conductivity of heterogeneous two-component systems. *Ind. Eng. Chem. Fundam.* **1962**, *1*, 187–191. [CrossRef]

4

Effects MHD and Heat Generation on Mixed Convection Flow of Jeffrey Fluid in Microgravity Environment over an Inclined Stretching Sheet

4

Iskander Tlili

Department of Mechanical and Industrial Engineering, College of Engineering, Majmaah University, Al-Majmaah 11952, Saudi Arabia; l.tlili@mu.edu.sa

Abstract: In this paper, Jeffrey fluid is studied in a microgravity environment. Unsteady two-dimensional incompressible and laminar g-Jitter mixed convective boundary layer flow over an inclined stretching sheet is examined. Heat generation and Magnetohydrodynamic MHD effects are also considered. The governing boundary layer equations together with boundary conditions are converted into a non-similar arrangement using appropriate similarity conversions. The transformed system of equations is resolved mathematically by employing an implicit finite difference pattern through quasi-linearization method. Numerical results of temperature, velocity, local heat transfer, and local skin friction coefficient are computed and plotted graphically. It is found that local skin friction and local heat transfer coefficients increased for increasing Deborah number when the magnitude of the gravity modulation is unity. Assessment with previously published results showed an excellent agreement.

Keywords: Jeffrey fluid; laminar g-Jitter flow; inclined stretching sheet; heat source/sink

1. Introduction

Nanoparticles composed of nanosized metals, oxides, and carbon nanotubes form fluid suspensions called nanofluids. Nanofluids are widely used in many practical applications including nano-electro-mechanical systems and in the industrial, manufacturing and medical sectors because compared to the conventional liquids, they are characterized by great thermal conductivities leading to an enhanced heat transfer rate. In this context, several researches were carried out in the last few decades to analyze theoretically and experimentally transport phenomena related to heat and nanofluid flow while considering diverse geometries, velocity, and temperature slip boundary conditions [1–6].

The gravity impact is abolished significantly in space, but the microgravity environment is able to reduce convective flows. g-Jitter which refers to the inertia impacts caused by the residual, oscillatory or transitory accelerations originating after squad waves, mechanical pulsations atmospheric slog and microgravity environments was investigated showing that microgravity is correlated to the frequency and magnitude of the periodical gravity modulation [7–11].

One of the major concerns of gravity is the study of the controversial impact of g-Jitter convective flow in various aspects. The impact of a gravitational field representative of g-Jitter was investigated [12–15]. It was demonstrated that g-Jitter has a great influence on the configuration of the three-dimensional boundary layer and the flow properties especially the skin friction and the heat rate. It was demonstrated that the skin friction coefficient is decreased when the magnetic field parameter rises [12,16]. Khoshnevis [17] studied the effect of residual and g-Jitter accelerations and reported that the diffusion procedure is more probably to be affected by the low-frequency involvement of g-Jitter. The effect of the gravity variations falls down against the rise of the forced flow velocity.

However, fuel injection velocity has the opposite effect when it augments [16]. Kumar et al. [18] studied the impact of gravity inflection in a couple stress liquid by using Ginzburg–Landau equations. It was underlined that to improve the rate of heat and mass transfer, the influence of the Prandtl number; concentration Rayleigh number; Lewis number and couple stress parameter on the Nusselt number and Sherwood number has to be increased. Mass transfer has the opposite behavior against Lewis quantity [15].

Magnetohydrodynamic convective boundary layer flow over an exponentially inclined permeable stretching surface was investigated by many researchers [9–11]. Reddy numerically studied [10] the scheme of joined boundary conditions for nonlinear ordinary differential equations. It was revealed that the momentum boundary layer thickness is reduced when the Casson fluid parameter augments. The thermal boundary layer thickness is improved when augmenting rates of the non-uniform heat source or sink parameters [12].

An extensive investigation has been conducted considering the different fluid physical conditions. Both the gravitational field and temperature gradient generate convective flows in porous and clear media. Greater temperature can be observed in the pure field as associated to the porous field in Couette, Poiseuille and widespread to Couette flows of an incompressible magneto hydrodynamic Jeffrey fluid among similar platters over homogeneous porous field employing slip boundary circumstances [18]. Convection flows of nanofluids in porous media have attracted great concern driven by material treating and solar energy gatherer uses. Bhadauria [19] focused on the impacts of flow and G-Jitter on chaotic convection in an anisotropic porous field. It was concluded that heat transfer is greater in the modulated system compared to the unmodulated system. Ghosh et al. [15] reported that sinusoidal g-Jitter leads to a flow streaming inside the porous cavity and time-dependent rolls privileged the encloser because of variances in thermal diffusivities among the solid matrix, wall, and fluid. The temperature profile has an opposite behavior for the Prandtl number [9,10,20] because of the influence of the slip parameter, non-Newtonian parameter, and Hartmann number [21] whereas it augments in parallel with the radiation parameter [4,11,12], Brinkman number, and Dufour and Soret numbers [14]. Compared to Maxwell and Oldroyd-B nanofluids, the Jeffrey nanofluid has superior heat transfer performance [22]. The heat transfer profile was studied in different conditions. Sandeep et al. analyzed numerically the momentum and heat transfer profile of Jeffrey, Maxwell, and Oldroyd-B nanofluids over a stretching surface to determine the impact of the transverse magnetic field, thermal radiation, non-uniform heat source/sink, and suction effects. Heat transfer rate was improved when the Biot number and suction parameter increase. The Jeffrey nanofluid has superior heat transfer performance than the Maxwell and Oldroyd-B nanofluids [23–30].

Some interesting studies concerning the flow of Jeffrey fluid have been conducted. Hayat et al. [8] considered a homogeneous–heterogeneous reaction in a nonlinear radiative flow of Jeffrey fluid between two stretchable rotating disks. The velocities augment in parallel with Deborah number. The thermal field and heat transfer rate are improved for the temperature ratio parameter. Khan et al. [14] analytically studied the Jeffrey liquid flow related to thermal-diffusion and diffusion-thermo characteristics using the homotopic method [31–34].

In the present work, we investigate the effects of Jeffrey fluid in a microgravity environment. Unsteady two-dimensional incompressible and laminar g-Jitter mixed convective boundary layer flow over an inclined stretching sheet is taken into account. The governing boundary layer equations together with boundary conditions are converted into a non-similar arrangement using appropriate similarity conversions. The transformed system of equations is solved numerically by using an implicit finite difference structure with quasi-linearization technique. Numerical results of velocity, temperature, local skin friction, and local heat transfer coefficient are computed and illustrated graphically.

2. Mathematical Formulation

The objective of this study is to investigate the unsteady incompressible flow of Jeffrey fluid past an inclined stretching sheet. In this problematic, the x-axis is ranging lengthways the surface with penchant angle γ to the perpendicular in the upward direction and y-axis is perpendicular to the surface. The plate is characterized with a linear speed $u_w(x)$ in x-direction. The temperature and flow of the platter varies linearly with the stretch x along the platter, anywhere $T_w(x) > T_\infty$ by means of $T_w(x)$ represent the temperature of the plate and T_∞ representing the uniform temperature of the ambient nanofluids.

The incessant stretching sheet is supposed to require the temperature and velocity in the arrangement of $u_w(x) = cx$ and $T_w(x) = T_\infty + ax$ where c and a are factors with $c > 0$. Mutually circumstances of reheating $(T_w(x) > T_\infty)$ and cooling $(T_w(x) < T_\infty)$ of the piece are considered, which settle to assisting flow for $a > 0$ and opposing flow for $a < 0$, correspondingly. The fluid conduct electricity in the control of a variable magnetic field $B(x) = \frac{B_0}{\sqrt{x}}$. And finally the heat generation effect is also considered.

In the typical boundary layer and Boussinesq calculations, the basic governing equations including the conservation of momentum, mass thermal energy equation of Jeffrey fluid can be written as [5,6,35],

$$\frac{\partial u}{\partial x} + \frac{\partial v}{\partial y} = 0 \tag{1}$$

$$
\begin{aligned}
\frac{\partial u}{\partial t} + u\frac{\partial u}{\partial x} + v\frac{\partial u}{\partial y} &= \frac{v}{1+\lambda_1}\left[\frac{\partial^2 u}{\partial y^2}\right] + \frac{v\lambda_2}{1+\lambda_1}\left[\frac{\partial^3 u}{\partial t\partial y^2} + u\frac{\partial^3 u}{\partial x\partial y^2} - \frac{\partial u}{\partial x}\frac{\partial^2 u}{\partial y^2}\right] \\
&+ \frac{v\lambda_2}{1+\lambda_1}\left[\frac{\partial u}{\partial y}\frac{\partial^2 u}{\partial x\partial y} + v\frac{\partial^3 u}{\partial y^3}\right] + g^*(t)\beta_T(T - T_\infty)\cos\alpha - \frac{\sigma B(x)^2}{\rho}u
\end{aligned}
\tag{2}
$$

$$\frac{\partial T}{\partial t} + u\frac{\partial T}{\partial x} + v\frac{\partial T}{\partial y} = \frac{k}{\rho c_p}\frac{\partial^2 T}{\partial y^2} + \frac{Q(x)}{\rho c_p}(T - T_\infty). \tag{3}$$

The suitable initial and boundary circumstances are prescribed as:

$$t = 0 : u = v = 0, T = T_\infty \text{ for any } x, y,$$

$$t > 0 : u_w(x) = ax, v = 0, T = T_w = T_\infty + bx \text{ at } y = 0, \tag{4}$$

$$u \to 0, \frac{\partial u}{\partial y} \to 0, T \to T_\infty \text{ as } y \to \infty,$$

where u and v are the speed modules lengthwise x and y axes, t represents time, and T characterizes the temperature of Jeffrey fluid. Meanwhile, ρ is the density, v is the kinematic viscosity, β_T is the volumetric coefficient of thermal expansion, C_p is the specific heat at constant pressure, k is the real thermal conductivity, λ_1 and λ_2 are two parameters related to the Jeffrey fluid which are, respectively, the fraction of relaxation to retardation times. It is worth mentioning that, for $\lambda_1 = \lambda_2 = 0$, and in the absence of MHD and heat generation terms, the problem reduces to the case of a regular viscous fluid, which is the same problem studied by Sharidan et al. (2006) [3]. The difficulty of the problematic is abridged by involving the subsequent similarity changes [3].

$$\tau = \omega t, \ \eta = \left(\frac{a}{v}\right)^{\frac{1}{2}}y, \ \psi = (av)^{\frac{1}{2}}xf(\tau,\eta), \ \theta(\tau,\eta) = \frac{(T - T_\infty)}{(T_w - T_\infty)}, \ g(\tau) = \frac{g(t)}{g_0}. \tag{5}$$

By employing the similarity transformations (5), Equation (1) is similarly fulfilled, and in addition, the following transformed governing equations are obtained:

$$\Omega\frac{\partial^2 f}{\partial\tau\partial\eta} + \left(\frac{\partial f}{\partial\eta}\right)^2 - f\frac{\partial^2 f}{\partial\eta^2} = \frac{1}{1+\lambda_1}\frac{\partial^3 f}{\partial\eta^3} + \frac{\beta}{1+\lambda_1}\Omega\left(\frac{\partial^4 f}{\partial\tau\partial\eta^3}\right) + \frac{\beta}{1+\lambda_1}\left(\left(\frac{\partial^2 f}{\partial\eta^2}\right)^2 - f\frac{\partial^4 f}{\partial\eta^4}\right)$$

$$+\lambda(1-\varepsilon\cos(\pi\tau))\cos\alpha\cdot\theta(\eta) - M\frac{\partial f}{\partial\eta}, \tag{6}$$

$$\frac{1}{\text{Pr}}\frac{\partial^2\theta}{\partial\eta^2} + f\frac{\partial\theta}{\partial\eta} - \theta\frac{\partial f}{\partial\eta} = \Omega\frac{\partial\theta}{\partial t} - Q_0\theta. \tag{7}$$

where

$$\Omega = \frac{\omega}{a}, \ M = \frac{\sigma B_0^2}{a\rho}, \ Q_0 = \frac{Q}{a\rho c_p}, \beta = a\lambda_2, \lambda = \frac{g_0\beta_T(T_w - T_\infty)\frac{x^3}{v^2}}{[u_w(x)\frac{x}{v}]^2} = \frac{Gr_x}{Re_x^2},$$

$$Gr_x = g_0\beta_T(T_w - T_\infty)\frac{x^3}{v^2}, \ Re_x = u_w(x)\frac{x}{v},$$

$$\tau = 0: \ \frac{\partial f}{\partial\eta}(\tau,\eta) = 0, \ f(\tau,\eta) = 0, \ \theta(\tau,\eta) = 0 \text{ for any } x, y,$$

$$\eta > 0: \ \frac{\partial f}{\partial\eta}(\tau,\eta) = 1, \ f(\tau,\eta) = 0, \ \theta(\tau,\eta) = \frac{b}{a} \text{ at } \eta = 0, \tag{8}$$

$$\eta > 0: \ f(\tau,\eta) \to 0, \ \frac{\partial^2\theta}{\partial\eta^2}(\tau,\eta) \to 0, \ \theta(\tau,\eta) \to 0 \text{ as } \eta \to \infty,$$

where

$$\text{Pr} = \frac{\mu c_p}{k},$$

The real amounts of main attention, such as the local Nusselt number, Nu_x and the skin friction coefficient, C_f, are defined as

$$C_f = \frac{\tau_w(x)}{\rho u_w^2} \text{ and } Nu_x = \frac{q_w(x)\,x}{k(T_w - T_\infty)} \tag{9}$$

where the $\tau_w(x)$ is the wall shear stress and $q_w(x)$ is the wall heat flux given by:

$$\tau_w = \frac{\mu}{1+\lambda_1}\left[\left(\frac{\partial u}{\partial y}\right) + \lambda_2\left(\frac{\partial^2 u}{\partial y\partial t} + u\frac{\partial^2 u}{\partial x\partial y} + v\frac{\partial^2 u}{\partial y^2}\right)\right]_{y=0} \text{ and } q_w(x) = -k\left(\frac{\partial T}{\partial y}\right)_{y=0} \tag{10}$$

The following skin friction coefficient and local Nusselt number are obtained as follows:

$$C_f Re_x^{1/2} = \frac{1}{(1+\lambda_1)}\left[\frac{\partial^2 f}{\partial\eta^2}(\tau,0) + \beta\left(\frac{\partial f}{\partial\eta}(\tau,0)\frac{\partial^2 f}{\partial\eta^2}(\tau,0) - f(\tau,0)\frac{\partial^3 f}{\partial\eta^3}(\tau,0)\right)\right],$$

$$\frac{Nu_x}{Re_x^{1/2}} = -\frac{\partial\theta}{\partial\eta}(\tau,0). \tag{11}$$

3. Numerical Method

The difficulty of the problematic is abridged by involving the subsequent similarity changes (Sharidan et al. (2006)) [3]. By employing the similarity transformations (5), Equations (1)–(3) with boundary conditions (4) transform to Equations (6)–(8). This model will be resolved numerically using Runge–Kutta–Fehlberg method of the seventh order (RKF7) joined with a shooting method. In the RKF7 method, several assessments are tolerable for each stage separately. For minor precision, this technique delivers the greatest effective outputs. A phase size $\Delta\eta = 0.001$ and a convergence condition of 10^{-6} were employed in the numerical calculations. The asymptotic boundary conditions given by Equation (11) were substituted using a value of 10 for the similarity variable $\eta_{\max}$ as follows:

$$f'(\eta, 10) = \theta'(\eta, 10) = 0$$

The choice of $\eta_{\max} = 10$ guarantees that all mathematical solutions move toward the asymptotic values appropriately. The other details of this method can be found in [24]. To check the accuracy of the current technique, the found results are compared in special cases with the results obtained by Hayat et al. [5]. These comparisons are presented clearly in Table 1 in terms of the Nu_m at the heat source. A very good arrangement is established among the results.

Table 1. Assessment of the Nu_m.

Ra	Hayat et al. [5]	Current
1000	5.321	5.332
10,000	5.487	5.496
100,000	7.212	7.223
1,000,000	13.946	14.101

4. Results and Discussion

Unsteady two-dimensional incompressible and laminar g-Jitter mixed convective boundary layer flow over an inclined stretching sheet in the presence of heat source/sink, viscous dissipation and magnetic field for Jeffrey fluid has been explored numerically by means of a shooting scheme based Runge–Kutta–Fehlberg-integration algorithm.

Figure 1a,b present the variation of dimensionless velocity with pertinent parameters in which Figure 1a illustrates the effect of magnetic field and g-Jitter frequency period on the dimensionless velocity and Figure 1b shows the effect of thermal expansion coefficient and mixed convection parameter on the dimensionless velocity. As anticipated, the velocity profile is higher for dimensionless boundary layer thickness $\eta = 0$ then decreases significantly along the inclined stretching sheet to reach zero, this effect is mathematically noticeable in Equation (8) which fulfills the assigned boundary condition. Furthermore, it can be seen that the velocity profile decreases with both magnetic field and thermal expansion coefficient while the g-Jitter frequency period and mixed convection parameter have a negligible effect on velocity profile. Consequently, boundary layer thickness has a similar effect and behavior as the dimensionless velocity. It is important to note that in the absence of a magnetic field the velocity profile is maximum and declines with the application of magnetic field, this can be elucidated by the fact that a magnetic field generates Lorentz force which leads to opposing the flow and, therefore, reduces the velocity profile. Similarly, increasing a thermal expansion coefficient leads to generating more thermal buoyancy that affect the velocity profile but remains a small alteration compared to the effect of a magnetic field. Both of Figure 1a,b refer to the case of assisting flow ($\lambda > 0$) except that in Figure 1b the mixed convection parameter increases.

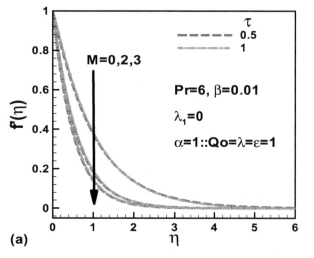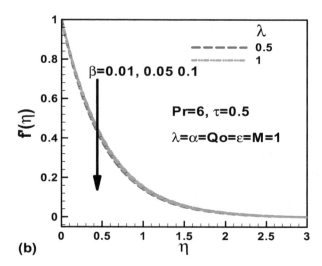

Figure 1. Variation of dimensionless velocity with (**a**) magnetic field and g-Jitter frequency period and with (**b**) thermal expansion coefficient and mixed convection parameter.

Figure 2a,b depict the variation of dimensionless temperature with pertinent parameters in which Figure 2a illustrates the effect of a magnetic field and g-Jitter frequency period on the dimensionless temperature and Figure 2b shows the effect of thermal expansion coefficient and mixed convection parameter on the dimensionless velocity. As expected, the temperature profile is higher for dimensionless boundary layer thickness $\eta = 0$ which then decreases significantly along the inclined stretching sheet to reach zero. This behavior is approved by the considered boundary condition and it is accurately harmonized with Equation (8). Moreover, tt can be seen that the dimensionless temperature variance of the velocity profile shown in Figure 1 augment slightly with both magnetic field and thermal expansion coefficient. Additionally, the thermal expansion coefficient and g-Jitter frequency have an insignificant effect on the temperature profile. Therefore, the thickness of the thermal boundary layer remains uniform in respect of the studied parameters. It should be pointed out that the overall effect of even very great magnetic fields, thermal expansion, and g-Jitter amplitudes on the temperature profile is very insignificant.

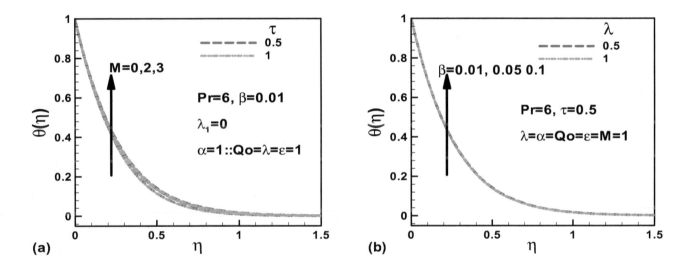

Figure 2. Variation of dimensionless temperature with (**a**) magnetic field and g-Jitter frequency period and with (**b**) thermal expansion coefficient and mixed convection parameter.

The effects of magnetic field and g-Jitter frequency on dimensionless skin friction are illustrated in Figure 3a,b. It is perceived that the dimensionless skin friction increases with both magnetic field and thermal expansion whereas it lessens with g-Jitter frequency. We realize that the velocity profile decreases with magnetic field; this effect will reduce the Reynold number. It can be interpreted from this fact that viscous force will be reduced compared to inertial force which, in turn, leads to augment dimensionless skin friction. Similarly, and for the same reason, it can be explained in Figure 1 that thermal expansion slightly reduces the velocity and consequently augments the dimensionless skin friction.

It is worthwhile to note that the skin friction coefficient Cf Rex-1/2 represents the velocity gradient at the surface; therefore, the velocity gradient at the surface will grow with thermal expansion and magnetic fields. It is very important and intriguing to note that the g-Jitter frequency period reduces skin friction. This effect can be physically elucidated by the fact that g-Jitter generates flow creating buoyancy forces due to the effect of vibration frequency distribution and density gradient which results to increase the acceleration of the fluid flow, this later owing to an augmented Reynolds number and consequently lessens skin friction.

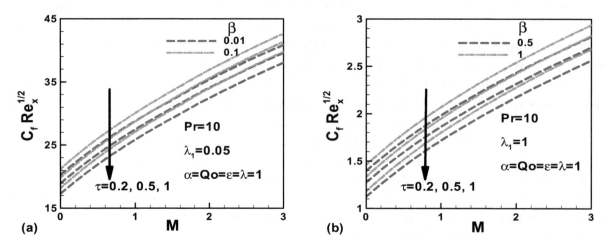

Figure 3. Variation of dimensionless skin friction with magnetic field and g-Jitter frequency period. (**a**) Assisting flow; (**b**) Opposing flow.

The effects of magnetic field, thermal expansion coefficient, and g-Jitter frequency period on the local Nusselt number are shown in Figure 4a,b. It is perceived that the local Nusselt number increases with both the thermal expansion coefficient and g-Jitter frequency period. Whereas, it decreases with magnetic field. It is clear that in the absence of a magnetic field, the Nusselt number is maximum and declines significantly until zero when the magnetic field augments, it can be interpreted on this fact that the increase of the magnetic field significantly enhances conduction heat transfer which, in turn, leads to lessening the local Nusselt number since it is well recognized that Nusselt numbers are the ratio of convection to conduction heat transfer. It is worthwhile to note that the local Nusselt number Nux Rex-1/2 represents the rate of heat transfer at the surface, consequently, heat transfer at the surface will decrease with magnetic field and increase with both the thermal expansion coefficient and g-Jitter frequency period. Applying a g-Jitter effect and increasing the thermal expansion coefficient, respectively, leads to acceleration of fluid flow and increases the thermal bouyancy effect, and both of them lead to increased convection heat transfer. Therefore, the local Nusselt number augments. Finally, it is realized that the Nusselt number at the inclined sheet surface with Jeffrey fluid and in a microgravity environment lessens with magnetic field and augments with the thermal expansion coefficient and g-Jitter frequency period. Thus, convection heat transfer will be enhanced by g-Jitter frequency and thermal expansion.

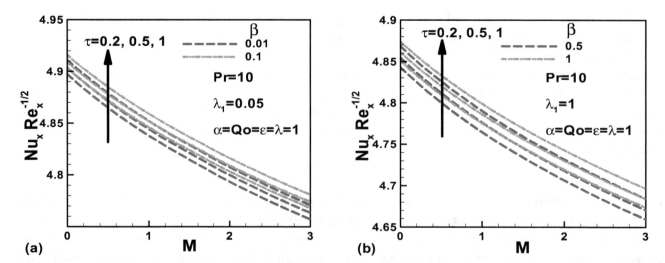

Figure 4. Variation of dimensionless heat transfer rate with several parameters in the presence of heat source for (**a**) assisting flow and (**b**) opposing flow.

5. Conclusions

Unsteady two-dimensional incompressible and laminar g-Jitter mixed convective boundary layer flow over an inclined stretching sheet in the presence of heat source/sink, viscous dissipation and magnetic field for Jeffrey fluid has been explored numerically by means of a shooting scheme based Runge–Kutta–Fehlberg-integration algorithm.

The results drawn from this study are mentioned as follows:

1. The velocity profile is higher for dimensionless boundary layer thickness $\eta = 0$ then decreases significantly along the inclined stretching sheet to reach zero. In the absence of a magnetic field, the velocity profile is maximum and declines with the application of a magnetic field

2. The temperature profile is higher for dimensionless boundary layer thickness $\eta = 0$ then decreases significantly along the inclined stretching sheet to reach zero. The thermal expansion coefficient and g-Jitter frequency have an insignificant effect on the temperature profile. Therefore, the thickness of the thermal boundary layer remains uniform in respect to the studied parameters.

3. The dimensionless skin friction increases with both magnetic field and thermal expansion whereas it lessens with g-Jitter frequency. The g-Jitter frequency period reduces skin friction; this effect can be physically elucidated by the fact that g-Jitter generates flow creating buoyancy forces due to the effect of the vibration frequency distribution and density gradients which results to increase the acceleration of the fluid flow.

4. The local Nusselt number increases with both the thermal expansion coefficient and g-Jitter frequency period. Whereas, it decreases with magnetic field. The Nusselt number at the inclined sheet surface with Jeffrey fluid and in microgravity environment lessens with magnetic field and augments with the thermal expansion coefficient and g-Jitter frequency period.

Acknowledgments: Iskander Tlili thanks the Deanship of Scientific Research and the Deanship of Community Service at Majmaah University, Kingdom of Saudi Arabia, for supporting this work. Iskander Tlili thanks Dana Ibrahim Alrabiah, Renad Fahad Aba Hussein, Ritaj Al Anazi, Maryam Khaled Al Mizani and Wateen Satam Al-Mutairi.

Nomenclature

Bi	Biot number
C	Nanoparticle volume fraction
C_p	Specific heat at constant pressure
C_f	Local skin-friction coefficient
Dn	Diffusivity of the microorganisms
D_B	Brownian diffusion coefficient
D_T	Thermophoretic diffusion coefficient of the microorganisms
f'	Dimensionless velocity
g	Gravitational acceleration
Gr_{x*}	Local Grashof number
h_f	Convective heat transfer coefficient
k	Thermal conductivity
Le	Lewis number
Lb	Bioconvection Lewis number
Nb	Brownian motion number
Nr	Buoyancy ratio parameter
Nt	Thermophoresis number
Nu_{x*}	Local Nusselt number
n	Density of motile microorganisms

Nn_x	Local density number
Pe	Bioconvection Péclet number
Pe_x	Local Péclet number
Pr	Prandtl number, ν/α
q_w	Wall heat flux
r	Local radius of the truncated cone
Ra_x	Modified Rayleigh number
Rb	Bioconvection Rayleigh number
Sh_x	Local Sherwood number
T	Temperature
u	Velocity component in the x-direction
U_r	Reference velocity
v	Velocity component in the y-direction
w_c	Maximum cell swimming speed
x	Streamwise coordinate
x_o	Distance of the leading edge of truncated cone measured from the origin
x^*	Distance measured from the leading edge of the truncated cone, x-x$_o$
y	Transverse coordinate
α	Thermal diffusivity
β	Coefficient of thermal expansion
γ	Average volume of a microorganism
σ	Motile parameter
η	Pseudo-similarity variable
θ	Dimensionless temperature
ϕ	Dimensionless nanoparticle volume fraction
ψ	Stream function
χ	Dimensionless density of motile microorganisms
ξ	Dimensionless distance
μ	Dynamic viscosity
ν	Kinematic viscosity
Ω	Half angle of the truncated cone
ρ_f	Density of the fluid
$\rho_{f\infty}$	Density of the base fluid
ρ_p	Density of the particles
$\rho_{m\infty}$	Density of the microorganism
$(\rho c)_f$	Heat capacity of the fluid
$(\rho c)_p$	Effective heat capacity of the nanoparticle material
ρ	Density
ψ	Stream function
w	Condition at the wall
∞	Condition at infinity

References

1. Rawi, N.A.; Kasim, A.R.M.; Isa, M.; Shafie, S. G-Jitter induced mixed convection flow of heat and mass transfer past an inclined stretching sheet. *J. Teknol.* **2014**, *71*. [CrossRef]

2. Shafie, S.; Amin, N.S.; Pop, I. G-Jitter free convection boundary layer flow of a micropolar fluid near a three-dimensional stagnation point of attachment. *Int. J. Fluid Mech. Res.* **2005**, *32*, 291–309. [CrossRef]

3. Sharidan, S.; Amin, N.; Pop, I. G-Jitter mixed convection adjacent to a vertical stretching sheet. *Microgravity-Sci. Technol.* **2006**, *18*, 5–14. [CrossRef]

4. Ramesh, K. Effects of viscous dissipation and Joule heating on the Couette and Poiseuille flows of a Jeffrey fluid with slip boundary conditions. *Propuls. Power Res.* **2018**, *7*, 329–341. [CrossRef]

5. Hayat, T.; Waqas, M.; Khan, M.I.; Alsaedi, A. Impacts of constructive and destructive chemical reactions in magnetohydrodynamic (MHD) flow of Jeffrey liquid due to nonlinear radially stretched surface. *J. Mol. Liq.* **2017**, *225*, 302–310. [CrossRef]

6. Selvi, R.K.; Muthuraj, R. MHD oscillatory flow of a Jeffrey fluid in a vertical porous channel with viscous dissipation. *Ain Shams Eng. J.* **2018**, *9*, 2503–2516. [CrossRef]

7. Dalir, N. Numerical study of entropy generation for forced convection flow and heat transfer of a Jeffrey fluid over a stretching sheet. *Alex. Eng. J.* **2014**, *53*, 769–778. [CrossRef]

8. Hayat, T.; Qayyum, S.; Imtiaz, M.; Alsaedi, A. Homogeneous-heterogeneous reactions in nonlinear radiative flow of Jeffrey fluid between two stretchable rotating disks. *Results Phys.* **2017**, *7*, 2557–2567. [CrossRef]

9. Khan, M.; Shahid, A.; Malik, M.Y.; Salahuddin, T. Thermal and concentration diffusion in Jeffery nanofluid flow over an inclined stretching sheet: A generalized Fourier's and Fick's perspective. *J. Mol. Liq.* **2018**, *251*, 7–14. [CrossRef]

10. Reddy, P.B.A. Magnetohydrodynamic flow of a Casson fluid over an exponentially inclined permeable stretching surface with thermal radiation and chemical reaction. *Ain Shams Eng. J.* **2016**, *7*, 593–602. [CrossRef]

11. Ramzan, M.; Bilal, M.; Chung, J.D. Effects of thermal and solutal stratification on jeffrey magneto-nanofluid along an inclined stretching cylinder with thermal radiation and heat generation/absorption. *Int. J. Mech. Sci.* **2017**, *131–132*, 317–324. [CrossRef]

12. Sandeep, N.; Sulochana, C. Momentum and heat transfer behaviour of Jeffrey, Maxwell and Oldroyd-B nanofluids past a stretching surface with non-uniform heat source/sink. *Ain Shams Eng. J.* **2018**, *9*, 517–524. [CrossRef]

13. Ahmad, K.; Ishak, A. Magnetohydrodynamic (MHD) Jeffrey fluid over a stretching vertical surface in a porous medium. *Propuls. Power Res.* **2017**, *6*, 269–276. [CrossRef]

14. Khan, M.I.; Waqas, M.; Hayat, T.; Alsaedi, A. Soret and Dufour effects in stretching flow of Jeffrey fluid subject to Newtonian heat and mass conditions. *Results Phys.* **2017**, *7*, 4183–4188. [CrossRef]

15. Ghosh, P.; Ghosh, M.K. Streaming flows in differentially heated square porous cavity under sinusoidal g-jitter. *Int. J. Therm. Sci.* **2009**, *48*, 514–520. [CrossRef]

16. Rouvreau, S.; Cordeiro, P.; Torero, J.L.; Joulain, P. Influence of g-jitter on a laminar boundary layer type diffusion flame. *Proc. Combust. Inst.* **2005**, *30*, 519–526. [CrossRef]

17. Khoshnevis, A.; Ahadi, A.; Saghir, M.Z. On the influence of g-jitter and prevailing residual accelerations onboard International Space Station on a thermodiffusion experiment. *Appl. Therm. Eng.* **2014**, *68*, 36–44. [CrossRef]

18. Kumar, A.; Gupta, V.K. Study of heat and mass transport in Couple-Stress liquid under G-jitter effect. *Ain Shams Eng. J.* **2018**, *9*, 973–984. [CrossRef]

19. Bhadauria, B.S.; Singh, A. Through flow and G-jitter effects on chaotic convection in an anisotropic porous medium. *Ain Shams Eng. J.* **2018**, *9*, 1999–2013. [CrossRef]

20. Shafie, S.; Amin, N.; Pop, I. G-Jitter free convection flow in the stagnation-point region of a three-dimensional body. *Mech. Res. Commun.* **2007**, *34*, 115–122. [CrossRef]

21. Asif, M.; Haq, S.U.; Islam, S.; Khan, I.; Tlili, I. Exact solution of non-Newtonian fluid motion between side walls. *Results Phys.* **2018**, *11*, 534–539. [CrossRef]

22. Agaie, B.G.; Khan, I.; Yacoob, Z.; Tlili, I. A novel technique of reduce order modelling without static correction for transient flow of non-isothermal hydrogen-natural gas mixture. *Results Phys.* **2018**, *10*, 532–540. [CrossRef]

23. Tlili, I.; Hamadneh, N.N.; Khan, W.A. Thermodynamic Analysis of MHD Heat and Mass Transfer of Nanofluids Past a Static Wedge with Navier Slip and Convective Boundary Conditions. *Arab. J. Sci. Eng.* **2018**, 1–13. [CrossRef]

24. Khalid, A.; Khan, I.; Khan, A.; Shafie, S.; Tlili, I. Case study of MHD blood flow in a porous medium with CNTS and thermal analysis. *Case Stud. Therm. Eng.* **2018**, *12*, 374–380. [CrossRef]

25. Khan, I.; Abro, K.A.; Mirbhar, M.N.; Tlili, I. Thermal analysis in Stokes' second problem of nanofluid: Applications in thermal engineering. *Case Stud. Therm. Eng.* **2018**, *12*, 271–275. [CrossRef]

26. Afridi, M.I.; Qasim, M.; Khan, I.; Tlili, I. Entropy generation in MHD mixed convection stagnation-point flow in the presence of joule and frictional heating. *Case Stud. Therm. Eng.* **2018**, *12*, 292–300. [CrossRef]

27. Khan, Z.A.; Haq, S.U.; Khan, T.S.; Khan, I.; Tlili, I. Unsteady MHD flow of a Brinkman type fluid between two side walls perpendicular to an infinite plate. *Results Phys.* **2018**, *9*, 1602–1608. [CrossRef]

28. Aman, S.; Khan, I.; Ismail, Z.; Salleh, M.Z.; Tlili, I. A new Caputo time fractional model for heat transfer enhancement of water based graphene nanofluid: An application to solar energy. *Results Phys.* **2018**, *9*, 1352–1362. [CrossRef]

29. Khan, Z.; Khan, I.; Ullah, M.; Tlili, I. Effect of thermal radiation and chemical reaction on non-Newtonian fluid through a vertically stretching porous plate with uniform suction. *Results Phys.* **2018**, *9*, 1086–1095. [CrossRef]

30. Imran, M.A.; Miraj, F.; Khan, I.; Tlili, I. MHD fractional Jeffrey's fluid flow in the presence of thermo diffusion, thermal radiation effects with first order chemical reaction and uniform heat flux. *Results Phys.* **2018**, *10*, 10–17. [CrossRef]

31. Rees, D.A.S.; Pop, I. The effect of g-jitter on vertical free convection boundary-layer flow in porous media. *Int. Commun. Heat Mass Transf.* **2000**, *27*, 415–424. [CrossRef]

32. Pandey, A.K.; Kumar, M. Natural convection and thermal radiation influence on nanofluid flow over a stretching cylinder in a porous medium with viscous dissipation. *Alex. Eng. J.* **2017**, *56*, 55–62. [CrossRef]

33. Ashorynejad, H.R.; Sheikholeslami, M.; Pop, I.; Ganji, D.D. Nanofluid flow and heat transfer due to a stretching cylinder in the presence of magnetic field. *Heat Mass Transf.* **2013**, *49*, 427–436. [CrossRef]

34. Buongiorno, J. Convective transport in nanofluids. *ASME Trans. J. Heat Transf.* **2006**, *128*, 240–250. [CrossRef]

35. Hayat, T.T.; Bibi, A.; Yasmin, H.H.; Alsaadi, F.E. Magnetic Field and Thermal Radiation Effects in Peristaltic Flow with Heat and Mass Convection. *ASME. J. Therm. Sci. Eng. Appl.* **2018**, *10*, 051018. [CrossRef]

5

MHD Stagnation Point Flow of Nanofluid on a Plate with Anisotropic Slip

Muhammad Adil Sadiq

Department of Mathematics, DCC-KFUPM, KFUPM Box 5084, Dhahran 31261, Saudi Arabia;
adilsadiq@kfupm.edu.sa

Abstract: In this article, an axisymmetric three-dimensional stagnation point flow of a nanofluid on a moving plate with different slip constants in two orthogonal directions in the presence of uniform magnetic field has been considered. The magnetic field is considered along the axis of the stagnation point flow. The governing Naiver–Stokes equation, along with the equations of nanofluid for three-dimensional flow, are modified using similarity transform, and reduced nonlinear coupled ordinary differential equations are solved numerically. It is observed that magnetic field M and slip parameter λ_1 increase the velocity and decrease the boundary layer thickness near the stagnation point. Also, a thermal boundary layer is achieved earlier than the momentum boundary layer, with the increase in thermophoresis parameter N_t and Brownian motion parameter N_b. Important physical quantities, such as skin friction, and Nusselt and Sherwood numbers, are also computed and discussed through graphs and tables.

Keywords: stagnation point flow; numerical solution; magnetic field; nanofuid

1. Introduction

The phenomenon of stagnation point flow has various uses in and aerodynamic industries. Such flows mainly compact with the movement of fluid close to the stagnated region of a rigid surface flowing in the fluid material, or retained with dynamics of fluid. Stagnation point has been studied by many researchers in the past because of its wide range of applications in engineering. Initially, stagnation point flow was analyzed by Hiemenz in 1911. He studied the two-dimensional stagnation point flow on a stationary plate. Stagnation point flow applications include cooling of electronic devices by fans, cooling of nuclear reactors, polymer extrusion, wire drawing, drawing of plastic sheets, and many hydrodynamic processes in engineering applications. Stagnation point flow possesses much physical significance, as it is used to calculate the velocity gradients and the rate of heat and mass transfer abutting to stagnation area of frames in high-speed flows, cooling of transpiration, rustproof designs of bearings, etc.

Recently, Borrelli et al. [1] deliberated over the impact of buoyancy on three-dimensional (3D) stagnation point flow. They stated that the buoyancy forces tend to favor an opposite flow. Later, Lok et al. [2] expanded on the work of Weidman [3] with buoyancy forces. They observed the discrete results for free convection and forced convection due to a singularity rising in the convection term. Steady oblique stagnation point flow of a viscous fluid was studied by Grosan et al. [4]. They solved the nonlinear coupled differential equation numerically using the Runge–Kutta method. It is observed that the location of the stagnation point depends strongly on the value of the shear parameter and magnetic parameter. Wang [5–7] discussed the three-dimensional stagnation flow in the absence of MHD and nanofluids on a flat plate, shrining disk, and rotating disk. Two-dimensional (2D) stagnation flow was discussed by Nadeem et al. [8] using HAM on a stretchable surface.

A fluid, heated by electric current in the presence of strong magnetic field, for example crystal growth in melting, has relevance in manufacturing industries. During the fluid motion, the association of electric current and magnetic field produces a divergence of Lorentz forces. This phenomenon prevents the convective motion of fluid and heat transfer characteristic changes accordingly. Ariel [9] investigated the flow near the stagnation point numerically for small magnetic fields; for large magnetic numbers, the perturbation technique was used. Raju and Sundeep [10] proved that with an increase in the magnetic number, there is an increase in the heat and mass transfer rates. They studied numerically the MHD flow of non-Newtonian fluid over a rotating cone or plate.

Generally, the size of nanoparticles is (1–100 nm). Currently, nanofluids are used for drug delivery in infected areas of the human body. Self-propagating objects containing drugs are used to remove blood clots in sensitive areas such as the brain, eye, heart, etc. Kleinstreuer [11] discussed the drug delivery system in humans at normal body temperature under the influence of some physical parameters such as nanoparticle length, artery diameter, and velocity of fluid. Recently, a mathematical model of nanofluid was developed by Choi [12]. Later, a contribution to heat transfer analysis in nanofluid was made by Buongiorno [13]. His mathematical model dealt with the non-homogeneous model for transport phenomena and heat transfer in nanofluids with applications to turbulence. Saleem et al. [14] discussed the effects of Brownian diffusion and thermophoresis on non-Newtonian fluid models, using HAM in the domain of a vertical rotating cone. Bachok et al. [15] studied the three-dimensional stagnation flow of a viscous fluid numerically, analyzed the velocity and heat transfer for different physical parameters, and compared three nanoparticles, namely C_u, Al_2O_3, T_iO_3. In [16] Ellahi et al. explored the heat and mass transfer of non-Newtonian fluid in an annulus in a porous medium using HAM. Recently, Sheikholeslami et al. [17] studied the effects of thermal radiation on steady viscous nanofluid in the presence of MHD numerically. Khan [18] explored Brownian diffusion and thermophoresis on stagnation point flow. He considered dual solutions for shrinking/stretching parameters and heat transfer in the presence of buoyancy forces on a stretchable surface. Mustafa et al. [19] investigated 3D nanofluid flow and heat transfer in two opposite directions on a plane horizontal stretchable surface. Thermal and momentum boundary layers were discussed using physical parameters such as Brownian motion and thermophoretic forces. Some more useful studies related to nanofluids can be found in [20–29].

In this article, an axisymmetric 3D stagnation point flow of a nanofluid on a moving plate with different slip constants in two orthogonal directions in the presence of uniform magnetic field has been considered and solved numerically.

2. Mathematical Formulation

Consider a stagnation point flow of a nanofluid over a plate with anisotropic slip in a Cartesian coordinate system, so that the x-axis is taken along the corrugations of plates, the y-axis is normal to the corrugations, and the z-axis is considered with the axis of stagnation flow. The velocities of the moving plate are (u, v) in (x, y) directions, respectively. A constant magnetic field is applied perpendicular to the corrugation along the axis of the stagnation flow in such a way that the magnetic Reynolds number is small. According to Wang [5], the potential flow far from the plate is defined as:

$$uu_x + vu_y + wu_z = -\frac{p_x}{\rho} + v\left(u_{xx} + u_{yy} + u_{zz}\right) - \frac{B_0^2}{\rho}u, \tag{1}$$

$$uv_x + vv_y + wv_z = -\frac{p_y}{\rho} + v\left(v_{xx} + v_{yy} + v_{zz}\right) - \frac{B_0^2}{\rho}v, \tag{2}$$

$$uw_x + vw_y + ww_z = -\frac{p_z}{\rho} + v\left(w_{xx} + w_{yy} + w_{zz}\right), \tag{3}$$

$$u\frac{\partial T}{\partial x} + v\frac{\partial T}{\partial y} + w\frac{\partial T}{\partial z} = \alpha_m \left(\frac{\partial^2 T}{\partial x^2} + \frac{\partial^2 T}{\partial y^2} + \frac{\partial^2 T}{\partial z^2} \right)$$

$$+ \frac{(\rho C)_p}{(\rho C)_f} \left[D_B \left(\frac{\partial C}{\partial x}\frac{\partial T}{\partial x} + \frac{\partial C}{\partial y}\frac{\partial T}{\partial y} + \frac{\partial C}{\partial z}\frac{\partial T}{\partial z} \right) \right]$$

$$+ \frac{D_T}{T_\infty} \left[\left(\frac{\partial T}{\partial x} \right)^2 + \left(\frac{\partial T}{\partial y} \right)^2 + \left(\frac{\partial T}{\partial z} \right)^2 \right], \tag{4}$$

$$u\frac{\partial C}{\partial x} + v\frac{\partial C}{\partial y} + w\frac{\partial C}{\partial z} = D_B \left[\frac{\partial^2 C}{\partial x^2} + \frac{\partial^2 C}{\partial y^2} + \frac{\partial^2 C}{\partial z^2} \right]$$

$$+ \frac{D_T}{T_\infty} \left[\frac{\partial^2 T}{\partial x^2} + \frac{\partial^2 T}{\partial y^2} + \frac{\partial^2 T}{\partial z^2} \right]. \tag{5}$$

and the boundary conditions are:

$$u - U = N_1 \mu \frac{\partial u}{\partial z} \quad , v - V = N_2 \mu \frac{\partial v}{\partial z}, \quad T = T_w, \quad C = C_\infty \quad at \quad z = 0,$$

$$u \to ax, v \to ay, T \to T_\infty, C \to C_\infty \quad at \quad z \to \infty. \tag{6}$$

where (u, v) are the velocity components in the (x, y) directions, v is the kinematic viscosity, T is the temperature, α_m is the thermal diffusivity, C is the volume of nanoparticles, $(\rho C)_p$ is the effective heat capacity of nanoparticles, $(\rho C)_f$ is the heat capacity of fluid, D_B is the Brownian diffusion coefficient and D_T is the thermophoretic diffusion coefficient. For the non-dimensionalization, we use the following similarity variables:

$$u = axf'(\eta) + Uh(\eta),$$

$$v = ayg'(\eta) + Vk(\eta),$$

$$w = -\sqrt{av} \left[f(\eta) + g(\eta) \right]. \tag{7}$$

where $\eta = \sqrt{a/v}\, z$. Using Equation (7) in Equations (5) and (6) finally we get:

$$f''' + f''(f + g) - (f')^2 - M^2 f' = -(1 + M^2) \tag{8}$$

$$g''' + g''(f + g) - (g')^2 - M^2 g' = -(1 + M^2) \tag{9}$$

$$h'' + h'(f + g) - hf' - M^2 h = 0 \tag{10}$$

$$k'' + k'(f + g) - kg' - M^2 k = 0 \tag{11}$$

$$\theta'' + P_r(f + g)\theta' + P_r \left[N_t \theta' \phi' + N_b (\theta')^2 \right] = 0 \tag{12}$$

$$\phi'' + S_c(f + g)\phi' + \frac{N_t}{N_b}\theta'' = 0. \tag{13}$$

and boundary conditions are:

$$f'(0) = \lambda_1 f''(0), \; g'(0) = \lambda_2 g''(0), \; h(0) = 1 + \lambda_1 h'(0), \; k(0) = 1 + \lambda_2 k'(0), \; f(0) = 0, g(0) = 0,$$

$$f'(\infty) \to 1, \; g'(\infty) \to 1, \; h(\infty) \to 1, \; k(\infty) \to 1, \; \theta(0) = 1, \; \theta(\infty) \to 0, \; \phi(0) = 1, \; \phi(\infty) \to 0. \tag{14}$$

here λ_1 and λ_2 are the slip parameters, P_r the prantle number, S_c the Schmidt number, N_t and N_b are thermophoresis parameter, Brownian motion parameters, respectively.

The expression for the skin friction coefficient, the local Nusselt number, and Sherwood number for second-grade fluid are defined as:

$$Re_x^{1/2} C_f = f''(0), \quad Nu_x Re_x^{-1/2} = -\theta'(0), \quad Sh_x Re_x^{-1/2} = -\varphi'(0), \tag{15}$$

where $Re_x = \frac{U_w x}{\nu}$ is the local Reynolds number. The solution of above coupled nonlinear differential equations are found numerically and discussed in the following section.

3. Result and Discussion

A system of nonlinear ordinary differential Equations (8)–(13) subject to the boundary conditions of Equation (14) are solved numerically using the Richardson extrapolation enhancement method. Richardson extrapolation is generally faster, and capable of handling BVP systems with unknown parameters. The values of these parameters can be determined under the presence of a sufficient number of boundary conditions. The solutions are discussed through graphs from Figures 1–10, and values of physical quantities, such as skin friction and Nusselt and Sherwood numbers, are presented in Tables 1–3.

Figures 1 and 2 show the variation of velocity profile f' and g' against η for different values of magnetic field M and slip parameter λ_1. It was observed that increasing in the values of M and λ_1 causes increase in the velocity profile, while boundary layer thickness reduces. Thus, these parameters cause a reduction in the momentum boundary layer. Analysis shows that increasing the values of these parameters to a sufficiently large level shows the monotonic behavior of velocity throughout the whole domain. Figures 3 and 4 shows the opposite behavior of h and k with the increment of M and λ_1, such that with the increase in value of these parameters, h and k decreases.

The temperature profile for the nanofluid against different values of thermophoresis parameter N_t and Brownian motion parameter N_b are plotted in Figures 5 and 6. As the temperature increase within the boundary layer, the values of these parameters increase. The thermal boundary layer is achieved earlier than the momentum boundary layer. The variation of nanoconcentration for different values of Schmidt number S_c and N_t is presented in Figures 7 and 8, respectively. It is observed that nanoconcentration ϕ decreases as the increase in S_c and boundary layer thickness decreases. Also, with the increase in N_t, the nanoconcentration decreases. Figures 9 and 10 show the velocity profile for different values of magnetic parameter $M = 0$ and for $M = 2$. It is observed that in the absence of magnetic parameter M, the boundary layer thickness is larger than while M is present. $M = 0$ in Figures 11 and 12 represents the results of Wang [5]. The slip parameter ratio can be defined as $\gamma = \frac{\lambda_2}{\lambda_1}$. Figures 13 and 14 describe the $f'(\eta), g'(\eta)$ for $\gamma = 0.5$. The range of γ varies from 0.2 to 10. $\gamma = 1$ represents the isotropic case where $f'(\eta) = g'(\eta)$ and $h(\eta) = k(\eta)$.

Table 1 shows local Nusselt number Nu_x and local Sherwood number Sh_x for the variation of P_r and thermophoresis parameter N_b. Here we see that with the increase of P_r, the local Nusselt number decreases, while local Sherwood number gives opposite results, meaning Sh_x increases. Moreover, with the increase of N_b, the results are again the opposite for Nu_x and Sh_x. Table 2 shows local Nusselt number and local Sherwood number for variations of slip parameter λ_1 and Brownian motion Nb. Here it is observed that with the increase of λ_1 both Nusselt number and local Sherwood number increase. Table 3 shows the skin friction coefficient C_f for different values of λ_1 and magnetic parameter M. Note that with the increment in λ_1, the value of skin friction decreases. A high value of M gives larger values of skin friction.

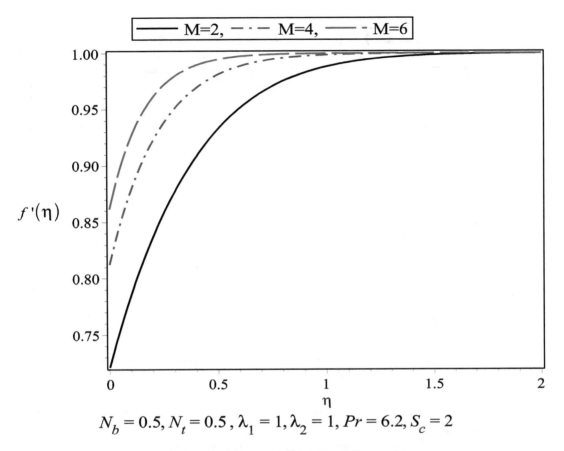

$N_b = 0.5, N_t = 0.5, \lambda_1 = 1, \lambda_2 = 1, Pr = 6.2, S_c = 2$

Figure 1. Variation of $f'(\eta)$ for different M.

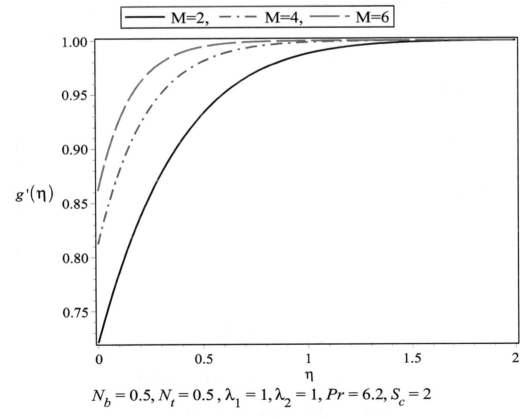

$N_b = 0.5, N_t = 0.5, \lambda_1 = 1, \lambda_2 = 1, Pr = 6.2, S_c = 2$

Figure 2. Variation of $g'(\eta)$ for different M.

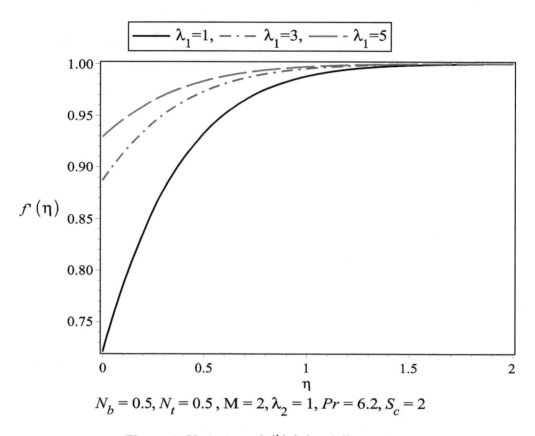

$N_b = 0.5, N_t = 0.5, M = 2, \lambda_2 = 1, Pr = 6.2, S_c = 2$

Figure 3. Variation of $f'(\eta)$ for different λ_1.

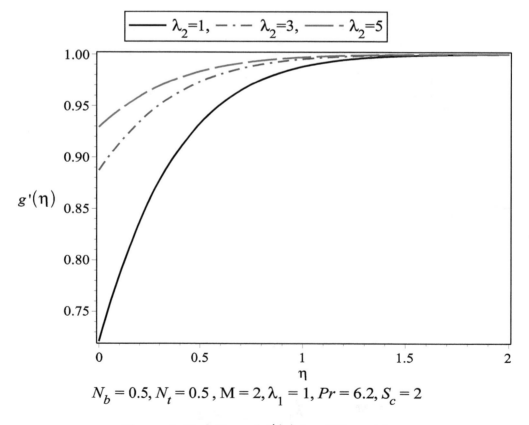

$N_b = 0.5, N_t = 0.5, M = 2, \lambda_1 = 1, Pr = 6.2, S_c = 2$

Figure 4. Variation of $g'(\eta)$ for different λ_2.

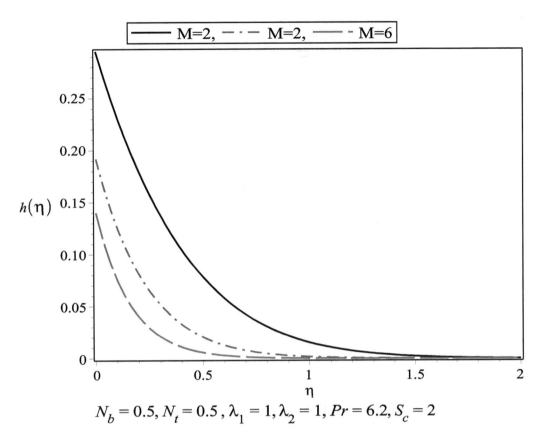

$N_b = 0.5, N_t = 0.5, \lambda_1 = 1, \lambda_2 = 1, Pr = 6.2, S_c = 2$

Figure 5. Variation of $h(\eta)$ for different M.

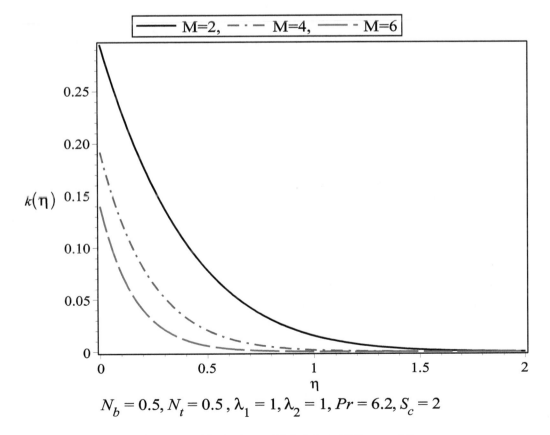

$N_b = 0.5, N_t = 0.5, \lambda_1 = 1, \lambda_2 = 1, Pr = 6.2, S_c = 2$

Figure 6. Variation of $k(\eta)$ for different M.

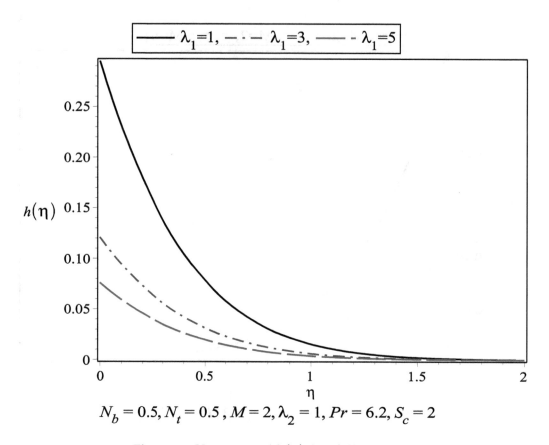

$$N_b = 0.5, N_t = 0.5, M = 2, \lambda_2 = 1, Pr = 6.2, S_c = 2$$

Figure 7. Variation of $h(\eta)$ for different λ_1.

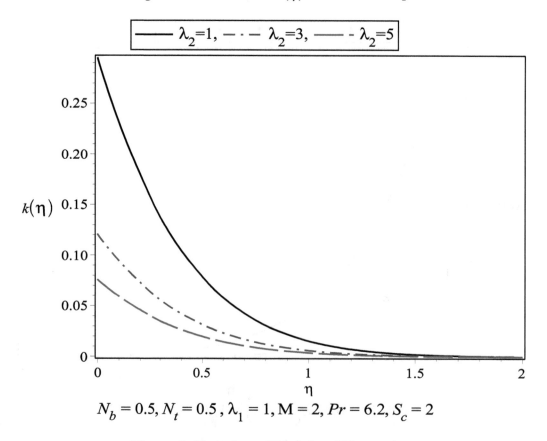

$$N_b = 0.5, N_t = 0.5, \lambda_1 = 1, M = 2, Pr = 6.2, S_c = 2$$

Figure 8. Variation of $k(\eta)$ for different λ_2.

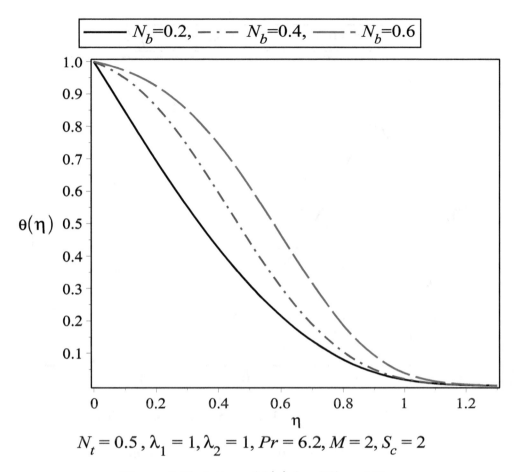

$$N_t = 0.5\,,\lambda_1 = 1, \lambda_2 = 1, Pr = 6.2, M = 2, S_c = 2$$

Figure 9. Variation of $\theta(\eta)$ for different N_b.

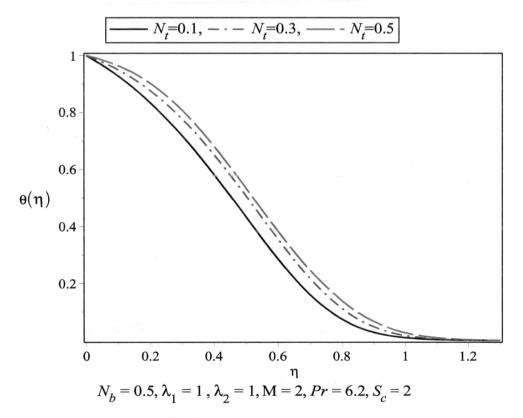

$$N_b = 0.5, \lambda_1 = 1\,, \lambda_2 = 1, M = 2, Pr = 6.2, S_c = 2$$

Figure 10. Variation of $\theta(\eta)$ for different N_t.

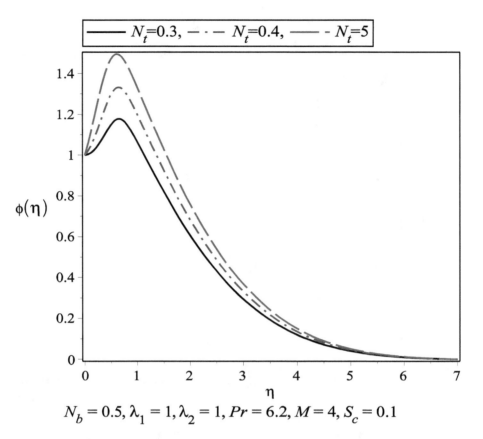

$N_b = 0.5, \lambda_1 = 1, \lambda_2 = 1, Pr = 6.2, M = 4, S_c = 0.1$

Figure 11. Variation of $\phi(\eta)$ for different N_t.

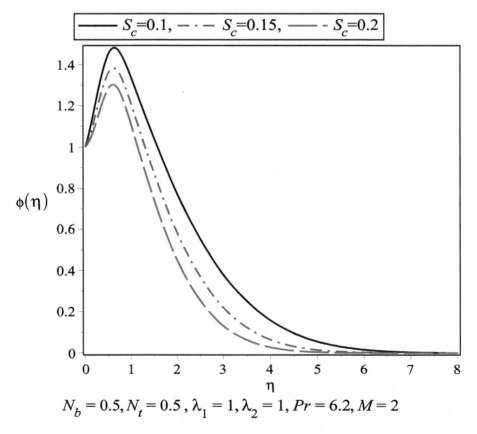

$N_b = 0.5, N_t = 0.5, \lambda_1 = 1, \lambda_2 = 1, Pr = 6.2, M = 2$

Figure 12. Variation of $\phi(\eta)$ for different S_c.

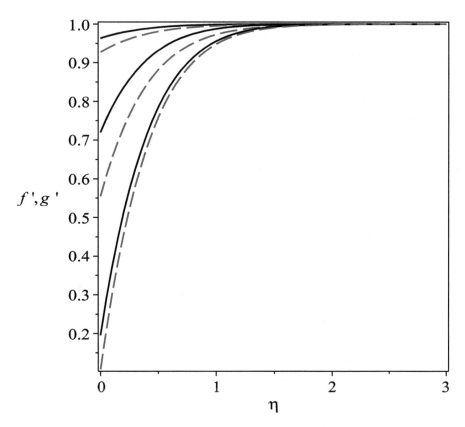

Figure 13. $f'(\eta)$ solid curves and $g'(\eta)$ dashed curves for $\gamma = \frac{\lambda_2}{\lambda_1} = 0.5$. From top: $\lambda_1 = 10, 1, 0.1$.

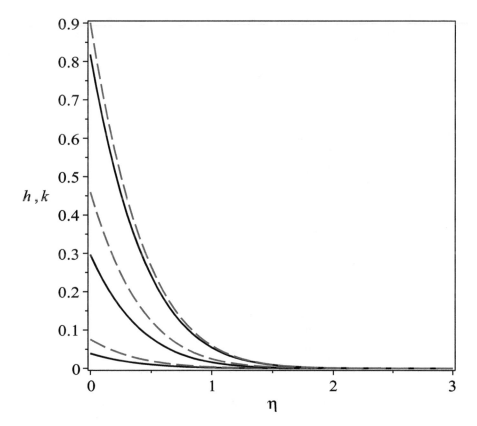

Figure 14. $h(\eta)$ solid curves and $k(\eta)$ dashed curves for $\gamma = \frac{\lambda_2}{\lambda_1} = 0.5$. From top: $\lambda_1 = 10, 1, 0.1$.

Table 1. Variation of Local Nusselt number Nu_x and Sherwood number Sh_x for different N_b and P_r.

| | $\lambda_1 = 1, \lambda_2 = 1, S_c = 2, M = 2, N_t = 0.5$ | | | | | |
| | $N_b = 0.1$ | | $N_b = 0.3$ | | $N_b = 0.5$ | |
P_r	Nu_x	Sh_x	Nu_x	Sh_x	Nu_x	Sh_x
5.5	0.67084	1.47332	0.45477	1.58078	0.30641	1.77542
5.6	0.66527	1.47440	0.44833	1.58438	0.29987	1.78168
5.7	0.65976	1.47546	0.44197	1.58792	0.29346	1.78781
5.8	0.65431	1.47651	0.43570	1.59139	0.28716	1.79381
5.9	0.64890	1.47754	0.42950	1.59481	0.28098	1.79968
6.0	0.64355	1.47856	0.42343	1.59816	0.27492	1.80543
6.1	0.63826	1.47956	0.41742	1.60145	0.26897	1.81105
6.2	0.63302	1.48055	0.41150	1.60468	0.26314	1.81655
6.3	0.62784	1.48152	0.40566	1.60785	0.25742	1.82192
6.4	0.62272	1.48248	0.39992	1.61096	0.25181	1.82718
6.5	0.61766	1.48342	0.39425	1.61401	0.24631	1.83232

Table 2. Variation of Local Nusselt number Nu_x and Sherwood number Sh_x for different M and λ_1.

| | $\lambda_2 = 1, S_c = 2, N_b = 0.5, N_t = 0.5, P_r = 6.2$ | | | | | |
| | $M = 2$ | | $M = 4$ | | $M = 6$ | |
λ_1	Nu_x	Sh_x	Nu_x	Sh_x	Nu_x	Sh_x
0.5	0.25129	1.78860	0.27363	1.85616	0.28527	1.88559
0.6	0.25465	1.79657	0.27603	1.86128	0.28698	1.88892
0.7	0.25737	1.80301	0.27791	1.86529	0.28829	1.89145
0.8	0.25962	1.80831	0.27943	1.86850	0.28933	1.89350
0.9	0.26152	1.81276	0.28068	1.87114	0.29018	1.89513
1.0	0.26314	1.81655	0.28172	1.87335	0.29087	1.89648
1.1	0.26453	1.81980	0.28261	1.87522	0.29146	1.89762
1.2	0.26575	1.82263	0.28337	1.87683	0.29196	1.89859
1.3	0.26682	1.82511	0.28403	1.87822	0.29239	1.89942
1.4	0.26776	1.82731	0.28461	1.87944	0.29277	1.90015
1.5	0.26861	1.82926	0.28513	1.88052	0.29310	1.90079

Table 3. Variation of Skin friction coefficient for different M and λ_1.

| | $\lambda_2 = 1, S_c = 2, N_b = 0.5, N_t = 0.5, P_r = 6.2$ | | |
| | $M = 2$ | $M = 4$ | $M = 6$ |
λ_1	C_f	C_f	C_f
0.5	1.12177	1.36687	1.51354
0.6	1.00998	1.20285	1.31469
0.7	0.91823	1.07391	1..16200
0.8	0.84163	0.96991	1.04108
0.9	0.77675	0.88425	0.94294
1.0	0.72109	0.81248	0.86171
1.1	0.67283	0.75148	0.79336
1.2	0.63060	0.69899	0.73505
1.3	0.59334	0.65335	0.68473
1.4	0.56022	0.61330	0.64086
1.5	0.53059	0.57788	0.60227

4. Conclusions

The current paper investigated the effects of uniform magnetic field of axisymmetric three-dimensional stagnation point flow of a nanofluid on a moving plate with different slip constants.

The governing equations were made dimensionless and then solved using the Richardson extrapolation enhancement method. The following are the findings of the above work:

- An increase in the magnetic field M and slip parameter λ_1 causes an increase in the velocity profile and decrease in the boundary layer thickness near the stagnation point.
- It is observed that in the absence of magnetic parameter M the boundary layer thickness is larger than while M is present.
- The thermal boundary layer increases with an increase in the thermophoresis parameter N_t and Brownian motion parameter N_b. It is observed that the thermal boundary layer is achieved earlier compared to the momentum boundary layer.
- It is observed that with the increase in S_c and N_t the nanoconcentration ϕ decreases and vice versa.

Acknowledgments: The author wishes to express his thanks to King Fahd University of Petroleum and Minerals and reviewers to improve the manuscript.

Abbreviations

The following abbreviations are used in this manuscript:

(u, v)	velocity Components
ν	kinematic viscosity
N_1, N_2	slip coefficient
T	temperature
α_m	thermal diffusivity
C	volume of nano particles
$(\rho C)_f$	heat capacity of fluid
D_B	Brownian diffusion coefficient
D_T	thermophoretic diffusion coefficient
λ_1, λ_2	slip parameters
N_t	thermophoresis parameter
N_b	browning motion parameter
C_f	skin friction coefficient
Nu_x	local Nusselt number
Sh_x	Sherwood number
Re_x	local Reynolds number
S_c	Schmidt number
Pr	prantle number
γ	ratio of slip parameters
ϕ	nano concentration
M	magnetic parameter

References

1. Borrelli, A.; Giantesio, G.; Patria, M.C. Numerical simulations of three-dimensional MHD stagnation-point flow of a micropolar fluid. *Comput. Math. Appl.* **2013**, *66*, 472–489. [CrossRef]
2. Lok, Y.Y.; Amin, N.; Pop, I. Non-orthogonal stagnation point flow towards a stretching shee. *Int. J. Non Linear Mech.* **2006**, *41*, 622–627. [CrossRef]
3. Tilley, B.S.; Weidman, P.D. Oblique two-fluid stagnation-point flow. *Eur. J. Mech. B Fluids* **1998**, *17*, 205–217. [CrossRef]
4. Grosan, T.; Pop, I.; Revnic, C.; Ingham, D.B. Magnetohydrodynamic oblique stagnation-point flow. *Meccanica* **2009**, *44*, 565. [CrossRef]
5. Wang, C.Y. Stagnation flow on a plate with anisotropic slip. *Eur. J. Mech. B Fluids* **2013**, *38*, 73–77. [CrossRef]
6. Wang, C.Y. Off-centered stagnation flow towards a rotating disc. *Int. J. Eng. Sci.* **2008**, *46*, 391–396. [CrossRef]

7. Wang, C.Y. Stagnation flow towards a shrinking sheet. *Int. J. Non Linear Mech.* **2008**, *43*, 377–382. [CrossRef]
8. Nadeem, S.; Hussain, A.; Khan, M. HAM solutions for boundary layer flow in the region of the stagnation point towards a stretching sheet. *Commun. Nonlinear Sci. Numer. Simul.* **2010**, *15*, 475–481. [CrossRef]
9. Ariel, P.D. Hiemenz flow in hydromagnetics. *Acta Mech.* **1994**, *103*, 31–43. [CrossRef]
10. Raju, C.S.; Sandeep, N. Heat and mass transfer in MHD non-Newtonian bio-convection flow over a rotating cone/plate with cross diffusion. *J. Mol. Liq.* **2016**, *215*, 115–126. [CrossRef]
11. Kleinstreuer, C.; Li, J.; Koo, J. Microfluidics of nano-drug delivery. *Int. J. Heat Mass Trans.* **2008**, *51*, 5590–5597. [CrossRef]
12. Choi, S.U.S. Enhancing thermal conductivity of fluids with nanoparticles. *ASME Publ. Fed* **1995**, *231*, 99–106.
13. Buongiorno, J. Convective transport in nanofluids. *J. Heat Transf.* **2006**, *128*, 240–250. [CrossRef]
14. Nadeem, S.; Saleem, S. Analytical study of third grade fluid over a rotating vertical cone in the presence of nanoparticles. *Int. J. Heat Mass Transf.* **2015**, *85*, 1041–1048. [CrossRef]
15. Bachok, N.; Ishak, A.; Nazar, R.; Pop, I. Flow and heat transfer at a general three-dimensional stagnation point in a nanofluid. *Phys. B Condens. Matter* **2010**, *405*, 4914–4918. [CrossRef]
16. Ellahi, R.; Aziz, S.; Zeeshan, A. Non Newtonian nanofluids flow through a porous medium between two coaxial cylinders with heat transfer and variable viscosity. *J. Porous Media* **2013**, *16*, 205–216. [CrossRef]
17. Sheikholeslami, M.; Ganji, D.; Javed, M.Y.; Ellahi, R. Effect of thermal radiation on nanofluid flow and heat transfer using two phase model. *J. Magn. Magn. Mater.* **2015**, *374*, 36–43. [CrossRef]
18. Makinde, O.D.; Khan, W.A.; Khan, Z.H. Buoyancy effects on MHD stagnation point flow and heat transfer of a nanofluid past a convectively heated stretching/shrinking sheet. *Int. J. Heat Mass Transf.* **2013**, *62*, 526–533. [CrossRef]
19. Junaid Ahmad Khan , M.; Mustafa, T.; Hayat, A.; Alsaedi, A. Three-dimensional flow of nanofluid over a non-linearly stretching sheet: An application to solar energy. *Int. J. Heat Mass Transf.* **2015**, *86*, 158–164. [CrossRef]
20. Upadhya, M.; Mahesha, S.; Raju, C.S.K. Unsteady Flow of Carreau Fluid in a Suspension of Dust and Graphene Nanoparticles With Cattaneo–Christov Heat Flux. *J. Heat Transf.* **2018**, *140*, 092401. [CrossRef]
21. Li, Z.; Sheikholeslami, M.; Ahmad Shafee, S.; Ali J Chamkha, S. Effect of dispersing nanoparticles on solidification process in existence of Lorenz forces in a permeable media. *J. Mol. Liq.* **2018**, *266*, 181–193. [CrossRef]
22. Raju, C.S.K.; Saleem, S.; Mamatha, S.U. Iqtadar Hussain, Heat and mass transport phenomena of radiated slender body of three revolutions with saturated porous: Buongiorno's model. *Int. J. Therm. Sci.* **2018**, *132*, 309–315. [CrossRef]
23. Ram, P.; Kumar, A. Analysis of Heat Transfer and Lifting Force in a Ferro-Nanofluid Based Porous Inclined Slider Bearing with Slip Conditions. *Nonlinear Eng.* **2018**. [CrossRef]
24. Soomro, F.A.; Hammouch, Z. Heat transfer analysis of CuO-water enclosed in a partially heated rhombus with heated square obstacle. *Int. J. Heat Mass Transf.* **2018**, *118*, 773–784.
25. Hayat, T.; Qayyum, S.; Alsaedi, A.; Ahmad, B. Results in Physics, Significant consequences of heat generation/absorption and homogeneous-heterogeneous reactions in second grade fluid due to rotating disk. *Results Phys.* **2018**, *8*, 223–230. [CrossRef]
26. Hussain, S.; Aziz, A.; Aziz, T.; Khalique, C.M. Slip Flow and Heat Transfer of Nanofluids over a Porous Plate Embedded in a Porous Medium with Temperature Dependent Viscosity and Thermal Conductivity. *Appl. Sci.* **2016**, *6*, 376. [CrossRef]
27. Anuar, N.; Bachok, N.; Pop, I. A Stability Analysis of Solutions in Boundary Layer Flow and Heat Transfer of Carbon Nanotubes over a Moving Plate with Slip Effect. *Energies* **2018**, *11*, 3243. [CrossRef]
28. Fetecau, C.; Vieru, D.; Azhar, W.A. Natural Convection Flow of Fractional Nanofluids Over an Isothermal Vertical Plate with Thermal Radiation. *Appl. Sci.* **2017**, *7*, 247. [CrossRef]
29. Khan, N.S.; Gul, T.; Islam, S.; Khan, I.; Alqahtani, A.M.; Alshomrani, A.S. Alqahtani and Ali Saleh Alshomrani, Magnetohydrodynamic Nanoliquid Thin Film Sprayed on a Stretching Cylinder with Heat Transfer. *Appl. Sci.* **2017**, *7*, 271. [CrossRef]

6

Analytical Study of the Head-On Collision Process between Hydroelastic Solitary Waves in the Presence of a Uniform Current

Muhammad Mubashir Bhatti [1,2] **and Dong Qiang Lu** [1,2,*]

1   Shanghai Institute of Applied Mathematics and Mechanics, Shanghai University, Yanchang Road, Shanghai 200072, China; muhammad09@shu.edu.cn
2   Shanghai Key Laboratory of Mechanics in Energy Engineering, Yanchang Road, Shanghai 200072, China
*   Correspondence: dqlu@shu.edu.cn

Abstract: The present study discusses an analytical simulation of the head-on collision between a pair of hydroelastic solitary waves propagating in the opposite directions in the presence of a uniform current. An infinite thin elastic plate is floating on the surface of water. The mathematical modeling of the thin elastic plate is based on the Euler–Bernoulli beam model. The resulting kinematic and dynamic boundary conditions are highly nonlinear, which are solved analytically with the help of a singular perturbation method. The Poincaré–Lighthill–Kuo method is applied to obtain the solution of the nonlinear partial differential equations. The resulting solutions are presented separately for the left- and right-going waves. The behavior of all the emerging parameters are presented mathematically and discussed graphically for the phase shift, maximum run-up amplitude, distortion profile, wave speed, and solitary wave profile. It is found that the presence of a current strongly affects the wavelength and wave speed of both solitary waves. A graphical comparison with pure-gravity waves is also presented as a particular case of our study.

Keywords: nonlinear hydroelastic waves; uniform current; thin elastic plate; solitary waves; PLK method

---

1. Introduction

The interaction between a deformable body and a moving fluid has received remarkable attention due to its numerous applications in offshore, polar engineering and industrial problems. Some applications in transportation systems can be observed in the cold region, where the ice sheet is treated as runways and roads, while air-cushion vehicles are very helpful in breaking the ice. These kinds of problems involve various mathematical challenges and present significant difficulties in the mathematical modeling of wave motion and ice deformation. Most of the previous theoretical and numerical results based on linear wave theories are not applicable to large amplitude waves. Hydroelasticity is associated with the deformation of elastic bodies due to hydrodynamics excitations, and together, these excitations are a result of body deformation. In hydroelastic problems, the elastic body and the fluid motion are coupled, which indicates that the deformation of the elastic body relies on the hydrodynamic forces and vice versa. Hydroelastic problems are difficult to analyze numerically and theoretically because, on the surface of the elastic body, hydrodynamic forces actively depend on the accelerations of the surface displacements.

In the past few years, various theoretical and numerical studies have been presented with the help of the Kirchhoff–Love plate theory to examine hydroelastic wave problems. For instance, Xia and Shen [1] analyzed the nonlinear interaction between hydroelastic solitary waves covered with ice. They used a simple perturbation method to obtain the solution for the nonlinear equations. They found that the wavelength, shape and celerity of nonlinear solitary waves depend on the wave amplitude. The wave speed is less than the wave speed in an open water region. Milewski et al. [2] discussed hydroelastic waves in deep water using a numerical method. They used a nonlinear model for an elastic plate and particularly discussed the dynamics of unforced and forced waves. Vanden–Broeck and Părău [3] investigated the generalized form of hydroelastic periodic and solitary waves in a two-dimensional channel. They derived weakly nonlinear solutions using a perturbation scheme, and fully nonlinear solutions were obtained with the help of a numerical method. Deike et al. [4] experimentally examined the behavior of nonlinear and linear waves propagating beneath an elastic sheet in the presence of flexural waves and surface tension. By using an optical method to derive a full space–time wave field, Deike et al. [4] observed that nonlinear shift occurs due to tension in a sheet by transverse motion of the fundamental mode of an elastic plate. They further noticed that the separation between associated timescales is satisfactory at each scale of a turbulent cascade which coincides with theoretical results. Wang and Lu [5] studied nonlinear hydroelastic waves traveling under an infinite elastic plate on the surface of deep water through the homotopy analysis method.

In the studies mentioned above, less attention has been given to hydroelastic waves in the presence of a uniform current. There are different reasons, i.e., thermal, wind, and tidal effects and the rotation of the earth, why ocean currents are often produced. According to engineering applications, it is essential to determine the behavior of current when it is required to perform refraction calculations, examine the water particle acceleration and velocities for force calculations on ocean structures [6] and calculate the wave height from subsurface pressure recordings. The presence of a current influences the wave speed and affects the observed wave period and the relationship between wavelengths. Physically, when the wave travels from one region to another region in the presence of a current, not only will the wavelength and wave speed change but also, probably, current-induced refraction will occur; furthermore, the wave height will be affected. Schulkes et al. [7] analyzed hydroelastic waves in the presence of a uniform current using linear potential flow theory. Bhattacharjee and Sahoo [8] addressed the interaction of flexural gravity waves with the wave current. They also used a linear approach to discuss the physical features of a floating elastic plate under the impact of a current. Later, Bhattacharjee and Sahoo [9] examined the effect of a uniform current on flexural-gravity waves that occur due to an initial disturbance at a point. Mohanty et al. [10] explored the simultaneous effects of compressive forces and a current on time-dependent hydroelastic waves with both single- and double-layer fluids propagating through a finite and infinite depth in a two-dimensional channel. They presented the asymptotic results for the Green function and the deflection of the elastic plate using the stationary phase method. Lu and Yeung [11] examined the unsteady flexural-gravity waves that occur due to the interaction of a fixed concentrated line load with the impact of a uniform current. They observed that the flexural-gravity wave motion depends on the ratio of the current speed to the group or phase speeds.

In recent decades, various authors have investigated the collision between solitary waves using different methodologies from different geometrical aspects [12–15]. Gardner et al. [16] introduced the inverse scattering transform method to determine the exact solution of the Korteweg–de Vries (KdV) equation and discussed various engrossing characteristics of the collision between solitary waves. According to this technique, one can easily obtain the solution for overtaking solitary waves, but this technique is not suitable for determining the solution of a head-on collision process between solitary waves. When two solitary waves come close to each other, they collide and transfer their positions and energies with each other. After separating, they regain their original shapes and positions. During this entire process of interaction, both solitary waves are very stable and preserve their identities. The features of solitary waves such as striking and colliding, can only be maintained in a conservative system. Su and Mirie [17] studied the head-on collision between two solitary waves with the help of the Poincaré–Lighthill–Kuo (PLK) method. Later, Mirie and Su [18] again numerically studied the head-on collision between solitary waves and observed that after the collision of solitary waves, they recovered their original amplitudes and positions; however, a difference of less than 2% was observed. Dai [19] investigated solitary waves at the interface of a two-layer fluid and considered a rigid bottom and surface of the channel. Mirie and Su [20] examined the head-on collision between internal solitary waves using a perturbation method. With the third-order solution, they observed that the amplitude and energy of the wave train diminish with time. Later, Mirie and Su [21] considered a similar mathematical modeling [20] with a different asymptotic expansion and derived a modified form of the KdV solution. They concluded that the collision process is inelastic, and a dispersive wave train occurs behind each emerging solitary wave. Recently, Ozden and Demiray [22] explored the work of Su and Mirie [17] with a different asymptotic assumption of the trajectory functions. The order of the trajectory functions considered by Su and Mirie [17] is ε, where ε is the perturbation parameter related to the wave amplitude. Ozden and Demiray [22] considered the order of trajectory functions to be ε^2 with a similar definition for ε. Marin and Öchsner [23] discussed the initial boundary value problem for a dipolar medium using the Green–Naghdi thermoelastic theory. Some more relevant studies on the head-on collision in single and two-layer fluids can be found in Refs. [24–28].

According to the previously published results, less attention has been given to hydroelastic solitary waves, and no attempt has been made to analyze the head-on collision mechanism between hydroelastic solitary waves in the presence of a uniform current. Recently, Bhatti and Lu [29] examined the head-on collision between two hydroelastic solitary waves using the Euler–Bernoulli beam model in the presence of compression. Therefore, the present study aims to discuss the head-on collision between two hydroelastic solitary waves under uniform current and surface tension effects. We apply a singular perturbation method to obtain the analytic results for the highly nonlinear coupled partial differential equations. The PLK method is the most appropriate technique to determine the collision properties, i.e., the head-on collision, wave speed, phase shift, distortion profile, and maximum run-up amplitude. The resulting series solutions are presented up to the third-order approximation. A graphical comparison with previously published results is also presented.

2. Mathematical Formulation

Consider a pair of nonlinear hydroelastic solitary waves propagating in the opposite directions through a finite channel. A Cartesian coordinate is selected to formulate the mathematical model, i.e., the x-axis is proposed to lie along the horizontal direction, and the z-axis is considered to lie along the vertical direction as shown in Figure 1. A thin elastic plate is floating on the surface of water at $z = H(x, t)$, and the horizontal bottom is located at $z = 0$. Let U_c be the intensity of an underlying uniform current moving from left to right ($U_c > 0$). An opposing current is defined as that moving from the right to left ($U_c < 0$). The normal velocity of the governing fluid is taken as zero. The fluid is supposed to be incompressible, homogenous and inviscid, and the motion be irrotational. The velocity field in terms of potential function $\phi(x, z, t)$ satisfies

$$\nabla^2 \phi = 0, \qquad\qquad (0 < z < H). \qquad\qquad (1)$$

The bottom boundary condition at $z = 0$ is written as

$$\frac{\partial \phi}{\partial z} = 0. \qquad\qquad (2)$$

The kinematic boundary condition at the water–plate interface ($z = H(x, t)$) is defined as [11,29]

$$\frac{\partial H}{\partial t} + U_c \frac{\partial H}{\partial x} + \nabla \phi \cdot \nabla H = \frac{\partial \phi}{\partial z}. \qquad\qquad (3)$$

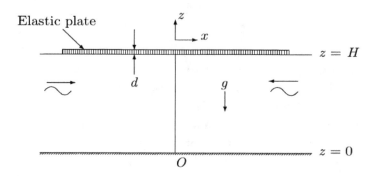

Figure 1. Schematic diagram.

The dynamic boundary condition reads [11,29]

$$\frac{\partial \phi}{\partial t} + \frac{1}{2}\left(U_c + |\nabla \phi|\right)^2 + gH - \frac{T}{\rho}\frac{\partial^2 H}{\partial x^2} + \frac{P_e}{\rho} = B_c(t). \tag{4}$$

In the above equation, U_c is a uniform current, g the gravitational acceleration, ρ the density of the fluid, and $B_c(t)$ the Bernoulli constant which is considered to be zero. T is coefficient of surface tension of the fluid. The expression for pressure P_e consists on the Euler–Bernoulli beam theory which can be written as

$$P_e = D\frac{\partial^4 H}{\partial x^4} + M\frac{\partial^2 H}{\partial t^2}, \tag{5}$$

where $D = Ed^3/[12(1-v^2)]$ is the flexural rigidity of the plate, $M = \rho_e d$, E Young's modulus, d constant thickness, v Poisson's ratio, and ρ_e the uniform mass density of the elastic plate.

With the help of Equations (1) and (2), the potential function $\phi(x,z,t)$ can be describe using the Taylor series expansion at $z = 0$, we have [17]

$$\phi(x,z,t) = \sum_{i=0}^{\infty}(-1)^i\frac{z^{2i}}{(2i)!}\nabla^{2i}\Phi, \tag{6}$$

where

$$\phi(x,0,t) = \Phi(x,t). \tag{7}$$

Using Equation (6), the kinematic and dynamic boundary condition can be obtained as

$$\frac{\partial H}{\partial t} + U_c\frac{\partial H}{\partial x} + \nabla \cdot \left[\sum_{i=0}^{\infty}(-1)^i\frac{H^{2i+1}}{(2i+1)!}\nabla^{2i}(\nabla\Phi)\right] = 0, \tag{8}$$

$$\begin{aligned}
&\frac{\partial \Phi}{\partial t} + \frac{1}{2}\left(U_c + |\nabla\phi|\right)^2 + gH + \frac{P_e}{\rho} - \frac{T}{\rho}\frac{\partial^2 H}{\partial x^2} \\
&+ \sum_{i=1}^{\infty}(-1)^i\frac{H^{2i}}{(2i)!}\left[\nabla^{2i}\Phi_t + U_c\nabla^{2j+1}\Phi + \frac{1}{2}\sum_{j=0}^{2i}(-1)^jC_j^{2i}\nabla^{j+1}\Phi * \nabla^{2i-j+1}\Phi\right],
\end{aligned} \tag{9}$$

where

$$C_j^{2i} = \binom{2i}{j} = \frac{2i!}{j!(2i-j)!}, \tag{10}$$

is a binomial coefficient. The asterisk in Equation (9) indicates an inner vector product for the multiplication of even j and odd i.

Equations (8) and (9) can be simplified in the following form as

$$\frac{\partial H}{\partial t} + U_c \frac{\partial H}{\partial x} + \frac{\partial}{\partial x}\left[HU + \sum_{i=1}^{\infty}(-1)^i \frac{H^{2i+1}}{(2i+1)!}\frac{\partial^{2i}U}{\partial x^{2i}} \right] = 0, \tag{11}$$

$$\begin{aligned}\frac{\partial U}{\partial t} &+ U_c \frac{\partial U}{\partial x} + \frac{\partial}{\partial x}\left[gH - \frac{T}{\rho}\frac{\partial^2 H}{\partial x^2} + \frac{U^2}{2} + \frac{P_e}{\rho} + \sum_{i=1}^{\infty}(-1)^i \frac{H^{2i}}{(2i)!}\left(\frac{\partial^{2i}U}{\partial t\partial x^{2i-1}}\right.\right.\\ &\left.\left.+ U_c \frac{\partial^{2i}U}{\partial x^{2i}} + \frac{1}{2}\sum_{j=0}^{\infty}(-1)^j C_j^{2i}\frac{\partial^{2i-j}U}{\partial x^{2i-j}}\frac{\partial^j U}{\partial x^j}\right)\right], \end{aligned}\tag{12}$$

where $U = \partial\Phi/\partial x$ is the tangential velocity at the bottom of the channel.

3. Solution Methodology

We will employ the PLK method in the ensuing section. Let us introduce the following coordinate transformations in the wave frame, we have

$$\xi_0 = \varepsilon^\delta k(x - Ct), \qquad\qquad \eta_0 = \varepsilon^\delta \bar{k}(x + \bar{C}t), \tag{13}$$

where k and $\bar{k}$ are the wave numbers of order unity for the right- and left-going waves, respectively; ε with $0 < \varepsilon \ll 1$ is a dimensionless parameter that represents the order of magnitude of the wave amplitude. C and $\bar{C}$ are the wave speeds of the right- and left-going solitary waves. Using the method of strained coordinates, we introduce the following transformation of wave frame coordinates with phase functions:

$$\xi_0 = \xi - \varepsilon k\theta(\xi,\eta), \qquad\qquad \eta_0 = \eta - \varepsilon\bar{k}\varphi(\xi,\eta), \tag{14}$$

where $\theta(\xi,\eta)$ and $\varphi(\xi,\eta)$ are the phase functions to be deduced in the perturbation analysis. The purpose of these functions is to obtain the asymptotic approximations which acquiesce us to analyze the phase changes due to a collision.

According to Ursell's theory of shallow water waves, we consider the scaling of the horizontal wavelength as $\delta = 1/2$. Although other values of δ are also possible, i.e., $0, 1/4, 1/8, 1$, these values have some restriction in our present case. However, these values are used by different authors [21,24,25] to derive various forms of the KdV equation in a two-layer fluid model but fail to give a KdV equation in our case. Later, Dai et al. [27] used $\delta = 0$ to discuss the head-on collision among solitary waves propagating in a compressible Mooney–Rivlin elastic rod. The value of δ plays a significant role and mainly depends upon physical and mathematical assumptions of the governing problem.

Let

$$H = H_0(1+\zeta), \tag{15}$$

where ζ is the non-dimensional elevation of the plate–fluid interface, and H_0 is the undisturbed depth of the fluid. Considering the linear part of Equations (11) and (12), and assuming the linear solutions of U and ζ take form of $\exp^{i(kx-\omega t)}$, where ω is the wave frequency, then we have the phase speed for the right-going wave as

$$C_c = \frac{c_0}{\chi_2}\left(F + \sqrt{F^2 + \chi_1\chi_2}\right), \tag{16}$$

and for the left-going wave it reads

$$\bar{C}_c = \frac{c_0}{\chi_2}\left(-F + \sqrt{F^2 + \chi_1\chi_2}\right), \tag{17}$$

where

$$\chi_1 = 1 + \Gamma(kH_0)^4 + \tau(kH_0)^2 - F^2, \tag{18}$$

$$\chi_2 = 1 + \sigma(kH_0)^2, \tag{19}$$

$$c_0 = \sqrt{gH_0}, \quad \Gamma = \frac{D}{\rho g H_0^4}, \quad \sigma = \frac{M}{\rho H_0}, \quad \tau = \frac{T}{\rho g H_0^2}, \quad F = \frac{U_c}{c_0}. \tag{20}$$

The value c_0 is the phase speed for pure gravity waves in shallow water of finite depth. The hydroelastic phase speed can be reduced for pure-gravity waves [17] by taking $D \to 0$, $M \to 0$, $T \to 0$, and $U_c \to 0$.

For convenience, let us introduce a column vector $\mathbf{T}$ defined as

$$\mathbf{T} = \begin{pmatrix} \zeta \\ U \end{pmatrix}. \tag{21}$$

New variables are introduced in the form of the following power series:

$$\theta(\xi, \eta) = \theta_0(\eta) + \varepsilon\theta_1(\xi, \eta) + \varepsilon^2\theta_2(\xi, \eta) + \dots \tag{22}$$

$$\varphi(\xi, \eta) = \varphi_0(\xi) + \varepsilon\varphi_1(\xi, \eta) + \varepsilon^2\varphi_2(\xi, \eta) + \dots \tag{23}$$

$$C = C_c \left(1 + \varepsilon a c_1 + \varepsilon^2 a^2 c_2 + \dots\right), \tag{24}$$

$$\overline{C} = \overline{C}_c \left(1 + \varepsilon b \bar{c}_1 + \varepsilon^2 b^2 \bar{c}_2 + \dots\right), \tag{25}$$

$$\mathbf{T} = \varepsilon\mathbf{T}_1(\xi, \eta) + \varepsilon^2\mathbf{T}_2(\xi, \eta) + \varepsilon^3\mathbf{T}_3(\xi, \eta) + \dots, \tag{26}$$

where c_n and $\bar{c}_n$ (i.e., $n = 1, 2, \dots$) are beneficial for removing the secular terms during the solution procedure; a and b are the amplitude factors.

4. Perturbation Analysis

Substituting Equation (22) into the resulting nonlinear partial differential equations, we obtain a set of coupled differential equations with coefficients in the form of $\varepsilon^{\frac{3}{2}}, \varepsilon^{\frac{5}{2}}, \varepsilon^{\frac{7}{2}}, \dots$, which are expressed in a sequence as follows.

4.1. Coefficients of $\varepsilon^{3/2}$

The system of first-order equations reduces to the following form as

$$k\mathbf{N}\frac{\partial \mathbf{T}_1}{\partial \xi} + \bar{k}\overline{\mathbf{N}}\frac{\partial \mathbf{T}_1}{\partial \eta} = 0, \tag{27}$$

where

$$\mathbf{N} = \begin{pmatrix} C_c\beta_- & 1 \\ c_0^2 & C_c\beta_- \end{pmatrix}, \quad \overline{\mathbf{N}} = \begin{pmatrix} \overline{C}_c\beta_+ & 1 \\ c_0^2 & \overline{C}_c\beta_+ \end{pmatrix}, \tag{28}$$

$$\beta_- = -1 + F\sqrt{\chi}, \quad \beta_+ = 1 + F\sqrt{\overline{\chi}}, \quad \chi = \frac{c_0^2}{C_c^2}, \quad \overline{\chi} = \frac{c_0^2}{\overline{C}_c^2}. \tag{29}$$

We define a matrix system to determine the solutions of the first order equations and higher order equations. Su and Mirie [17] introduced a new form of transformation to obtain the solutions, but the transformation fails to provide a solution for the hydroelastic wave speed (see Equation (16)). Zhu and Dai [24] successfully used a similar methodology for a two-layer fluid model, and Dai et al. [27] used a matrix system to examine the solutions for a single-layer fluid model.

The right and left characteristic vectors of $\mathbf{N}$ and $\overline{\mathbf{N}}$ are

$$\mathbf{R} = \begin{pmatrix} 1 \\ -C_c \beta_- \end{pmatrix}, \quad \overline{\mathbf{R}} = \begin{pmatrix} 1 \\ -\overline{C}_c \beta_+ \end{pmatrix}, \tag{30}$$

$$\mathbf{L} = \begin{pmatrix} 1, & -\dfrac{1}{C_c \beta_-} \end{pmatrix}, \quad \overline{\mathbf{L}} = \begin{pmatrix} 1, & -\dfrac{1}{\overline{C}_c \beta_+} \end{pmatrix}. \tag{31}$$

The right and left characteristic vectors in Equations (30) and (31) are introduced to determine the solution at each order. The right characteristic vectors are helpful for obtaining the solution at each order of approximation, whereas the left characteristic vectors are beneficial for solving the coupled equations at a higher order. In higher-order approximations, the resulting equations are more complex and highly coupled. It is impossible to solve the equations directly. Therefore, left characteristic vectors are beneficial for making the coupled equations into one equation at each order of approximation.

Let us consider the solution of Equation (27) in the following form

$$\mathbf{T}_1 = aA(\xi)\mathbf{R} + bB(\eta)\overline{\mathbf{R}}, \tag{32}$$

where $A(\xi)$ and $B(\eta)$ are arbitrary functions to be determined in the next order.

The first-order solution can be written as

$$\zeta_1 = aA(\xi) + bB(\eta), \tag{33}$$

$$U_1 = -\left(\overline{C}_c \beta_+ bB(\eta) + C_c \beta_- aA(\xi)\right). \tag{34}$$

In the above equation, by taking $F = 0$, the present results reduce to results similar to those obtained by Bhatti and Lu [29] for hydroelastic solitary waves.

4.2. Coefficients of $\varepsilon^{5/2}$

The system of second-order equations reduces to the following form as

$$\begin{aligned}
&\mathbf{N}k\frac{\partial \mathbf{T}_2}{\partial \xi} + aC_c k\left(\mathbf{E}_1 A' + \mathbf{E}_2 AA' + \mathbf{E}_3 A''' + \mathbf{E}_4 A'\right) \\
&+\overline{\mathbf{N}}k\frac{\partial \mathbf{T}_2}{\partial \eta} + b\overline{C}_c k\left(\overline{\mathbf{E}}_1 B' + \overline{\mathbf{E}}_2 BB' + \overline{\mathbf{E}}_3 B''' + \overline{\mathbf{E}}_4 B'\right) = 0,
\end{aligned} \tag{35}$$

where $\mathbf{E}_n$ and $\overline{\mathbf{E}}_n$ i.e., $(n = 1, 2, 3, 4)$ are presented in Appendix A.

Let us assume a general solution of the following form

$$\mathbf{T}_2 = X(\xi, \eta)\mathbf{R} + Y(\xi, \eta)\overline{\mathbf{R}}. \tag{36}$$

Using Equation (36) in Equation (35), and multiplying it by $\mathbf{L}$ and $\overline{\mathbf{L}}$, we obtain

$$\begin{aligned}
&\mathbf{L}\overline{\mathbf{N}}\mathbf{R}k\frac{\partial X}{\partial \eta} + aC_c k\left(\mathbf{L}\mathbf{E}_1 A' + \mathbf{L}\mathbf{E}_2 AA' + \mathbf{L}\mathbf{E}_3 A''' + \mathbf{L}\mathbf{E}_4 A'\right) \\
&+b\overline{C}_c k\left(\mathbf{L}\overline{\mathbf{E}}_1 B' + \mathbf{L}\overline{\mathbf{E}}_2 BB' + \mathbf{L}\overline{\mathbf{E}}_3 B''' + \mathbf{L}\overline{\mathbf{E}}_4 B'\right) = 0,
\end{aligned} \tag{37}$$

$$\overline{\mathbf{LNR}}k\frac{\partial Y}{\partial \overline{\xi}} + aC_ck\left(\overline{\mathbf{LE}}_1 A' + \overline{\mathbf{LE}}_2 AA' + \overline{\mathbf{LE}}_3 A''' + \overline{\mathbf{LE}}_4 A'\right)$$
$$+ b\overline{C}_c\overline{k}\left(\overline{\mathbf{LE}}_1 B' + \overline{\mathbf{LE}}_2 BB' + \overline{\mathbf{LE}}_3 B''' + \overline{\mathbf{LE}}_4 B'\right) = 0, \tag{38}$$

where $\mathbf{L\overline{N}R}$, $\overline{\mathbf{LNR}}$, $\mathbf{LE}_n$ and $\mathbf{L\overline{E}}_n$ ($n = 1, 2, 3, 4$) represent the inner products $\mathbf{L} \cdot \overline{\mathbf{N}} \cdot \mathbf{R}$, $\overline{\mathbf{L}} \cdot \mathbf{N} \cdot \overline{\mathbf{R}}$, $\mathbf{L} \cdot \mathbf{E}_n$ and $\mathbf{L} \cdot \overline{\mathbf{E}}_n$, respectively.

The above equation is further divided into three parts namely, (i) secular terms, (ii) local terms and (iii) non-local terms.

4.2.1. Secular Terms

Secular terms in Equation (37) are those terms which do not depend on η. The other terms cannot be treated as secular terms, because if we integrate these terms with respect to η, then these terms become unbounded in space and time and show a secular behavior. The secular terms are

$$\mathbf{LE}_1 A' + \mathbf{LE}_2 AA' + \mathbf{LE}_3 A''' = 0. \tag{39}$$

Let

$$c_1 = \frac{1}{2}, \quad k^2 H_0^2 = 3a, \tag{40}$$

then Equation (39) reduces to the following form

$$\gamma A''' - 3\beta_- AA' - A' = 0, \tag{41}$$

where

$$\gamma = -\frac{1}{\beta_-}\left(\beta_-^2 + 3\alpha\right), \quad \alpha = \sigma - \chi\tau. \tag{42}$$

The solution of the above KdV equation is found as

$$A = -\frac{1}{\beta_-}\text{sech}^2\frac{\xi}{2\sqrt{\gamma}}. \tag{43}$$

Similarly,

$$\overline{\mathbf{LE}}_1 B' + \overline{\mathbf{LE}}_2 BB' + \overline{\mathbf{LE}}_3 B''' = 0. \tag{44}$$

Let

$$\bar{c}_1 = \frac{1}{2}, \quad \bar{k}^2 H_0^2 = 3b. \tag{45}$$

Then Equation (44) reduces to the following form

$$\overline{\gamma}B''' + 3\beta_+ BB' - B' = 0, \tag{46}$$

$$B = \frac{1}{\beta_+}\text{sech}^2\frac{\eta}{2\sqrt{\overline{\gamma}}}, \tag{47}$$

where

$$\overline{\gamma} = \frac{1}{\beta_+}\left(\beta_+^2 + 3\alpha\right). \tag{48}$$

We have obtained a third-order KdV equation for the hydroelastic wave profile. A third-order KdV profile was also presented by Bhatti and Lu [29] in the absence of a uniform current, i.e., $F = 0$. Furthermore, the stiffness of the plate appears in the third-order approximation.

4.2.2. Non-Local Terms

The non-local terms are not secular. However, these terms are helpful for determining the phase shifts. Therefore, we will leave these terms as they are. The non-local terms appearing in Equation (37) are

$$\mathbf{LE_4} = 0. \tag{49}$$

It follows that

$$\theta_0 = \frac{b}{kY} \int_{-\infty}^{\eta} B d\eta_1, \tag{50}$$

where

$$Y = \frac{2\beta_+ \beta_- - \sqrt{\chi \overline{\chi}} - S\beta_-^2}{\beta_-(2\beta_+ + S\beta_-)}, \quad S = \frac{C_c}{\overline{\overline{C_c}}}. \tag{51}$$

Similarly, we have

$$\overline{\mathbf{LE_4}} = 0. \tag{52}$$

It follows that

$$\varphi_0 = \frac{a}{k\overline{Y}} \int_{+\infty}^{\xi} A d\xi_1, \tag{53}$$

where

$$\overline{Y} = \frac{2\beta_- \beta_+ - \sqrt{\chi \overline{\chi}} - \overline{S}\beta_+^2}{\beta_+(2\beta_- + \overline{S}\beta_+)}, \quad \overline{S} = \frac{\overline{C_c}}{C_c}. \tag{54}$$

4.2.3. Local Terms

The local terms are those terms that are helpful in examining the wave speed for the left- and right-going solitary waves. The local terms appearing in Equation (37) can be written in the following form

$$\mathbf{L\overline{N}R\overline{k}} \frac{\partial X}{\partial \eta} + b\overline{C_c}\overline{k} \left(\mathbf{L\overline{E}_1} B' + \mathbf{L\overline{E}_2} BB' + \mathbf{L\overline{E}_3} B''' + \mathbf{L\overline{E}_4} B' \right) = 0. \tag{55}$$

Integrating the above equation with respect to η, we obtain the resulting equation after simplification

$$X(\xi, \eta) = C_1 b^2 B + \frac{C_2 b^2}{2} B^2 + \frac{C_3 b^2}{\overline{\gamma}} \left(B - \frac{3\beta_+}{2} B^2 \right) + C_4 ab AB + a^2 A_1(\xi), \tag{56}$$

where C_n $(n = 1 \ldots 4)$ are presented in Appendix B.
Similarly

$$\overline{\mathbf{LN}}\overline{R}k \frac{\partial Y}{\partial \xi} + aC_c k \left(\overline{\mathbf{LE}_1} A' + \overline{\mathbf{LE}_2} AA' + \overline{\mathbf{LE}_3} A''' + \overline{\mathbf{LE}_4} A' \right). \tag{57}$$

Integrating the above equation with respect to ξ, we get the resulting equation after simplification

$$Y(\xi, \eta) = \overline{C}_1 a^2 A + \frac{a^2 \overline{C}_2}{2} A^2 + \frac{\overline{C}_3 a^2}{\gamma} \left(A + \frac{3\beta_-}{2} A^2 \right) + \overline{C}_4 ab AB + b^2 B_1(\eta), \tag{58}$$

where $\overline{C}_n$ $(n = 1 \ldots 4)$ are presented in Appendix C. The unknown arbitrary functions $A_1(\xi)$ and $B_1(\eta)$ will be determined in the next order.

4.3. Coefficients of $\varepsilon^{7/2}$

In the third-order system, we obtain the following equation

$$\begin{aligned}
&Nk \frac{\partial T_3}{\partial \xi} + aC_c k \left(F_1 A' + F_2 AA' + F_3 A' A^2 + F_4 \left(A' A_1 + AA_1' \right) + F_5 A_1' + F_6 A_1''' \right. \\
&\left. + F_7 A' + \overline{F}_8 A' \right) + \overline{N}k \frac{\partial T_3}{\partial \eta} + b\overline{C}_c \overline{k} \left(\overline{F}_1 B' + \overline{F}_2 BB' + \overline{F}_3 B' B^2 \right. \\
&\left. + \overline{F}_4 \left(B' B_1 + B_1' B \right) + \overline{F}_5 B_1' + \overline{F}_6 B_1''' + \overline{F}_7 B' + \overline{F}_8 B' \right) = 0,
\end{aligned} \tag{59}$$

where F_n and $\overline{F}_n$ i.e., $(n = 1 \ldots 8)$ are presented in Appendix A.

Let us assume a general solution of the following form

$$T_3 = X_1(\xi, \eta) \mathbf{R} + Y_1(\xi, \eta) \overline{\mathbf{R}}. \tag{60}$$

Using Equation (60) in Equation (59), and multiplying it by $\mathbf{L}$ and $\overline{\mathbf{L}}$, we obtain

$$\begin{aligned}
&\mathbf{L\overline{N}R}\overline{k} \frac{\partial X_1}{\partial \eta} + aC_c k \left[(\mathbf{LF}_1 + \mathbf{LF}_2 A + \mathbf{LF}_3 A^2 + \mathbf{LF}_7 + \mathbf{L\overline{F}}_8) A' + \mathbf{LF}_5 A_1' \right. \\
&+ \mathbf{LF}_6 A_1''' + \mathbf{LF}_4 \left(A' A_1 + AA_1' \right) \right] + b\overline{C}_c \overline{k} \left[(\mathbf{LF}_8 + \mathbf{L\overline{F}}_1 + \mathbf{L\overline{F}}_7 + \mathbf{L\overline{F}}_2 B \right. \\
&\left. + \mathbf{L\overline{F}}_3 B^2) B' + \mathbf{L\overline{F}}_4 \left(B' B_1 + B_1' B \right) + \mathbf{L\overline{F}}_5 B_1' + \mathbf{L\overline{F}}_6 B_1''' \right] = 0,
\end{aligned} \tag{61}$$

$$\begin{aligned}
&\mathbf{\overline{L}NR}k \frac{\partial Y_1}{\partial \xi} + aC_c k \left[(\mathbf{\overline{L}F}_1 + \mathbf{\overline{L}F}_2 A + \mathbf{\overline{L}F}_3 A^2 + \mathbf{\overline{L}F}_7 + \mathbf{\overline{L}F}_8) A' + \mathbf{\overline{L}F}_5 A_1' \right. \\
&+ \mathbf{\overline{L}F}_6 A_1''' + \mathbf{\overline{L}F}_4 \left(A' A_1 + AA_1' \right) \right] + b\overline{C}_c \overline{k} \left[(\mathbf{\overline{L}F}_8 + \mathbf{\overline{L}F}_1 + \mathbf{\overline{L}F}_7 + \mathbf{\overline{L}F}_2 B \right. \\
&\left. + \mathbf{\overline{L}F}_3 B^2) B' + \mathbf{\overline{L}F}_4 \left(B' B_1 + B_1' B \right) + \mathbf{\overline{L}F}_5 B_1' + \mathbf{\overline{L}F}_6 B_1''' \right] = 0,
\end{aligned} \tag{62}$$

where $\mathbf{L\overline{N}R}$, $\mathbf{\overline{L}N\overline{R}}$, $\mathbf{LF}_n$ and $\mathbf{L\overline{F}}_n$ $(n = 1 \ldots 8)$ represent the inner products $\mathbf{L} \cdot \overline{\mathbf{N}} \cdot \mathbf{R}$, $\overline{\mathbf{L}} \cdot \mathbf{N} \cdot \overline{\mathbf{R}}$, $\mathbf{L} \cdot \mathbf{F}_n$ and $\mathbf{L} \cdot \overline{\mathbf{F}}_n$, respectively.

The above equation is further divided into the following three parts: (i) secular terms, (ii) local terms and (iii) non-local terms.

4.3.1. Secular Terms

The secular terms appearing in this order are found as

$$(\mathbf{LF}_1 + \mathbf{LF}_2 A + \mathbf{LF}_3 A^2) A' + \mathbf{LF}_4 \left(A' A_1 + AA_1' \right) + \mathbf{LF}_5 A_1' + \mathbf{LF}_6 A_1'''. \tag{63}$$

The above equation is simplified as

$$A_1''' - A_1' - 3\beta_- \left(A_1 A' + A_1' A \right) + \left(-2c_2 + C_6 \right) A' + C_7 AA' + C_8 A^2 A'. \tag{64}$$

Upon integrating the above equation we get

$$A_1'' - A_1 - 3\beta_- A_1 A + \left(-2c_2 + C_6 \right) A + \frac{C_7}{2} A^2 + \frac{C_8}{3} A^3, \tag{65}$$

where C_6, C_7 and C_8 are presented in Appendix B.

Let

$$c_2 = \frac{C_6}{2}. \tag{66}$$

Then the solution of Equation (64) can be written as

$$A_1 = C_9 A + C_{10} A^2, \tag{67}$$

where C_9 and C_{10} are presented in Appendix B.

Similarly, we have

$$(\overline{\mathbf{LF}}_1 + \overline{\mathbf{LF}}_2 B + \overline{\mathbf{LF}}_3 B^2) B' + \overline{\mathbf{LF}}_4 (B' B_1 + B_1' B) + \overline{\mathbf{LF}}_5 B_1' + \overline{\mathbf{LF}}_6 B_1'''. \tag{68}$$

The above equation is simplified as

$$B_1''' - B_1' + 3\beta_+ (B_1 B' + B_1' B) - (2\bar{c}_2 + \overline{C}_6) B' - \overline{C}_7 A A' - \overline{C}_8 B^2 B'. \tag{69}$$

Upon integrating the above equation, we obtain

$$B_1'' - B_1 + 3\beta_+ B_1 B - (2\bar{c}_2 + \overline{C}_6) B - \frac{\overline{C}_7}{2} B^2 - \frac{\overline{C}_8}{3} B^3, \tag{70}$$

where $\overline{C}_6, \overline{C}_7$ and $\overline{C}_8$ are presented in Appendix C.

Let

$$\bar{c}_2 = -\frac{\overline{C}_6}{2}. \tag{71}$$

Then the solution of Equation (69) can be written as

$$B_1 = \overline{C}_9 B + \overline{C}_{10} B^2, \tag{72}$$

where $\overline{C}_9$ and $\overline{C}_{10}$ are presented in Appendix C.

This completes the solutions for Equations (56) and (58).

4.3.2. Non-Local Terms

The non-local terms appearing in this order are found as

$$\mathbf{LF}_7 A' = 0. \tag{73}$$

The above equation can be written as

$$\theta_1 = \bar{\theta}_{1,0} \int_{-\infty}^{\eta} B d\eta_1 + \bar{\theta}_{1,1} \int_{-\infty}^{\eta} B^2 d\eta_1, \tag{74}$$

where

$$\bar{\theta}_{1,0} = \frac{b}{\bar{\theta}_{1,2}} \left[C_{11} + a C_{12} A - \frac{C_{14} + a C_{15} A}{\beta_-} \right], \tag{75}$$

$$\bar{\theta}_{1,1} = \frac{b^2}{\bar{\theta}_{1,2}} \left[C_{13} - \frac{C_{16}}{\beta_-} \right], \tag{76}$$

$$\bar{\theta}_{1,2} = -\bar{k}\left[\frac{2\beta_+\beta_- - \sqrt{\chi\bar{\chi}} - S\beta_-^2}{\beta_-}\right],\tag{77}$$

and C_n ($n = 11\ldots 16$) are presented in Appendix B.

Similarly, we have

$$\mathbf{L\bar{F}}_7 B' = 0.\tag{78}$$

The above equation reduces to

$$\varphi_1 = \bar{\varphi}_{1,0}\int_{+\infty}^{\xi} A\,d\xi_1 + \bar{\varphi}_{1,1}\int_{+\infty}^{\xi} A^2\,d\xi_1,\tag{79}$$

where

$$\bar{\varphi}_{1,0} = \frac{a}{\bar{\varphi}_{1,2}}\left[\bar{C}_{11} + b\bar{C}_{12}B - \frac{\bar{C}_{14} + b\bar{C}_{15}B}{\beta_-}\right],\tag{80}$$

$$\bar{\varphi}_{1,1} = \frac{a^2}{\bar{\varphi}_{1,2}}\left[\bar{C}_{13} - \frac{\bar{C}_{16}}{\beta_-}\right],\tag{81}$$

$$\bar{\varphi}_{1,2} = -k\left[\frac{2\beta_-\beta_+ - \sqrt{\chi\bar{\chi}} - \bar{S}\beta_+^2}{\beta_+}\right],\tag{82}$$

and $\bar{C}_n$ ($n = 11\ldots 16$) are presented in Appendix C.

In Equation (74), all the terms appearing are similar to the first-order phase shift and show a simple phase shift behavior except for the third term in $\bar{\theta}_{1,0}$. Few terms in $\bar{\theta}_{1,0}$ depend on ξ when $\eta \to +\infty$; therefore, the wave profile is different before and after the collision process (see Figures 6 and 7) because θ_1 enters into the argument of function $A(\xi)$. A similar behavior has been observed for the left-going wave.

4.3.3. Local Terms

The local terms are found as

$$\mathbf{L\bar{N}R}\bar{k}\frac{\partial X_1}{\partial \eta} + b\bar{C}_c\bar{k}\left[(\mathbf{LF}_8 + \mathbf{L\bar{F}}_1 + \mathbf{L\bar{F}}_7 + \mathbf{L\bar{F}}_2 B + \mathbf{L\bar{F}}_3 B^2)B'\right.$$
$$\left. + \mathbf{L\bar{F}}_4(B'B_1 + B_1'B) + \mathbf{L\bar{F}}_5 B_1' + \mathbf{L\bar{F}}_6 B_1'''\right] + aC_c k\mathbf{L\bar{F}}_8 A' = 0.\tag{83}$$

Integrating the above equation, we obtain

$$X_1 = \frac{1}{C_5}\left(C_{17}B + C_{18}B^2 + C_{19}B^3\right) + a^3 A_2(\xi),\tag{84}$$

where C_{17}, C_{18} and C_{19} are presented in Appendix B.

Similarly, we have

$$\mathbf{\bar{L}NR}\bar{k}\frac{\partial Y_1}{\partial \xi} + aC_c k\left[(\mathbf{\bar{L}F}_1 + \mathbf{\bar{L}F}_2 A + \mathbf{\bar{L}F}_3 A^2 + \mathbf{\bar{L}F}_7 + \mathbf{\bar{L}F}_8)A' + \mathbf{\bar{L}F}_5 A_1'\right.$$
$$\left. + \mathbf{\bar{L}F}_6 A_1''' + \mathbf{\bar{L}F}_4(A'A_1 + AA_1')\right] + b\bar{C}_c\bar{k}\mathbf{\bar{L}F}_8 = 0.\tag{85}$$

Integrating the above equation, we obtain

$$Y_1 = \frac{1}{\overline{C}_5}(\overline{C}_{17}A + \overline{C}_{18}A^2 + \overline{C}_{19}A^3) + b^3 B_2(\eta), \tag{86}$$

where $\overline{C}_{17}, \overline{C}_{18}$ and $\overline{C}_{19}$ are presented in Appendix C.

In the above equation, $A_2(\xi)$ and $B_2(\eta)$ are the undetermined functions. For further analysis, we will end our calculations here, and the solutions for $A_2(\xi)$ and $B_2(\eta)$ are neglected.

5. Analytical Results

The series solutions in the preceding section are summarized in the following form.

The interfacial elevation at the water–plate interface can be written with the help of Equations (32) and (36), and we have

$$\zeta = \varepsilon(aA + bB) + \varepsilon^2 [X(\xi, \eta) + Y(\xi, \eta)], \tag{87}$$

where $X(\xi, \eta)$ and $Y(\xi, \eta)$ are defined in Equations (56) and (58).

The distortion profile can be calculated with the help of Equation (87). Therefore, the functions that are products of $B(\eta)$ and $A(\xi)$ must be removed. For this purpose, by taking $B(\eta) = 0$ in Equation (87), the distortion profile at the water–plate interface can be written as

$$\zeta = a\varepsilon A + \varepsilon^2 a^2 \left[\overline{C}_1 A + \frac{\overline{C}_2}{2}A^2 + \frac{\overline{C}_3}{\gamma}\left(A + \frac{3\beta_-}{2}A^2\right) + a^2 A_1(\xi)\right]. \tag{88}$$

The maximum run-up $\zeta_{\max}$ during the collision process at the water–plate interface can be obtained by taking $A = B = 1$ in Equation (87), namely

$$\zeta_{\max}\Big|_{A=B=1} = \zeta. \tag{89}$$

Following from Equations (32) and (36), the velocity at the bottom reads

$$U = -\varepsilon \left[\overline{C}_c\beta_+ bB + C_c\beta_- aA\right] - \varepsilon^2 \left[C_c\beta_- X(\xi, \eta) + \overline{C}_c\beta_+ Y(\xi, \eta)\right]. \tag{90}$$

Using Equations (40) and (65), the series solutions for the right- and left-going wave speeds are given by

$$C = C_c \left(1 + \frac{1}{2}\varepsilon a + \frac{C_6}{2}\varepsilon^2 a^2 + O(\varepsilon^3)\right), \tag{91}$$

$$\overline{C} = \overline{C}_c \left(1 + \frac{1}{2}\varepsilon b - \frac{\overline{C}_6}{2}\varepsilon^2 b^2 + O(\varepsilon^3)\right). \tag{92}$$

The phase shifts for the right- and left-going wave read

$$\theta = \theta_0 + \varepsilon\theta_1 + O(\varepsilon^2), \tag{93}$$

$$\varphi = \varphi_0 + \varepsilon\varphi_1 + O(\varepsilon^2), \tag{94}$$

where $\theta_0, \theta_1, \varphi_0$, and φ_1 are given in Equations (50), (53), (74) and (79), respectively.

6. Graphical Analysis

This section describes the graphical behaviors of all the physical parameters involved in the governing two-dimensional hydroelastic wave problem. To determine the results in a more significant manner, Figures 2–16 depict the water–plate interface, distortion profile, wave speed, phase shift, and maximum run-up amplitude during the collision process. We consider the physical parameters, unless otherwise stated, $E = 10^6\,\mathrm{N\,m^{-2}}$, $d = 0.05\,\mathrm{m}$, $F = 0.3\,\mathrm{m\,s^{-1}}$, $g = 9.8\,\mathrm{m\,s^{-2}}$, $\rho = 10^3\,\mathrm{kg\,m^{-3}}$, $T = 0.075\,\mathrm{N\,m^{-1}}$, $H_0 = 1\,\mathrm{m}$ and $\rho_e = 917\,\mathrm{kg\,m^{-3}}$ for the graphical results. It is worth mentioning here that by taking $\sigma = 0$, $\Gamma = 0$, $\tau = 0$, and $F = 0$ in Equations (3) and (4), the present results reduce to those obtained by Su and Mirie [17] for pure-gravity waves.

Figure 2 shows the wave profile for different values of Γ and σ. When the effects of the elastic plate are taken into account, then significantly changes in the wave profile are observed. The parameter Γ is directly proportional to the plate thickness d and Young's modulus E. When Γ and σ increase, the plate becomes significantly stiffer, which produces a reactive force that opposes the deformation of hydroelastic waves. Figure 3 is plotted to see the effect of the current on the solitary wave profile. We can see from this figure that when the current is moving from right to left, $F < 0$; then, the wavelength, amplitude and speed of the solitary waves are affected as shown in the region $x \in [-50,0]$. However, when $F > 0$, a similar and converse behavior is found for the second solitary waves in the region $x \in [0,50]$. Figure 4 shows that an increment in the surface tension parameter τ significantly diminishes the wave profile and that the wave profile before and after the collision process becomes narrower as the surface tension increases. Figure 5 shows a graphical comparison with previously published results presented by Su and Mirie [17]. We can observe that when $\Gamma = 0$, $\sigma = 0$, $\tau = 0$, and $F = 0$, our results are in excellent agreement with those of Su and Mirie [17] for pure gravity waves, which ensures the validity of the present results and the methodology used.

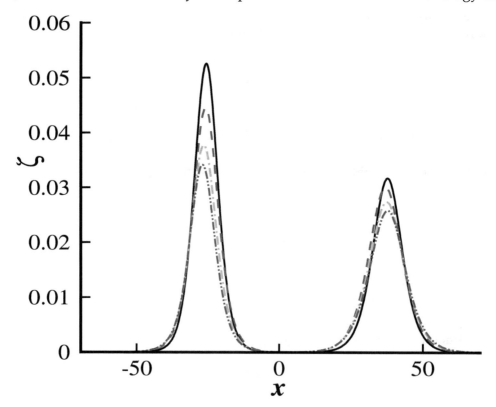

Figure 2. Head-on collision between two solitary waves for different values of Γ and σ. Solid line: $\Gamma = 0$, $\sigma = 0$; dashed line: $\Gamma = 0.07$, $\sigma = 0.5$; dot-dashed line: $\Gamma = 0.09$, $\sigma = 1.1$; dot-dot-dashed line: $\Gamma = 0.1$, $\sigma = 1.4$.

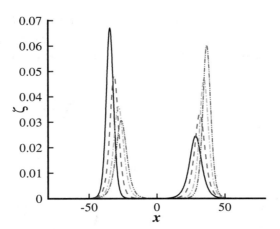

Figure 3. Head-on collision between two solitary waves for different values of F. Solid line: $F = -0.3$ (opposing current); dashed line: $F = -0.1$ (no current); dot-dashed line: $F = 0.3$ (following current); dot-dot-dashed line: $F = 0.5$ (following current).

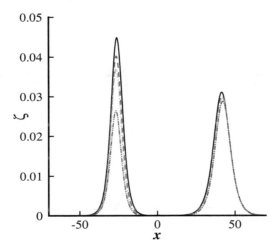

Figure 4. Head-on collision between two solitary waves for different values of τ. Solid line: $\tau = 0.5$; dashed line: $\tau = 0.7$; dot-dashed line: $\tau = 0.8$; dot-dot-dashed line: $\tau = 1$.

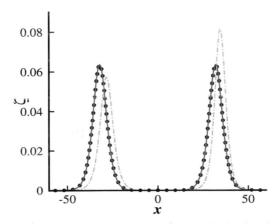

Figure 5. Comparison of head-on collision between two solitary waves with previously published results. Solid line (present results): $\Gamma = 0$, $\sigma = 0$, $\tau = 0$, $F = 0$; circles (Su and Mirie [17]): $\Gamma = 0$, $\sigma = 0$, $\tau = 0$, $F = 0$; dot-dashed line: $\Gamma = 0.11$, $\sigma = 0.09$, $\tau = 0.01$, $F = 0.3$.

Figures 6 and 7 show the distortion profile during the head-on collision process for $F > 0$ and $F < 0$. In both figures, we can see that before the collision process, the wave profile is similar for $F > 0$ and $F < 0$. We can observe from these figures that during the collision process, the wave profile tilts backward in the direction of wave propagation. However, the wave profile remains symmetric before the collision process. Further, we can see that the wave profile is less affected by the following current $F > 0$ compared with the opposing current $F < 0$. A similar behavior was also observed by Su and Mirie [17] for pure gravity waves and Bhatti and Lu [29] for nonlinear hydroelastic waves in the presence of a compressive force.

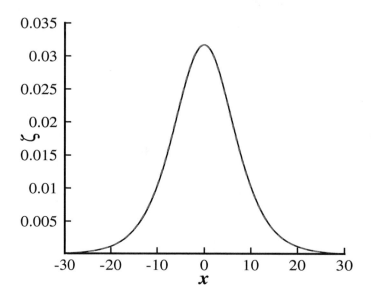

Figure 6. Distortion profile for $F > 0$ (following current). Solid line: before collision; dashed line: after collision.

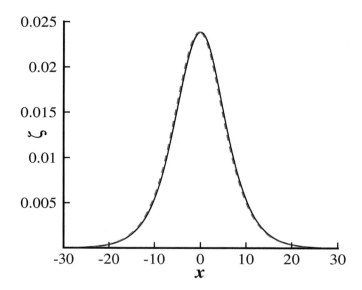

Figure 7. Distortion profile for $F < 0$ (opposing current). Solid line: before collision; dashed line: after collision.

Figures 8–10 are plotted for the phase shift profile against different values of Γ, σ, τ and F. It can be noted from Figure 8 that the initially phase shift profile increases due to the presence of the elastic plate, while its behavior becomes the opposite as the wave amplitude rises. Figure 9 shows that the phase shift profile increases for higher values of the following current ($F > 0$), whereas its behavior is converse for higher values of the opposing current ($F < 0$). Figure 10 shows the effects of the surface tension τ on the phase shift profile. From this figure, we observe that the surface tension results are uniform throughout the domain, whereas the phase shift remarkably decreases due to a strong influence of the surface tension.

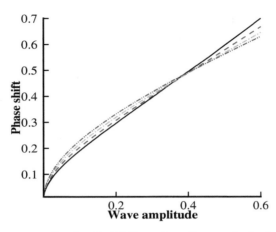

Figure 8. Phase shift *vs* wave amplitude. Solid line: $\Gamma = 0$, $\sigma = 0$; dashed line: $\Gamma = 0.11$, $\sigma = 0.09$; dot-dashed line: $\Gamma = 0.88$, $\sigma = 0.18$; dot-dot-dashed line: $\Gamma = 3$, $\sigma = 0.27$.

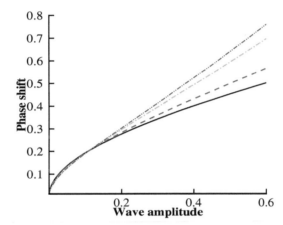

Figure 9. Phase shift *vs* wave amplitude. Solid line: $F = -0.3$ (opposing current); dashed line: $F = -0.1$ (opposing current); dot-dashed line: $F = 0.3$ (following current); dot-dot-dashed line: $F = 0.5$ (following current).

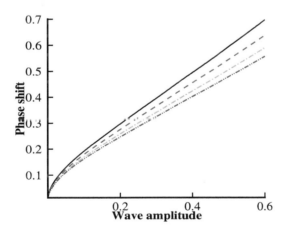

Figure 10. Phase shift *vs* wave amplitude. Solid line: $\tau = 0.1$; dashed line: $\tau = 0.5$; dot-dashed line: $\tau = 0.8$; dot-dot-dashed line: $\tau = 1$.

Figures 11–13 show the variation of the wave speed with multiple values of Γ, σ, τ and F. In Figure 11, we can observe that the wave speed decreases significantly when Γ increases. Furthermore, the behavior of the left-going wave speed will be opposite to that of the right-going one. Figure 12 is plotted to analyze the wave speed behavior for the left- and right-going solitary waves. In Figure 12, we can easily observe that the left-going wave speed tends to diminish with the following current $F > 0$, while the behavior is opposite for the right-going solitary wave. From Figure 13, we note that the surface tension strongly influences the wave speed. We can see that the wave speed decreases due to an increment in the surface tension parameter τ.

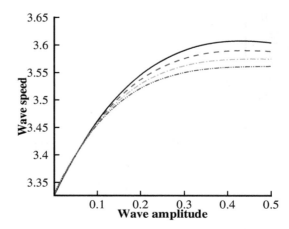

Figure 11. Wave speed *vs* wave amplitude. Solid line: $\Gamma = 0.3$; dashed line: $\Gamma = 0.4$; dot-dashed line: $\Gamma = 0.5$; dot-dot-dashed line: $\Gamma = 0.6$.

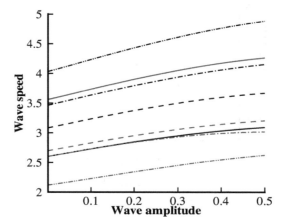

Figure 12. Wave speed *vs* wave amplitude. Solid line: $F = -0.4$ (opposing current); dashed line: $F = 0$ (no current); dot-dashed line: $F = 0.5$ (following current); dot-dot-dashed line: $F = 1$ (following current). Red line: left-going wave; black line: right-going wave.

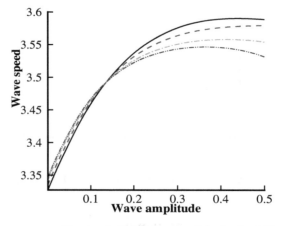

Figure 13. Wave speed *vs* wave amplitude. Solid line: $\tau = 0.1$; dashed line: $\tau = 0.3$; dot-dashed line: $\tau = 0.5$; dot-dot-dashed line: $\tau = 0.6$.

Figures 14–16 present the variations in the maximum run-up during the collision process. It can be noted from Figure 14 that the plate deflection creates an opposing force, which tends to resist the maximum run-up amplitude. Therefore, an increment of Γ or σ tends to diminish the maximum run-up amplitude during the collision process. However, in Figure 15, we can see that the maximum run-up amplitude increases for negative values of the current $F < 0$, whereas its behavior is similar for higher values of the current when $F > 0$. It can be observed from Figure 16 that the surface tension parameter τ significantly affects the wave speed compared with Γ, F and σ. An enhancement in the surface tension tends to reduce the maximum run-up amplitude.

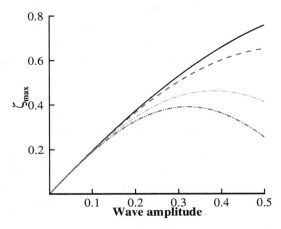

Figure 14. Maximum run-up vs wave amplitude. Solid line: $\Gamma = \sigma = 0$; dashed line: $\Gamma = 0.07, \sigma = 0.5$; dot-dashed line: $\Gamma = 0.09, \sigma = 1.1$; dot-dot-dashed line: $\Gamma = 0.1, \sigma = 1.4$.

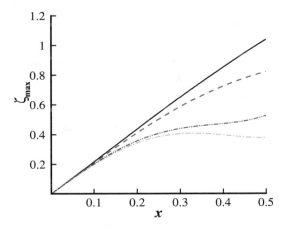

Figure 15. Maximum run-up vs wave amplitude. Solid line: $F = -0.5$ (opposing current); dashed line: $F = -0.4$ (opposing current); dot-dashed line: $F = 0.4$ (following current); dot-dot-dashed line: $F = 0.5$ (following current).

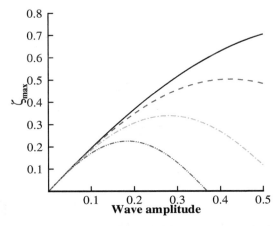

Figure 16. Maximum run-up vs wave amplitude. Solid line: $\tau = 0.5$; dashed line: $\tau = 1.0$; dot-dashed line: $\tau = 1.5$; dot-dot-dashed line: $\tau = 2$.

7. Concluding Remarks

In this article, we analytically studied the behavior of a current in the head-on collision process between a pair of hydroelastic solitary waves propagating in the opposite directions. The Poincaré–Lighthill–Kuo method is successfully applied to obtain the solution for the governing nonlinear partial differential equations. The resulting solutions are presented up to the third-order approximation for right- and left-going solitary waves. The impact of different essential parameters is discussed with the help of graphs and presented mathematically for the water–plate interface, wave speed, phase shift, distortion profile, and maximum run-up amplitude during the collision process.

A graphical comparison with previously published results is also presented, and it is found that the present results are in excellent agreement, which ensures that the results for the hydroelastic wave problem are correct. It is also found that the presence of a thin elastic plate significantly reduces the amplitude of the wave profile. Furthermore, we noted that the presence of a current not only affects the wavelength and wave amplitude but also produces a remarkable effect on the wave speed. The phase shift markedly decreases due to the more significant influence of the elastic plate. It is also noted that the phase shift tends to increase for the following current, whereas its behavior is converse for the opposing current. The maximum run-up amplitude increases due to the strong influence of the following and opposing currents. It is observed that very small tilting occurs in the distortion profile during the collision process for positive values of the current, whereas greater tilting in the wave profile is seen for negative current values.

Author Contributions: D.Q.L.; conceptualization, M.M.B.; methodology, validation, formal analysis and writing–original draft preparation, D.Q.L.; investigation, resources, supervision, project administration, funding acquisition and writing–review and editing.

Acknowledgments: This research was sponsored by the National Natural Science Foundation of China under Grant No. 11872239.

Appendix A

$$\mathbf{E}_1 = -ac_1 \begin{pmatrix} 1 \\ -C_c\beta_- \end{pmatrix}, \tag{A1}$$

$$\mathbf{E}_2 = -a\beta_- \begin{pmatrix} 2 \\ -C_c\beta_- \end{pmatrix}, \tag{A2}$$

$$\mathbf{E}_3 = 3a \begin{pmatrix} \dfrac{\beta_-}{6} \\ C_c \left(\dfrac{\beta_-^2}{2} + \alpha \right) \end{pmatrix}, \tag{A3}$$

$$\mathbf{E}_4 = \begin{pmatrix} -abk\left(\overline{S}\beta_+ + \beta_-\right)B + 2ak\bar{k}\left(\overline{S}\beta_+ - \beta_-\right)\dfrac{\partial\theta_0}{\partial\eta} \\ ak\overline{C}_c \left(b\beta_+\beta_-B + \bar{k}\dfrac{\partial\theta_0}{\partial\eta}\left(\sqrt{\chi\overline{\chi}} - \beta_-\beta_+\right) \right) \end{pmatrix}, \tag{A4}$$

$$\overline{\mathbf{E}}_1 = b\bar{c}_1 \begin{pmatrix} 1 \\ \overline{C}_c\beta_+ \end{pmatrix}, \tag{A5}$$

$$\overline{\mathbf{E}}_2 = b\beta_+ \begin{pmatrix} 2 \\ \overline{C}_c\beta_+ \end{pmatrix}, \tag{A6}$$

$$\overline{\mathbf{E}}_3 = 3b \begin{pmatrix} -\dfrac{\beta_+}{6} \\ \overline{C}_c \left(\dfrac{-\beta_+^2}{2} + \alpha \right) \end{pmatrix}, \tag{A7}$$

$$\overline{\mathbf{E}}_4 = \begin{pmatrix} -ab\bar{k}\left(\beta_+ + S\beta_-\right)A - bk\bar{k}\left(\beta_+ - S\beta_-\right)\dfrac{\partial\varphi_0}{\partial\xi} \\ b\bar{k}C_c \left(a\beta_+\beta_-A + k\dfrac{\partial\varphi_0}{\partial\xi}\left(\sqrt{\chi\overline{\chi}} - \beta_-\beta_+\right) \right) \end{pmatrix}, \tag{A8}$$

$$\mathbf{F}_1 = a^2 \begin{pmatrix} \dfrac{1}{2\gamma^2}\left(\overline{C}_3 S\beta_+ - \overline{C}_3\gamma + \overline{C}_1 S\beta_+\gamma - \gamma^2(\overline{C}_1 + 2c_2)\right) \\ C_c \left[c_1\overline{C}_1 S\beta_+ + c_2\beta_- + \dfrac{9\Gamma\chi}{\gamma^2} + \dfrac{3\overline{C}_3\sigma}{\gamma^2} - \dfrac{3S\overline{C}_3\beta_+\beta_-}{2\gamma^2} \right. \\ \left. -\dfrac{9\beta_-^2}{2\gamma^2} + \dfrac{3\overline{C}_1\sigma}{\gamma} + \dfrac{S\overline{C}_3\beta_+}{2\gamma} + \dfrac{3S\beta_+\beta_-\overline{C}_1}{2\gamma} - \dfrac{3\beta_-}{4\gamma} \right. \\ \left. -\dfrac{3\overline{C}_3\tau\chi}{\gamma^2} - \dfrac{3\overline{C}_1\tau\chi}{\gamma} \right] \end{pmatrix}, \tag{A9}$$

$$\mathbf{F}_2 = a^2 \begin{pmatrix} \left[-2\left\{ (S\beta_+ + \beta_-)\left(\overline{C}_1 + \dfrac{\overline{C}_3}{\gamma}\right)\right\} - c_1\left(\overline{C}_2 + \dfrac{3\overline{C}_3\beta_-}{\gamma}\right) \right. \\ \left. + \dfrac{3S\beta_+}{\gamma^2}\left(15\overline{C}_3\beta_- + 4\overline{C}_2\gamma + 3\overline{C}_1\gamma\beta_-\right) \right] \\ C_c \left[2S\beta_+\beta_-\left(\overline{C}_1 + \dfrac{\overline{C}_3}{\gamma}\right) + 3\left(\dfrac{\alpha - c_1 S\beta_+\beta_-}{\gamma^2}\right) \right. \\ \left. (15\overline{C}_3\beta_- + 4\overline{C}_2\gamma + 3\gamma\overline{C}_1\beta_-) + c_1 S\beta_+\left(\overline{C}_2 + \dfrac{3\overline{C}_3\beta_-}{\gamma}\right) \right. \\ \left. + \dfrac{135\Gamma\beta_-}{\gamma^2} - \dfrac{45\beta_-^3}{8\gamma^2} + \dfrac{\beta_-^2}{4\gamma} \right] \end{pmatrix}, \tag{A10}$$

$$\mathbf{F}_3 = a^2 \begin{pmatrix} -3c_1(S\beta_+ + \beta_-)\left(\overline{C}_2 + \dfrac{3\overline{C}_3\beta_-}{\gamma}\right) + 15c_1^2 S\beta_+\left(\dfrac{3\overline{C}_3\beta_-^2}{\gamma^2} + \dfrac{\overline{C}_2\beta_-}{\gamma}\right) \\ C_c\beta_- \left[\dfrac{45c_1(3\overline{C}_3\beta_- + \overline{C}_2\gamma)(2\alpha - S\beta_-\beta_+)}{\gamma^2} + 3c_1 S\beta_+ \right. \\ \left. \left(\overline{C}_2 + \dfrac{3\overline{C}_3\beta_-}{\gamma}\right) + \dfrac{405\chi\Gamma\beta_-}{\gamma^2} + \dfrac{135c_1\beta_-^3}{8\gamma^2} + \dfrac{27c_1\beta_-^2}{2\gamma} \right] \end{pmatrix}, \tag{A11}$$

$$\mathbf{F}_4 = \mathbf{E}_2, \ \ \mathbf{F}_5 = \mathbf{E}_1, \ \ \mathbf{F}_6 = \mathbf{E}_3, \tag{A12}$$

$$\mathbf{F}_7 = \begin{pmatrix} bC_{11}B + abC_{12}AB + b^2C_{13}B^2 + \bar{k}(\overline{S} - \beta_-)\dfrac{\partial \theta_1}{\partial \eta} \\[2mm] C_{\mathrm{c}}\left(bC_{14}B + abC_{15}AB + b^2C_{16}B^2 + \bar{k}\left(\sqrt{\overline{\chi}\chi} - \beta_-\beta_+\right)\dfrac{\partial \theta_1}{\partial \eta}\right) \end{pmatrix}, \tag{A13}$$

$$\mathbf{F}_8 = -k\overline{C}_{\mathrm{c}}\dfrac{\partial \theta_1}{\partial \xi}\begin{pmatrix} 0 \\[1mm] S^2\chi + \beta_+^2 \end{pmatrix}, \tag{A14}$$

$$\overline{\mathbf{F}}_1 = b^2 \begin{pmatrix} \dfrac{1}{2}C_1 + \bar{c}_2 + \dfrac{3C_3 S\beta_-}{6\overline{\gamma}^2} + \dfrac{C_3}{2\overline{\gamma}} + \dfrac{3C_1 S\beta_-}{6\overline{\gamma}} \\[3mm] \overline{C}_{\mathrm{c}}\left[-\bar{c}_2\beta_+ - \dfrac{1}{2}C_1 S\beta_- + \dfrac{9\overline{\chi}\Gamma}{\overline{\gamma}^2} + \dfrac{3C_3\sigma}{\overline{\gamma}^2} - \dfrac{9\beta_+^2}{24\overline{\gamma}^2} - \dfrac{3C_3 S\beta_-\beta_+}{2\overline{\gamma}}^2 \right. \\[3mm] \left. + \dfrac{3c_1\sigma}{\overline{\gamma}} + \dfrac{3\beta_+}{4\overline{\gamma}} - \dfrac{C_3 S\beta_-}{2\overline{\gamma}} + \dfrac{3c_1 S\beta_-\beta_+}{2\overline{\gamma}} - 3\overline{\chi}\tau\left(\dfrac{C_3}{\overline{\gamma}^2} + \dfrac{C_1}{\overline{\gamma}}\right)\right] \end{pmatrix}, \tag{A15}$$

$$\overline{\mathbf{F}}_2 = b^2 \begin{pmatrix} \left[-2\left\{(\beta_+ + S\beta_-)\left(C_1 + \dfrac{C_3}{\overline{\gamma}}\right)\right\} + \dfrac{1}{2}\left(C_2 - \dfrac{3\overline{C}_3\beta_+}{\overline{\gamma}}\right)\right. \\[3mm] \left. + \dfrac{3S\beta_-}{6\overline{\gamma}^2}\left(15C_3\beta_+ + 4C_2\overline{\gamma} + 3C_1\overline{\gamma}\beta_+\right)\right] \\[3mm] \overline{C}_{\mathrm{c}}\left[2S\beta_+\beta_-\left(C_1 + \dfrac{C_3}{\overline{\gamma}}\right) - 3\left(\dfrac{\alpha + c_1\beta_+\beta_-}{\overline{\gamma}^2}\right)\right. \\[3mm] \left(15\overline{C}_3\beta_- - 4\overline{C}_2\overline{\gamma} + 3\overline{\gamma}\overline{C}_1\beta_-\right) - \bar{c}_1 S\beta_+\left(C_2 + \dfrac{3C_3\beta_-}{\overline{\gamma}}\right) \\[3mm] \left. - \dfrac{135\Gamma S^2\beta_+}{\overline{\gamma}^2} + \dfrac{45S^2\beta_+^3}{8\overline{\gamma}^2} + \dfrac{\beta_+^2}{4\overline{\gamma}}\right] \end{pmatrix}, \tag{A16}$$

$$\overline{\mathbf{F}}_3 = b^2 \begin{pmatrix} -3c_1(\beta_+ + S\beta_-)\left(C_2 + \dfrac{3C_3\beta_-}{\overline{\gamma}}\right) + 15c_1^2 S\beta_-\left(\dfrac{3C_3\beta_+^2}{\overline{\gamma}^2} - \dfrac{C_2\beta_+}{\overline{\gamma}}\right) \\[3mm] \overline{C}_{\mathrm{c}}\left[\dfrac{45c_1\beta_+(3C_3\beta_- - C_2\overline{\gamma})(2\alpha + S\beta_-\beta_+)}{\overline{\gamma}^2} + 3c_1 S\beta_+\beta_- \right. \\[3mm] \left. \left(C_2 - \dfrac{3C_3\beta_+}{\overline{\gamma}}\right) + \dfrac{405\Gamma\chi\beta_+^2}{\overline{\gamma}^2} - \dfrac{135c_1 S^2\beta_+^4}{8\gamma^2} - \dfrac{27c_1\beta_+^3}{\overline{\gamma}}\right] \end{pmatrix}, \tag{A17}$$

$$\overline{\mathbf{F}}_4 = \overline{\mathbf{E}}_2, \ \ \overline{\mathbf{F}}_5 = \overline{\mathbf{E}}_1, \ \ \overline{\mathbf{F}}_6 = \overline{\mathbf{E}}_3, \tag{A18}$$

$$\overline{\mathbf{F}}_7 = \begin{pmatrix} a\overline{C}_{11}A + ab\overline{C}_{12}AB + a^2\overline{C}_{13}A^2 + k(\beta_+ + S\beta_-)\dfrac{\partial \varphi_1}{\partial \xi} \\[2mm] \overline{C}_{\mathrm{c}}\left(a\overline{C}_{14}A + ab\overline{C}_{15}AB + ab\overline{C}_{16}A^2 + k\left(\sqrt{\overline{\chi}\chi} - \beta_-\beta_+\right)\dfrac{\partial \varphi_1}{\partial \xi}\right) \end{pmatrix}, \tag{A19}$$

$$\overline{\mathbf{F}}_8 = \bar{k}\dfrac{\partial \varphi_1}{\partial \eta}\begin{pmatrix} -2\beta_- \\[1mm] C_{\mathrm{c}}(\beta_-^2 - \overline{S}\sqrt{\overline{\chi}}) \end{pmatrix}, \tag{A20}$$

where C_n and $\overline{C}_n$ ($n = 11 \ldots 16$) are defined in Appendixs B and C.

Appendix B

$$C_1 = -\frac{\overline{S}\beta_+ + \beta_-}{2C_5\beta_-}, \tag{A21}$$

$$C_2 = \frac{\beta_+(\overline{S}\beta_+ + 2\beta_-)}{C_5\beta_-}, \tag{A22}$$

$$C_3 = \frac{3\overline{S}}{C_5\beta_-}\left(3\beta_+^2 + 6\alpha - S\beta_-\beta_+\right), \tag{A23}$$

$$C_4 = \frac{1}{C_5}\left(2\beta_+ + \overline{S}\beta_- + \frac{C_c\overline{Y}(\chi - \beta_-^2)}{\beta_-}\right), \tag{A24}$$

$$C_5 = 2\beta_+ - \frac{S}{\beta_-} - \frac{\sqrt{\chi}\sqrt{\overline{\chi}}}{\beta_-}, \tag{A25}$$

$$C_6 = \frac{1}{2\gamma^2}\left(\overline{C}_3\overline{S}\beta_+ - \overline{C}_3\gamma + \overline{C}_1\overline{S}\beta_+\gamma - \gamma^2\overline{C}_1\right) - 2\beta_- A_2 - \frac{1}{\beta_-}\left[c_1\overline{C}_1\overline{S}\beta_+ + c_2\beta_- + \frac{9\Gamma\chi}{\gamma^2}\right.$$
$$+\frac{3\overline{C}_3\sigma}{\gamma^2} - \frac{3\overline{S}C_3\beta_+\beta_-}{2\gamma^2} - \frac{9\beta_-^2}{2\gamma^2} + \frac{3\overline{C}_1\sigma}{\gamma} + \frac{\overline{S}C_3\beta_+}{2\gamma} + \frac{3\overline{S}\beta_+\beta_-\overline{C}_1}{2\gamma} - \frac{3\beta_-}{4\gamma} - \frac{3\overline{C}_3\tau\chi}{\gamma^2}$$
$$\left.-\frac{3\overline{C}_1\tau\chi}{\gamma}\right], \tag{A26}$$

$$C_7 = \left[-2\left\{(\overline{S}\beta_+ + \beta_-)\left(\overline{C}_1 + \frac{\overline{C}_3}{\gamma}\right)\right\} - c_1\left(\overline{C}_2 + \frac{3\overline{C}_3\beta_-}{\gamma}\right) + (15\overline{C}_3\beta_- + 4\overline{C}_2\gamma\right.$$
$$+3\overline{C}_1\gamma\beta_-)\frac{3\overline{S}\beta_+}{\gamma^2}\right] - \frac{1}{\beta_-}\left[2\overline{S}\beta_+\beta_-\left(\overline{C}_1 + \frac{\overline{C}_3}{\gamma}\right) + 3\left(\frac{\alpha - c_1\overline{S}\beta_+\beta_-}{\gamma^2}\right)(15\overline{C}_3\beta_-\right.$$
$$+4\overline{C}_2\gamma + 3\gamma\overline{C}_1\beta_-) + c_1\overline{S}\beta_+\left(\overline{C}_2 + \frac{3\overline{C}_3\beta_-}{\gamma}\right) + \frac{135\Gamma\beta_-}{\gamma^2} - \frac{45\beta_-^3}{8\gamma^2} + \frac{\beta_-^2}{4\gamma}\right], \tag{A27}$$

$$C_8 = -3c_1(\overline{S}\beta_+ + \beta_-)\left(\overline{C}_2 + \frac{3\overline{C}_3\beta_-}{\gamma}\right) + 15c_1^2\overline{S}\beta_+\left(\frac{3\overline{C}_3\beta_-^2}{\gamma^2} + \frac{\overline{C}_2\beta_-}{\gamma}\right)$$
$$-\left[\frac{45c_1(3\overline{C}_3\beta_- + \overline{C}_2\gamma)(2\alpha - \overline{S}\beta_-\beta_+)}{\gamma^2} + 3c_1\overline{S}\beta_+\left(\overline{C}_2 + \frac{3\overline{C}_3\beta_-}{\gamma}\right) + \frac{405\chi\Gamma\beta_-}{\gamma^2}\right.$$
$$\left.+\frac{135c_1\beta_-^3}{8\gamma^2} + \frac{27c_1\beta_-^2}{2\gamma}\right], \tag{A28}$$

$$C_9 = -\frac{4C_8 + 3C_7\beta_-}{9\beta_-^2}, \tag{A29}$$

$$C_{10} = \frac{2C_8}{3\beta_-}. \tag{A30}$$

$$
\begin{aligned}
C_{11} = \ & -c_1 a (C_4 + \overline{C}_4) - (b\beta_- + 2a\overline{S}\beta_+)\left(\overline{C}_1 + \frac{\overline{C}_3}{\gamma}\right) - 3c_1(b+a)\left(\frac{\overline{C}_4\overline{S}\beta_+}{\overline{\gamma}} + \frac{C_4\beta_-}{\overline{\gamma}}\right) \\
& + c_1 a\overline{S}Y + \frac{c_1 a\beta_- Y}{\gamma} - \frac{c_1 b\overline{S}\beta_+ Y}{\overline{\gamma}} + \frac{c_1 a\beta_-}{\overline{\gamma}} + \frac{b\overline{S}\beta_+ Y}{\overline{\gamma}} + \frac{b\beta_- Y}{\overline{\gamma}} + \frac{a\beta_- Y}{\gamma},
\end{aligned} \tag{A31}
$$

$$
\begin{aligned}
C_{12} = \ & -2(C_4 + \overline{C}_4)\beta_- - 2(\overline{C}_4\overline{S}\beta_+ + C_4\beta_-) + \frac{3c_1\overline{S}C_4\beta_-\beta_+}{\gamma} + \frac{3c_1 C_4\beta_-^2}{\gamma} \\
& - 2\overline{S}\beta_+\left(\overline{C}_2 + \frac{3\overline{C}_3\beta_-}{\gamma}\right) - 2\beta_- Y + \frac{9c_1\beta_-^2 Y}{\gamma},
\end{aligned} \tag{A32}
$$

$$
\begin{aligned}
C_{13} = & -2(C_4+\overline{C}_4)\overline{S}\beta_+ - 2(\overline{C}_4\overline{S}\beta_+ + C_4\beta_-) + \frac{9c_1\overline{S}C_4\beta_+^2}{\overline{\gamma}} + \frac{9c_1 C_4\beta_-\beta_+}{\overline{\gamma}} - \frac{3\overline{S}\beta_+^2 \overline{Y}}{\overline{\gamma}} \\
& - 2\beta_-\left(\overline{C}_2 - \frac{3\overline{C}_3\beta_+}{\overline{\gamma}}\right) + 2(C_4+\overline{C}_4)\overline{S}Y\beta_+ - 2Y\left(\overline{S}\beta_+ + \beta_- + \overline{C}_4\overline{S}\beta_+ + C_4\beta_-\right) \\
& - \frac{3c_1\beta_+\beta_-\overline{Y}}{\overline{\gamma}},
\end{aligned} \tag{A33}
$$

$$
\begin{aligned}
C_{14} = & \ c_1 a(\overline{C}_4\overline{S}\beta_+ + C_4\beta_-) + (b\beta_-^2 + a\beta_+^2\overline{S}^2)\left(C_1 + \frac{C_3}{\overline{\gamma}}\right) + \sigma(3\overline{S}^2 b - 6b\overline{S} + 3a) \\
& \left(\frac{C_4}{\gamma} + \frac{\overline{C}_4}{\gamma}\right) + \frac{3b\beta_-}{2\overline{\gamma}}(\overline{S}C_4 + C_4) + \frac{3b\overline{S}\beta_+}{2\overline{\gamma}}(\overline{S}C_4 + C_4) + \frac{3a\beta_-}{2\gamma}(\overline{S}C_4 + C_4) \\
& - a\overline{S}^2 Y\beta_+\beta_-\left(\overline{C}_1 + \frac{\overline{C}_3}{\gamma}\right) + \frac{3a\beta_-^2 Y}{\gamma} + \frac{3a\beta_-\beta_+\overline{S}Y}{2\gamma} + \frac{3b\beta_+^2\overline{S}Y}{2\overline{\gamma}} + \frac{3b\beta_+\beta_-\overline{S}Y}{\overline{\gamma}} \\
& + \frac{3b\beta_+\beta_-\overline{S}Y}{2\overline{\gamma}} + \frac{3a\beta_+\beta_-\overline{S}Y}{2\overline{\gamma}} - \frac{4b\beta_+\overline{S}^2 Y}{\overline{\gamma}} + \frac{4a\beta_-\overline{S}Y}{\gamma} - \frac{7b\overline{S}\beta_+ \overline{Y}}{2\overline{\gamma}} + \frac{3b\overline{S}\beta_+\beta_-}{2\overline{\gamma}} \\
& - \frac{3a\overline{S}\beta_+\beta_-}{2\overline{\gamma}} - 3ab\chi\tau\left(\frac{b}{\overline{\gamma}} + \frac{a}{\gamma}\right)(C_4+\overline{C}_4) - c_1 a\overline{S}\beta_- Y + a\overline{S}\sqrt{\chi\overline{\chi}}Y\left(\overline{C}_1 + \frac{\overline{C}_3}{\gamma}\right) \\
& - \frac{a^2\sigma Y}{\gamma} - \frac{a^2\chi\tau Y}{\gamma} - \tau\chi\left(\frac{3b\overline{Y}}{\overline{\gamma}} + \frac{3aY}{\gamma}\right) - 3\sigma\left(\frac{b\overline{Y}\overline{S}}{\overline{\gamma}} + \frac{aY}{\gamma}\right) \\
& - 3\chi\tau\left(\frac{2b\overline{Y}}{\overline{\gamma}} + \frac{bY}{\overline{\gamma}} + \frac{aY}{\gamma}\right) + 3\sigma\left(-\frac{2b\overline{S}Y}{\overline{\gamma}} + \frac{b\overline{S}^2 Y}{\overline{\gamma}} + \frac{aY}{\gamma}\right),
\end{aligned} \tag{A34}
$$

$$
\begin{aligned}
C_{15} = & \ 2\beta_-(\overline{C}_4\overline{S}\beta_+ + C_4\beta_-) + \overline{S}^2\beta_+^2\left(\overline{C}_2 + \frac{3\overline{C}_3\beta_-}{\gamma}\right) + \frac{9\beta_-\alpha}{\gamma}(C_4+\overline{C}_4) \\
& - \frac{3ab^2\overline{S}\beta_+\beta_-^2}{2\gamma} + 2\beta_-^2 Y + \chi Y\left(\overline{C}_2 + \frac{3\overline{C}_3\beta_-}{\gamma}\right) - \frac{9\sigma\beta_- Y}{\gamma} - \frac{27\beta_-\tau\chi Y}{\gamma} \\
& + \frac{9\overline{S}\beta_-^2 Y}{4\gamma} - \frac{9\overline{S}\beta_-^3 Y}{\gamma} + \frac{9\overline{S}\beta_-^2\beta_+ Y}{\gamma} - \frac{3\beta_-^2}{\gamma}(\overline{S}C_4\beta_+ + C_4\beta_-) - \overline{S}^2\beta_+ Y\beta_- \\
& \left(\overline{C}_2 + \frac{\overline{C}_3}{\gamma}\right) + \frac{9\beta_-^3}{\gamma},
\end{aligned} \tag{A35}
$$

$$
\begin{aligned}
C_{16} = \ & 2c_1\overline{S}\beta_+(\overline{C}_4\overline{S}\beta_+ + C_4\beta_-) + c_1\beta_-^2\left(C_2 - \frac{3C_3\beta_+}{\overline{\gamma}}\right) + \frac{3\sigma\overline{S}c_1\beta_+}{\overline{\gamma}}(\overline{S}-2) \\
& (C_4+\overline{C}_4) - \frac{9\overline{S}\beta_+^2\beta_-}{4\overline{\gamma}} + \frac{27\chi\tau\beta_+}{2\gamma}(C_4+\overline{C}_4) - \frac{9\overline{S}\sigma\beta_+\overline{Y}}{2\overline{\gamma}} + \frac{9\chi\tau\overline{Y}}{2\overline{\gamma}} + Y \\
& (C_4+\overline{C}_4) + 2\overline{S}\beta_+\beta_-Y + \frac{9\beta_+\tau\chi}{2\overline{\gamma}}(2\overline{Y}+Y) + \frac{9\sigma\beta_+\overline{S}}{2\overline{\gamma}}(2\overline{Y}-\overline{S}Y) - \frac{9\overline{S}\beta_+^2\overline{Y}}{4\overline{\gamma}} \\
& + \frac{81\overline{S}^2\beta_+^2\overline{Y}}{4\overline{\gamma}} + \frac{63\overline{S}\beta_+^2\overline{Y}}{4\overline{\gamma}} + \frac{18\overline{S}\beta_+\beta_-Y}{\overline{\gamma}},
\end{aligned}
\tag{A36}
$$

$$
C_{17} = b\left(\mathbf{L\overline{F}}_8 + \mathbf{L\overline{F}}_1 + \mathbf{L\overline{F}}_7 + \frac{\mathbf{L\overline{F}}_6 C_8}{\gamma}\right) + Sa\mathbf{L\overline{F}}_8 A',
\tag{A37}
$$

$$
C_{18} = b\left[\frac{\mathbf{L\overline{F}}_2}{2} + \mathbf{L\overline{F}}_4\overline{C}_9 + \mathbf{L\overline{F}}_5\overline{C}_9 + \mathbf{L\overline{F}}_6\left(\frac{4\overline{C}_{10}}{\gamma} - \frac{3\beta_+}{2\gamma}\right)\right],
\tag{A38}
$$

$$
C_{19} = b\left(\frac{\mathbf{L\overline{F}}_3}{3} + \mathbf{L\overline{F}}_4\overline{C}_{10} + \mathbf{L\overline{F}}_5\overline{C}_{10} - \frac{5\mathbf{L\overline{F}}_6\overline{C}_{10}\beta_+}{\gamma}\right).
\tag{A39}
$$

Appendix C

$$
\overline{C}_1 = \frac{\beta_+ + S\beta_-}{2\overline{C}_5\beta_+},
\tag{A40}
$$

$$
\overline{C}_2 = \frac{\beta_-(2\beta_+ + S\beta_-)}{\overline{C}_5\beta_+},
\tag{A41}
$$

$$
\overline{C}_3 = \frac{3S}{6\overline{C}_5\beta_+}\left(6\alpha - 3\beta_-^2 - \overline{S}\beta_+\beta_-\right),
\tag{A42}
$$

$$
\overline{C}_4 = \frac{1}{\overline{C}_5}\left(\overline{S}\beta_+ + 2\beta_- + \frac{\overline{C}_c Y(\chi-\beta_+^2)}{\beta_+}\right),
\tag{A43}
$$

$$
\overline{C}_5 = 2\beta_- - \frac{\overline{S}}{\beta_+} - \frac{\sqrt{\chi\overline{\chi}}}{\beta_+}.
\tag{A44}
$$

$$
\begin{aligned}
\overline{C}_6 = \ & \frac{1}{2}C_1 + \frac{3C_3S\beta_-}{6\overline{\gamma}^2} + \frac{C_3}{2\overline{\gamma}} + \frac{3C_1S\beta_-}{6\overline{\gamma}} \\
& - \frac{1}{\beta_+}\left[-\frac{1}{2}C_1S\beta_- + \frac{9\overline{\chi}\Gamma}{\overline{\gamma}^2} + \frac{3C_3\sigma}{\overline{\gamma}^2} - \frac{9\beta_+^2}{24\overline{\gamma}^2} - \frac{3C_3S\beta_-\beta_+}{2\overline{\gamma}}^2\right. \\
& \left. + \frac{3c_1\sigma}{\overline{\gamma}} + \frac{3\beta_+}{4\overline{\gamma}} - \frac{C_3S\beta_-}{2\overline{\gamma}} + \frac{3c_1S\beta_-\beta_+}{2\overline{\gamma}} - 3\overline{\chi}\tau\left(\frac{C_3}{\overline{\gamma}^2} + \frac{C_1}{\overline{\gamma}}\right)\right],
\end{aligned}
\tag{A45}
$$

$$\overline{C}_7 = \left[-2\left\{ (\beta_+ + S\beta_-)\left(C_1 + \frac{C_3}{\overline{\gamma}}\right)\right\} + \frac{1}{2}\left(C_2 - \frac{3\overline{C}_3\beta_+}{\overline{\gamma}}\right) + (15C_3\beta_+ + 4C_2\overline{\gamma}\right.$$

$$\left. + 3C_1\overline{\gamma}\beta_+)\frac{3S\beta_-}{6\overline{\gamma}^2}\right] - \frac{1}{\beta_+}\left[2S\beta_+\beta_-\left(C_1 + \frac{C_3}{\overline{\gamma}}\right) - 3\left(\frac{\alpha + c_1\beta_+\beta_-}{\overline{\gamma}^2}\right)(15\overline{C}_3\beta_- \right. \tag{A46}$$

$$\left. - 4\overline{C}_2\overline{\gamma} + 3\overline{\gamma}\overline{C}_1\beta_-) - \overline{c}_1S\beta_+\left(C_2 + \frac{3C_3\beta_-}{\overline{\gamma}}\right) - \frac{135\Gamma S^2\beta_+}{\overline{\gamma}^2} + \frac{45S^2\beta_+^3}{8\overline{\gamma}^2} + \frac{\beta_+^2}{4\overline{\gamma}}\right],$$

$$\overline{C}_8 = -3c_1(\beta_+ + S\beta_-)\left(C_2 + \frac{3C_3\beta_-}{\overline{\gamma}}\right) + 15c_1^2 S\beta_-\left(\frac{3C_3\beta_+^2}{\overline{\gamma}^2} - \frac{C_2\beta_+}{\overline{\gamma}}\right) - \frac{1}{\beta_+}$$

$$\left[\frac{45c_1\beta_+(3C_3\beta_- - C_2\overline{\gamma})(2\alpha + S\beta_-\beta_+)}{\overline{\gamma}^2} + 3c_1 S\beta_+\beta_-\left(C_2 - \frac{3C_3\beta_+}{\overline{\gamma}}\right) + \frac{405\Gamma\chi\beta_+^2}{\overline{\gamma}^2}\right. \tag{A47}$$

$$\left. - \frac{135c_1 S^2\beta_+^4}{8\gamma^2} - \frac{27c_1\beta_+^3}{\overline{\gamma}}\right],$$

$$\overline{C}_9 = -\frac{4\overline{C}_8 - 3\overline{C}_7\beta_+}{9\beta_+^2}, \tag{A48}$$

$$\overline{C}_{10} = \frac{2\overline{C}_8}{3\beta_+}, \tag{A49}$$

$$\overline{C}_{11} = c_1 b(C_4 + \overline{C}_4) - 2(bS\beta_- + a\beta_+)\left(\overline{C}_1 + \frac{\overline{C}_3}{\gamma}\right) + \left(\frac{b}{\overline{Y}} + \frac{3a}{2Y}\right)(\overline{C}_4\beta_+ + C_4\beta_- S)$$

$$- \frac{bS\overline{Y}}{2} + \frac{b\beta_+\overline{Y}}{6\overline{\gamma}} + \frac{a\beta_- YS}{2\gamma} + \frac{a\beta_+\overline{Y}}{2\gamma} + \frac{aS\beta_- Y}{\gamma}, \tag{A50}$$

$$\overline{C}_{12} = 2\beta_+(C_4 + \overline{C}_4) - 2(\overline{C}_4\beta_+ + C_4 S\beta_-) - \beta_- S\left(C_2 - \frac{3C_3\beta_+}{\overline{\gamma}}\right) - 2\beta_+\overline{Y} - \frac{3\beta_+^2\overline{Y}}{2\overline{\gamma}}$$

$$- \frac{\beta_+}{2\overline{\gamma}}(\overline{C}_4\beta_+ + C_4 S\beta_-), \tag{A51}$$

$$\overline{C}_{13} = (C_4 + \overline{C}_4) - (C_4 S\beta_- + \overline{C}_4\beta_+) - 2\beta_+\left(\overline{C}_2 + \frac{3\overline{C}_3\beta_-}{\gamma}\right) + \frac{3\beta_-}{4\gamma}(\beta_+\overline{Y} + 2S\beta_- Y)$$

$$+ \frac{9\beta_-}{4\gamma}(\overline{C}_4\beta_+ + C_4 S\beta_-) + \beta_-(C_4 + \overline{C}_4)S\overline{Y} - (\beta_+ + S\beta_-) - \overline{Y}(\overline{C}_4\beta_+ + C_4 S\beta_-) \tag{A52}$$

$$+ \frac{3SY\beta_-^2}{4\gamma},$$

$$\overline{C}_{14} = -c_1 b(\overline{C}_4 \beta_+ + C_4 S\beta_-) + \left(b\beta_-^2 S^2 + a\beta_+^2\right)\left(\overline{C}_1 + \frac{\overline{C}_3}{\gamma}\right) + \frac{3a\beta_- YS^2}{\gamma} - \frac{3b\sigma\overline{Y}}{\overline{\gamma}}$$

$$3\sigma(\overline{C}_4 + C_4)\left[\frac{b}{\overline{\gamma}} + \frac{S^2 a}{\gamma} - \frac{2Sa}{\gamma}\right] - \frac{3\beta_+ \beta_- S}{2}\left(\frac{b}{\overline{\gamma}} - \frac{a}{\gamma}\right) + \frac{3a\beta_-^2 YS^2}{\gamma} + 3b\beta_+^2 S^2$$

$$\left(\frac{\overline{Y}}{\overline{\gamma}} + \frac{1}{\gamma}\right) - 3\tau\chi(C_4 + \overline{C}_4)\left(\frac{b}{\overline{\gamma}} + \frac{a}{\gamma}\right) + \frac{bS\beta_+ \overline{Y}}{2} + b\left(C_1 + \frac{C_3}{\overline{\gamma}}\right) - \frac{3b\tau\chi\overline{Y}}{\overline{\gamma}}$$

$$- 3\tau\chi\left(\frac{b\overline{Y}}{\overline{\gamma}} + \frac{a\overline{Y}}{\gamma} + \frac{2aY}{\gamma}\right) + 3\sigma\left(\frac{b}{\overline{\gamma}} + \frac{aS^2}{\gamma} - \frac{2aSY}{\gamma}\right) - 3\tau\sqrt{\chi\overline{\chi}}S\left(\frac{\overline{Y}}{\overline{\gamma}} + \frac{Y}{\gamma}\right) \quad \text{(A53)}$$

$$+ 3\left(\frac{b\overline{Y}}{\overline{\gamma}} + \frac{aSY}{\gamma}\right) + \frac{3}{2}S\left[\beta_+ \overline{Y}\left(\frac{b}{\overline{\gamma}} + \frac{a}{\gamma}\right) - \frac{\beta_- Y}{\gamma}(Sa - b)\right] + \frac{3b\beta_+^2 \overline{Y}}{2\overline{\gamma}}$$

$$+ 15S\beta_+ \overline{Y}\left(\frac{b}{\overline{\gamma}} + \frac{a}{\gamma}\right) + \frac{9b\beta_+ \overline{Y}}{\overline{\gamma}} + \frac{3aS^2 \beta_- Y}{\gamma} + (bc_1\beta_+ + 3c_1 a\beta_+ + aS\beta_-)$$

$$\frac{1}{\gamma}(\beta_+ + S\beta_-) - bS^2\overline{Y}\beta_-^2\left(C_1 + \frac{C_3}{\overline{\gamma}}\right) + 3c_1 S^2 \beta_+ \overline{Y}\beta_-\left(\frac{b}{\overline{\gamma}} + \frac{a}{\gamma}\right),$$

$$\overline{C}_{15} = 2\beta_+(\overline{C}_4 \beta_+ + C_4 S\beta_-) + S\beta_-^2\left(C_2 - \frac{3C_3\beta_+}{\overline{\gamma}}\right) - \frac{9\beta_+ \sigma}{Y} + \frac{27\beta_+ \chi\tau\overline{Y}}{\overline{\gamma}} + \frac{18\beta_+^2 \overline{Y}}{2\overline{\gamma}}$$

$$(C_4 + \overline{C}_4) + \frac{3S\beta_+^2 \beta_-}{2\gamma} - \frac{9\beta_+ \chi\tau}{\overline{\gamma}}(C_4 + \overline{C}_4) + 2\beta_+^2 \overline{Y} + \chi - S\beta_-^2 a^2 b\left(C_2 - \frac{3C_3\beta_+}{\overline{\gamma}}\right) \quad \text{(A54)}$$

$$\overline{Y}\left(C_2 - \frac{3C_3\beta_+}{\overline{\gamma}}\right) + \frac{3\sigma\beta_+ \overline{Y}}{\overline{Y}} - \frac{9\beta_+^2 \overline{Y}}{\overline{\gamma}}(1 + S) - \frac{9\beta_+}{2\overline{\gamma}}\left(\overline{C}_4 \beta_+^2 + C_4 S\beta_-\right) - \frac{27\beta_+^3 \overline{Y}}{2\overline{\gamma}},$$

$$\overline{C}_{16} = S\beta_-(\overline{C}_4 \beta_+ + C_4 S\beta_-) + c_1\beta_+^2\left(\overline{C}_2 + \frac{3\overline{C}_3\beta_-}{\gamma}\right) + \frac{3c_1\sigma S}{\gamma}(S - 2)(C_4 + \overline{C}_4)$$

$$+ \frac{9c_1 S\beta_+ \beta_-^2}{2\gamma} + \frac{9c_1 S^2 \chi\tau\beta_-}{\gamma}(C_4 + \overline{C}_4) + S^2\chi\overline{Y}(C_4 + \overline{C}_4) + 4S\beta_+ \beta_- \overline{Y}$$

$$+ \frac{3b\sigma\beta_- YS}{2\gamma} - \frac{3S^2\chi\tau\beta_- Y}{2\gamma} - \frac{9\tau\chi S^2 \beta_-}{2\gamma}(\overline{Y} + Y) + \frac{9S\sigma\beta_-}{2\gamma}(S\overline{Y} - 2Y) \quad \text{(A55)}$$

$$+ \frac{9\beta_-^2 YS^2}{4\gamma} + \frac{18\beta_+ \beta_- S\overline{Y}}{\gamma} - \frac{63S\beta_-^2 Y}{2\gamma}(1 + S) - S\overline{Y}\beta_-(\overline{C}_4 \beta_+ + C_4 S\beta_-) + \frac{9\beta_- \beta_+}{4\gamma}$$

$$(\overline{C}_4 \beta_+ + S\beta_- C_4) - \frac{9S\beta_-^2}{2\gamma}(\overline{C}_4 \beta_+ + S\beta_- C_4) + \frac{9\beta_+^2 S\beta_- \overline{Y}}{2\gamma} + \frac{9S^2\beta_-^2 Y}{4\gamma},$$

$$\overline{C}_{17} = a\left(\overline{\mathbf{LF}}_1 + \overline{\mathbf{LF}}_7 + \frac{\overline{\mathbf{LF}}_6 C_8}{\gamma} + \overline{\mathbf{LF}}_8\right) + \overline{S}b\overline{\mathbf{LF}}_8 B', \quad \text{(A56)}$$

$$\overline{C}_{18} = a\left[\frac{\overline{\mathbf{LF}}_2}{2} + \overline{\mathbf{LF}}_4 C_9 + \overline{\mathbf{LF}}_5 C_9 + \overline{\mathbf{LF}}_6\left(\frac{4C_{10}}{\gamma} + \frac{3\beta_-}{2\gamma}\right)\right], \quad \text{(A57)}$$

$$\overline{C}_{19} = a\left(\frac{\overline{\mathbf{LF}}_3}{3} + \overline{\mathbf{LF}}_4 C_{10} + \overline{\mathbf{LF}}_5 C_{10} + \frac{5\overline{\mathbf{LF}}_6 C_{10}\beta_-}{\gamma}\right). \quad \text{(A58)}$$

References

1.   Xia, X.; Shen, H.T. Nonlinear interaction of ice cover with shallow water waves in channels. *J. Fluid Mech.* **2002**, *467*, 259–268. [CrossRef]

2.   Milewski, P.A.; Vanden-Broeck, J.M.; Wang, Z. Hydroelastic solitary waves in deep water. *J. Fluid Mech.* **2011**, *679*, 628–640. [CrossRef]

3.   Vanden-Broeck, J.M.; Părău, E.I. Two-dimensional generalized solitary waves and periodic waves under an ice sheet. *Philos. Trans. R. Soc. A Math. Phys. Eng. Sci.* **2011**, *369*, 2957–2972. [CrossRef] [PubMed]

4.   Deike, L.; Bacri, J.C.; Falcon, E. Nonlinear waves on the surface of a fluid covered by an elastic sheet. *J. Fluid Mech.* **2013**, *733*, 394–413. [CrossRef]

5.   Wang, P.; Lu, D.Q. Analytic approximation to nonlinear hydroelastic waves traveling in a thin elastic plate floating on a fluid. *Sci. China Phys. Mech. Astron.* **2013**, *56*, 2170–2177. [CrossRef]

6.   Hedges, T.S. Combinations of waves and currents: An introduction. *Proc. Inst. Civ. Eng.* **1987**, *82*, 567–585. [CrossRef]

7.   Schulkes, R.M.S.M.; Hosking, R.J.; Sneyd, A.D. Waves due to a steadily moving source on a floating ice plate. part 2. *J. Fluid Mech.* **1987**, *180*, 297–318. [CrossRef]

8.   Bhattacharjee, J.; Sahoo, T. Interaction of current and flexural gravity waves. *Ocean Eng.* **2007**, *34*, 1505–1515. [CrossRef]

9.   Bhattacharjee, J.; Sahoo, T. Flexural gravity wave generation by initial disturbances in the presence of current. *J. Mar. Sci. Technol.* **2008**, *13*, 138–146. [CrossRef]

10.   Mohanty, S.K.; Mondal, R.; Sahoo, T. Time dependent flexural gravity waves in the presence of current. *J. Fluids Struct.* **2014**, *45*, 28–49. [CrossRef]

11.   Lu, D.Q.; Yeung, R.W. Hydroelastic waves generated by point loads in a current. *Int. J. Offshore Polar Eng.* **2015**, *25*, 8–12.

12.   Fenton, J.D.; Rienecker, M.M. A fourier method for solving nonlinear water-wave problems: Application to solitary-wave interactions. *J. Fluid Mech.* **1982**, *118*, 411–443. [CrossRef]

13.   Lin, P. A numerical study of solitary wave interaction with rectangular obstacles. *Coast. Eng.* **2004**, *51*, 35–51. [CrossRef]

14.   Khan, U.; Ellahi, R.; Khan, R.; Mohyud-Din, S.T. Extracting new solitary wave solutions of benny–luke equation and phi-4 equation of fractional order by using (g'/g)-expansion method. *Opt. Quantum Electron.* **2017**, *49*, 362. [CrossRef]

15.   Abdel-Gawad, H.; Tantawy, M. Mixed-type soliton propagations in two-layer-liquid (or in an elastic) medium with dispersive waveguides. *J. Mol. Liq.* **2017**, *241*, 870–874. [CrossRef]

16.   Gardner, C.S.; Greene, J.M.; Kruskal, M.D.; Miura, R.M. Method for solving the Korteweg-de Vries equation. *Phys. Rev. Lett.* **1967**, *19*, 1095–1097. [CrossRef]

17.   Su, C.H.; Mirie, R.M. On head-on collisions between two solitary waves. *J. Fluid Mech.* **1980**, *98*, 509–525. [CrossRef]

18.   Mirie, R.M.; Su, C.H. Collisions between two solitary waves. part 2. A numerical study. *J. Fluid Mech.* **1982**, *115*, 475–492. [CrossRef]

19.   Dai, S.Q. Solitary waves at the interface of a two-layer fluid. *Appl. Math. Mech.* **1982**, *3*, 777–788.

20.   Mirie, R.M.; Su, C.H. Internal solitary waves and their head-on collision. i. *J. Fluid Mech.* **1984**, *147*, 213–231. [CrossRef]

21.   Mirie, R.M.; Su, C.H. Internal solitary waves and their head-on collision. ii. *Phys. Fluids (1958–1988)* **1986**, *29*, 31–37. [CrossRef]

22.   Ozden, A.E.; Demiray, H. Re-visiting the head-on collision problem between two solitary waves in shallow water. *Int. J. Non-Linear Mech.* **2015**, *69*, 66–70. [CrossRef]

23.   Marin, M.; Öchsner, A. The effect of a dipolar structure on the hölder stability in Green–Naghdi thermoelasticity. *Contin. Mech. Thermodynam.* **2017**, *29*, 1365–1374. [CrossRef]

24.   Zhu, Y.; Dai, S.Q. On head-on collision between two gKdV solitary waves in a stratified fluid. *Acta Mech. Sin.* **1991**, *7*, 300–308.

25.   Zhu, Y. Head-on collision between two mKdV solitary waves in a two-layer fluid system. *Appl. Math. Mech.* **1992**, *13*, 407–417.

26. Huang, G.; Lou, S.; Xu, Z. Head-on collision between two solitary waves in a Rayleigh-Bénard convecting fluid. *Phys. Rev. E* **1993**, *47*, R3830. [CrossRef]
27. Dai, H.H.; Dai, S.Q.; Huo, Y. Head-on collision between two solitary waves in a compressible Mooney-Rivlin elastic rod. *Wave Motion* **2000**, *32*, 93–111. [CrossRef]
28. Bhatti, M.M.; Lu, D.Q. Head-on collision between two hydroelastic solitary waves with Plotnikov-Toland's plate model. *Theor. Appl. Mech. Lett.* **2018**, *8*, 384–392. [CrossRef]
29. Bhatti, M.M.; Lu, D.Q. Head-on collision between two hydroelastic solitary waves in shallow water. *Qual. Theory Dynam. Syst.* **2018**, *17*, 103–122. [CrossRef]

7

Unsteady Flow of Fractional Fluid between Two Parallel Walls with Arbitrary Wall Shear Stress using Caputo–Fabrizio Derivative

Muhammad Asif [1], **Sami Ul Haq** [2], **Saeed Islam** [1], **Tawfeeq Abdullah Alkanhal** [3], **Zar Ali Khan** [4], **Ilyas Khan** [5,*] **and Kottakkaran Sooppy Nisar** [6]

[1]   Department of Mathematics, Abdul Wali Khan University, Mardan 23200, Pakistan; asif_best1986@yahoo.com (M.A.); saeed.sns@gmail.com (S.I.)
[2]   Department of Mathematics, Islamia College University, Peshawar 25000, Pakistan; samiulhaqmaths@yahoo.com
[3]   Department of Mechatronics and System Engineering, College of Engineering, Majmaah University, Majmaah 11952, Saudi Arabia; t.alkanhal@mu.edu.sa
[4]   Department of Mathematics, University of Peshawar, Peshawar 25000, Pakistan; zaralikhangmk@gmail.com
[5]   Faculty of Mathematics and Statistics, Ton Duc Thang University, Ho Chi Minh City 72915, Vietnam
[6]   Department of Mathematics, College of Arts and Science, Prince Sattam bin Abdulaziz University, Wadi Al-Dawaser 11991, Saudi Arabia; n.sooppy@psau.edu.sa
*   Correspondence: ilyaskhan@tdt.edu.vn

Abstract: In this article, unidirectional flows of fractional viscous fluids in a rectangular channel are studied. The flow is generated by the shear stress given on the bottom plate of the channel. The authors have developed a generalized model on the basis of constitutive equations described by the time-fractional Caputo–Fabrizio derivative. Many authors have published different results by applying the time-fractional derivative to the local part of acceleration in the momentum equation. This approach of the fractional models does not have sufficient physical background. By using fractional generalized constitutive equations, we have developed a proper model to investigate exact analytical solutions corresponding to the channel flow of a generalized viscous fluid. The exact solutions for velocity field and shear stress are obtained by using Laplace transform and Fourier integral transformation, for three different cases namely (i) constant shear, (ii) ramped type shear and (iii) oscillating shear. The results are plotted and discussed.

Keywords: viscous fluid; Caputo–Fabrizio time-fractional derivative; Laplace and Fourier transformations; side walls; oscillating shear stress

1. Introduction

The branch of mathematics that studies derivatives and integrals is called calculus, i.e., discussing integer order derivatives and integrals. When the order of derivatives changes from integer order to real (non-integer) order a new branch of calculus comes into being, called fractional calculus. Fractional order derivatives occur in many physical problems for example, frequency-dependent damping behavior of objects, velocity of infinite thin plate in a viscous fluid, creeping and relaxation functions of viscoelastic materials, and the control of dynamical systems as mentioned in [1–4]. Fractional calculus provides more generalized derivatives, and therefore it has more applications as compared with the classical or integer order derivatives. Fractional differential equations also explain the phenomena in electrochemistry, acoustics, electromagnetics, viscoelasticity, and material science [5–10]. For the last twenty years, a lot of work has been done on fractional calculus. Some authors [11–13] have used the formal definitions of fractional calculus like Riemann–Liouville and Caputo operators. These definitions provide a strong basis for the modern approach of Caputo and Fabrizio who have presented the definition without singular kernel [14].

The Caputo–Fabrizio differential operator is used by many authors to obtain exact solutions concerning real life problems [15–17]. All the benefits of Reimann–Liouville and Caputo definitions are also included in Caputo–Fabrizio, which is a worthy point of this definition. The Caputo–Fabrizio definition has been used by different authors in the medical sciences for example, the cancer treatment model and the flow of blood through veins under the effect of magnetic field [18,19]. Shah et al. [20] investigated the exact solutions over an isothermal vertical plate of free convectional flow of viscous fluids by using a definition of the Caputo–Fabrizio time-fractional derivative. Free convectional time-fractional flow with Newtonian heating near a vertical plate including mass diffusion has been investigated by Vieru et al. [21].

The effect of side walls over the velocity of a non-Newtonian fluid while the motion is produced due to the oscillation of the lower plate has been investigated by Fetecau et al. [22]. In addition, Haq et al. [23] have analyzed the exact solution of viscous fluid over an infinite plate using the Caputo–Fabrizio fractional order derivatives. Most of the authors have discussed different fluids using the fractional order differential operator defined by Caputo, Caputo–Fabrizio etc. and published many interesting results by applying the fractional order definition only to the local part of acceleration. Henry et al. [24] and Hristov [25,26] have suggested a generalized Fourier law for the thermal heat flux. It is clear from their discussion that the fractional differential operator has been employed in the constitutive relation of energy equation, rather than directly using it in the governing equation. This approach is appealing to mathematical and physical aspects of fluid mechanics. Hameid et al. [27] applied the definition of the fractional order derivative to the convective part of a constitutive equation and explained their model in a very interesting way. Vieru et al. [28] have followed the discussion presented by Hameid et al. [27] by applying the fractional derivative definition in a constitutive equation.

Keeping in mind all the above discussions, we present this article exploring the effect of side walls on the motion of an incompressible fluid using generalized fractional constitutive equations and the Caputo–Fabrizio derivative through a rectangular channel.

2. Problem Formulation

Consider an incompressible fluid, which is viscous in nature, present over an infinite plate between two parallel side walls that are at right angles to the horizontal plate as shown in Figure 1. Initially both the plate and fluid are at rest for $t = 0$, after time $t > 0$, and the flow is generated by the shear stress given by $\tau_0\, f(t)$ which engender the velocity as

$$V = v(y, z, t)i, \tag{1}$$

where i stands for unit vector.

In the absence of the body forces and the pressure gradient in the flow direction, the linear momentum equation in the x-direction is:

$$\rho \frac{\partial u_x}{\partial t} + u_x \frac{\partial u_x}{\partial x} + u_y \frac{\partial u_x}{\partial y} + u_z \frac{\partial u_x}{\partial z} = \frac{\partial \tau_{xy}}{\partial y} + \frac{\partial \tau_{xz}}{\partial z} \tag{2}$$

Therefore, in our case, the velocity field is the advection terms in Equation (2) that are zero with initial and boundary conditions as follows:

$$v(y, z, 0) = 0, \; for \; y > 0 \; and \; z \in [0, l] \,,$$
$$v(y, 0, t) = v(y, l, t) = 0 \; for \; y, t > 0 \,, \tag{3}$$
$$\tau_{xy}(0, z, t) = \mu \; {}^{CF}D_t^\alpha \frac{\partial v(y, z, t)}{\partial y}\bigg|_{y=0} = \tau_0\, f(t) \,,$$

The above relation (Equation (3)), shows the shear stress which is non-trivial, where μ is dynamic viscosity, $v = \frac{\mu}{\rho}$, where τ_0 shows the constant parameter, and it is assumed that $f(\cdot)$ is a dimensionless, piecewise, continuous function such that $f(0) = 0$, $v(y,z,t)$ and $\frac{\partial v(y,z,t)}{\partial y} \to 0$ as $y \to \infty$, $z \in [0,l]$ for $t > 0$.

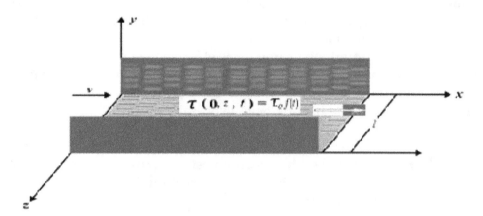

Figure 1. Geometry of the problem.

We have the Caputo–Fabrizio derivative operator of order α given by Zafar et al. [29]

$$^{CF}D_t^\alpha[h(t)] = \frac{1}{1-\alpha}\int_0^t h'(s)\exp\left[-\frac{\alpha(t-s)}{1-\alpha}\right]ds \text{ for } 0 \le \alpha < 1. \tag{4}$$

In the present paper, we consider the generalized constitutive equations with the Caputo–Fabrizio time-fractional derivative, namely:

$$\tau_{xy} = \mu \,^{CF}D_t^\alpha\left(\frac{\partial v}{\partial y}\right) \text{ and } \tau_{xz} = \mu \,^{CF}D_t^\alpha\left(\frac{\partial v}{\partial z}\right) \text{ for } \alpha \in [0,1). \tag{5}$$

It is known that any constitutive equation must satisfy the principle of material objectivity, and therefore it must be frame-invariant with respect to Euclidean transformations. Yang et al. ([30], Equation (3.1)) have formulated a constitutive equation with fractional derivatives for generalized upper-convected Maxwell fluids on the basis of the convected coordinate system. They have proven that the proposed constitutive equation is frame-indifferent and have studied some particular cases of the proposed equation.

By applying the Laplace transform, the constitutive Equation (5) for the shear stress τ_{xy} can be written in the following equivalent form:

$$\tau_{xy} = \frac{1}{1-\alpha}\frac{\partial v}{\partial y} - \frac{\alpha}{(1-\alpha)^2}\int_0^t \exp\left(\frac{-\alpha(t-\tau)}{1-\alpha}\right)\frac{\partial v(y,\tau)}{\partial y}d\tau, \ 0 < \alpha < 1. \tag{6}$$

Equation (6) is equivalent to the equation studied by Yang et al. ([30], Equation (4.5)), therefore, the proposed constitutive equations given by Equation (5) satisfy the principle of material objectivity.

3. Problem Solution

Using Equation (5) in Equation (2), applying Laplace transform to the obtained form and simplifying the result we get:

$$v(y,z,q) = \frac{v}{(1-\alpha)q+\alpha}\left[\frac{\partial^2 v(y,z,q)}{\partial y^2} + \frac{\partial^2 v(y,z,q)}{\partial z^2}\right], \tag{7}$$

where $v(y, z, q) = \int_0^\infty v(y, z, t) \exp(-qt)dt$ is the Laplace transform with respect to t.

Applying the Fourier cosine transform with respect to variable y namely $v(\xi, z, q) = \int_0^\infty v(y, z, q) \cos(y\xi)dy$ and finite Fourier sine transform with respect to variable z $v(\xi, n, q) = \int_0^l v(\xi, z, q) \sin\left(\frac{n\pi z}{l}\right)dz$, $n = 1, 2, \ldots$, we obtain:

$$v_{sc}(\xi, q) = \frac{v}{(1-\alpha)q + \alpha + v\left(\xi^2 + \lambda_n^2\right)} \frac{\tau_0}{\mu} \sqrt{\frac{2}{\pi}} f(q) \frac{(-1)^n - 1}{\lambda_n}, \tag{8}$$

where $\lambda_n = \frac{n\pi}{l}$ and subscript "sc" represents finite Fourier sine and inifinite cosine transforms.
For simplification, Equation (8) can be written as:

$$v_{sc}(\xi, q) = \frac{\tau_0}{\mu} \frac{1 - (-1)^n}{\lambda_n} \sqrt{\frac{2}{\pi}} \frac{f(q)}{(\xi^2 + \lambda_n^2)} - \frac{\frac{\tau_0}{\mu} \frac{1 - (-1)^n}{\lambda_n} \sqrt{\frac{2}{\pi}} f(q)}{\left[\frac{(1-\alpha)q + \alpha}{(\xi^2 + \lambda_n^2)\left[(1-\alpha)q + \alpha + v\left(\xi^2 + \lambda_n^2\right)\right]}\right]}. \tag{9}$$

Applying the inverse Laplace transformation, we get:

$$v_{sc}(\xi, t) = \frac{\tau_0}{\mu} \sqrt{\frac{2}{\pi}} \frac{1 - (-1)^n}{\lambda_n} \frac{f(t)}{(\xi^2 + \lambda_n^2)} + \frac{1}{(1-\alpha)(\xi^2 + \lambda_n^2)} e^{-A(\xi)t}$$
$$\times \left[\frac{\tau_0}{\mu} \sqrt{\frac{2}{\pi}} \frac{1 - (-1)^n}{\lambda_n}(1-\alpha)f^\bullet(t) + \frac{\tau_0}{\mu} \sqrt{\frac{2}{\pi}} \frac{1 - (-1)^n}{\lambda_n} \alpha f(t)\right], \tag{10}$$

where $A(\xi) = \frac{v\left(\xi^2 + \lambda_n^2\right) + \alpha}{1 - \alpha}$.
Applying the inverse Fourier transformation, we find:

$$v(y, z, t) = \frac{2}{\pi} \frac{2}{l} \frac{\tau_0}{\mu} \sum_{n=1}^\infty \frac{(-1)^n - 1}{\lambda_n} \sin(\lambda_n z) \left[\begin{array}{c} f(t) \int_0^\infty \frac{\cos(y\xi)}{\xi^2 + \lambda_n^2} d\xi + \quad [(1-\alpha)f^\bullet(t) + \alpha f(t)] \\ \int_0^\infty \frac{\cos(y\xi)}{(\xi^2 + \lambda_n^2)} e^{-A(\xi)t} d\xi \end{array}\right], \tag{11}$$

where $m = 2n - 1$ and $l = 2h$. Changing the origin by using $z = z^* + h$,

$$v(y, z, t) = \frac{2\tau_0}{\mu\pi h} \sum_{n=1}^\infty \frac{(-1)^{n+1?} \cos(\gamma_m z^*)}{\gamma_m} \left[\begin{array}{c} f(t) \int_0^\infty \frac{\cos(y\xi)}{\xi^2 + \gamma_m^2} d\xi + \quad [(1-\alpha)f^\bullet(t) + \alpha f(t)] \\ \int_0^\infty \frac{\cos(y\xi)}{(\xi^2 + \gamma_m^2)} e^{-A(\xi)t} d\xi \end{array}\right], \tag{12}$$

ignoring the * notation, keeping in view the following result:

$$\int_0^\infty \frac{\cos(ax)}{b^2 + x^2} dx = \frac{\pi}{2b} e^{-ab}, \ for \ a > 0 \ Re(b) > 0,$$

and putting in Equation (11) we get:

$$v(y, z, t) = \frac{2}{h} \sum_{n=1}^\infty (-1)^{n+1} \frac{\tau_0}{\mu} \frac{\cos\gamma_m z}{\gamma_m} f(t) \frac{e^{-y\gamma_m}}{\gamma_m} - \frac{4}{\pi h} \sum_{n=1}^\infty (-1)^{n+1} \frac{\tau_0}{\mu} \frac{\cos\gamma_m z}{\gamma_m}$$
$$\times \left[\int_0^\infty \int_0^t \frac{\cos(y\xi)}{\xi^2 + \gamma_m^2} f^\bullet(t - \tau)e^{-A(\xi)\tau} d\tau d\xi + \frac{\alpha}{1-\alpha} \int_0^\infty \int_0^t \frac{\cos(y\xi)}{\xi^2 + \gamma_m^2} f^\bullet(t - \tau)e^{-A(\xi)\tau} d\tau d\xi\right]. \tag{13}$$
where $\gamma_m = (2n - 1)\frac{\pi}{2h}$

4. Graphical Illustration and Discussions

After finding the general solution for the velocity of the fluid, we discuss three different cases which are very useful in engineering. The obtained results are presented graphically for the three cases. Figures 2–5 show different profiles of velocity by taking case I (constant shear) into consideration. Figures 6–9 show the behavior of fluid velocity for case II (ramped type shear), and Figures 10–12 show the same discussion for Case III (oscillating shear).

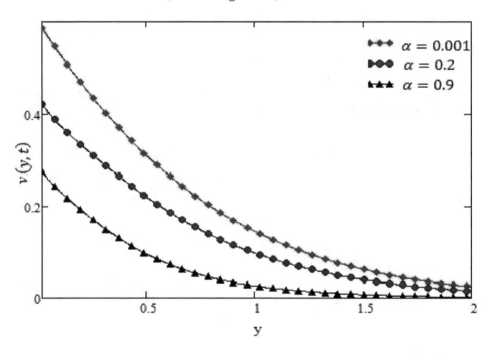

Figure 2. Profiles of velocity given by Equation (14) of constant shear for different values of α, for $t = 20$ s, $h = 0.6$ m, $\tau_0 = -1$ N/m^2, $\nu = 0.1$ m^2/s, and $\mu = 1.4$ kg·m/s.

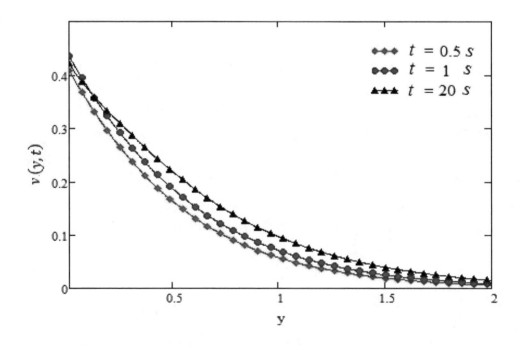

Figure 3. Profiles of velocity given by Equation (14) of constant shear for different values of t, for $\alpha = 0.2$, $h = 0.6$ m, $\tau_0 = -1$ N/m^2, $\nu = 0.1$ m^2/s, and $\mu = 1.4$ kg·m/s.

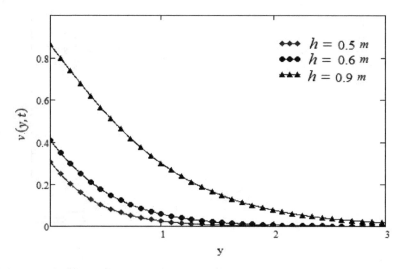

Figure 4. Profiles of velocity given by Equation (14) of constant shear for different values of h, for $t = 0.5$ s, $\alpha = 0.2$, $\tau_0 = -1$ N/m^2, $\nu = 0.1$ m^2/s, and $\mu = 1.4$ kg·m/s.

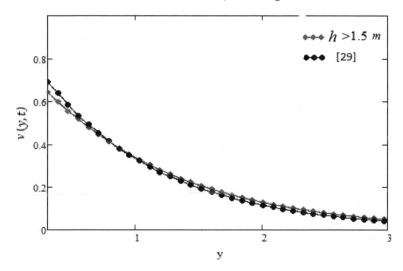

Figure 5. The value of h required for overlapping the velocity profiles of equation (14) and infinite plate [29] in constant shear for $t = 0.5$ s, $\alpha = 0.2$, $\tau_0 = -1$ N/m^2, $\nu = 0.1$ m^2/s, $\mu = 1.4$ kg·m/s, and $h > 1.5$ m.

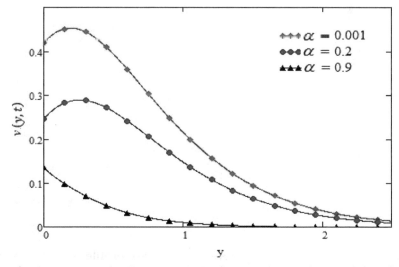

Figure 6. Profiles of velocity given by Equation (15) of ramped type shear velocity for different values of α, for $t = 0.5$ s, $h = 0.6$ m, $\tau_0 = -1$ N/m^2, $\nu = 0.1$ m^2/s, and $\mu = 1.4$ kg·m/s.

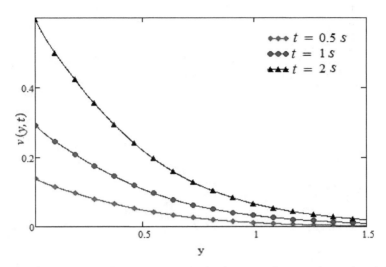

Figure 7. Profiles of velocity given by Equation (15) of ramped type shear for different values of t, for $\alpha = 0.9$, $h = 0.6$ m, $\tau_0 = -1$ N/m^2, $\nu = 0.1$ m^2/s, and $\mu = 1.4$ kg·m/s.

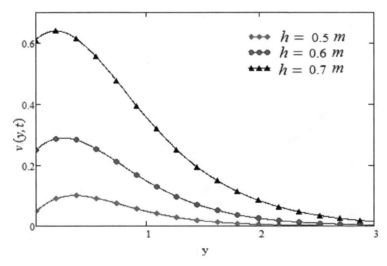

Figure 8. Profiles of velocity given by Equation (15) of ramped type shear for different values of h, for $t = 0.5$ s, $\alpha = 0.2$, $\tau_0 = -1$ N/m^2, $\nu = 0.1$ m^2/s, and $\mu = 1.4$ kg·m/s.

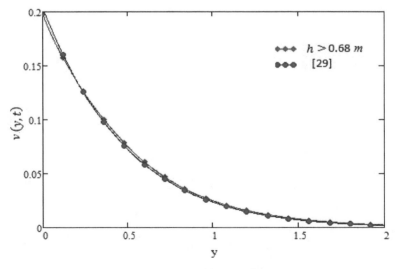

Figure 9. The value of h required for overlapping the velocity profiles of Equation (15) and infinite plate [29] in ramped type shear for $t = 0.4$ s, $\alpha = 0.9$, $\nu = 0.1$ m^2/s, $\tau_0 = -1.5$ N/m^2, $\mu = 1.4$ kg·m/s and $h \geq 0.68$.

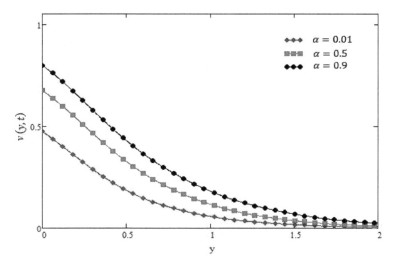

Figure 10. Profiles of velocity given by Equation (16) of oscillating shear for different values of α, for $t = 1$ s, $h = 0.6$ m, $\nu = 0.05$ m^2/s, $\tau_0 = -1$ N/m^2, $\mu = 1.4$ kg·m/s, and $\omega = 1$.

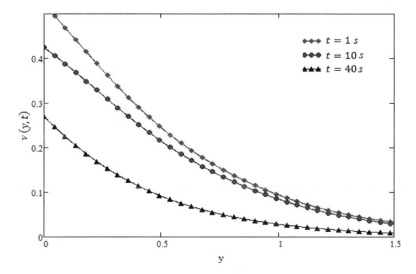

Figure 11. Profiles of velocity given by Equation (16) of oscillating shear for different values of t, for $\alpha = 0.5$, $h = 0.6$ m, $\tau_0 = -1$ N/m^2, $\nu = 0.1$ m^2/s, $\mu = 1.4$ kg·m/s, and $\omega = 1$.

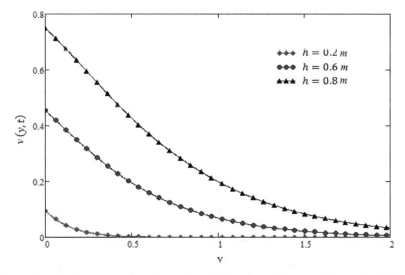

Figure 12. Profiles of velocity given by Equation (16) of oscillating shear for different values of h, for $t = 1$ s, $\alpha = 0.5$, $\tau_0 = -1$ N/m^2, $\nu = 0.1$ m^2/s, $\mu = 1.4$ kg·m/s, and $\omega = 1$.

4.1. Case I (Constant Shear)

Taking $f(t) = H(t)$, where $H(t)$ is Heaviside unit step function we find the velocity profile as:

$$
v(y,z,t) = \frac{2}{h}\sum_{n=1}^{\infty}(-1)^{n+1}\frac{\tau_0}{\mu}\frac{\cos\gamma_m z}{\gamma_m}\left[\begin{array}{c} e^{-y\,\gamma_m} + (1-\alpha)(t-\alpha) \\ \times e^{-yA}\frac{B_n\cos(yB_n)+A_n\sin(yB_n)}{c_n\left(\sqrt{b_n^4+c_n^2}\right)} \end{array}\right]
$$
$$
-\frac{4}{\pi h}\sum_{n=1}^{\infty}(-1)^{n+1}\frac{\tau_0}{\mu}\frac{\cos\gamma_m z}{\gamma_m}\left[\int_0^{\infty}\left(\frac{e^{-A(\xi)t}-1}{(A(\xi))^2} - \frac{\alpha e^{-A(\xi)t}}{(1-\alpha)A(\xi)}\right)\frac{\cos(y\xi)}{\xi^2+\gamma_m^2}d\xi\right], \tag{14}
$$

Taking the following identity into account:

$$
\int_0^{\infty}\frac{\cos(y\xi)}{\left(\xi^2-b_n^2\right)^2+c_n^2} = \frac{\pi}{2c_n}e^{-yA}\frac{B_n\cos(yB_n)+A_n\sin(yB_n)}{\sqrt{b_n^4+c_n^2}},
$$

where $2A_n^2 = \sqrt{b_n^4+c_n^2}+b_n^2$ and $2B_n^2 = \sqrt{b_n^4+c_n^2}-b_n^2$ in which: $b_n = \frac{v\gamma_m^2+\frac{\alpha}{2}}{v}$ and $c_n = \frac{v\gamma_m^4+\alpha\gamma_m^2}{v}-b_n^2$.

Figure 2 shows that as the value of α ($0 < \alpha < 1$) is increasing, the fluid velocity is decreasing. Figure 3 shows the velocity profiles for different times, which implies that as time passages the fluid velocity increases in constant case. Figure 4 shows the profiles of velocity for different values of h, if we increase the distance between side walls the fluid velocity will increase while keeping the other parameters constant. Figure 5 shows the curve of Equation (14) is overlapping as that of [29] constant case if we increase the distance between the side walls to $h \geq 1.5$, which gives our result for constant case more validity.

4.2. Case II (Ramped Type Shear)

Taking $f(t) = tH(t)$, we find the velocity profile as:

$$
v(y,z,t) = \frac{2}{d}\sum_{n=1}^{\infty}(-1)^{n+1}\frac{\tau_0}{\mu}\frac{\cos\gamma_m z}{\gamma_m}\frac{e^{-y\,\gamma_m}}{\gamma_m} + \frac{2}{h}\sum_{n=1}^{\infty}(-1)^{n+1}\frac{\tau_0}{\mu}\frac{\cos\gamma_m z}{\gamma_m}\frac{e^{-y\,A}}{c_n}
$$
$$
\times\frac{B_n\cos(yB_n)+A_n\sin(yB_n)}{\left(\sqrt{b_n^4+c_n^2}\right)}\left[t(2-\alpha)-t^2(1-\alpha)\right] - \frac{4}{\pi h}\sum_{n=1}^{\infty}(-1)^{n+1}
$$
$$
\times\frac{\tau_0}{\mu}\frac{\cos\gamma_m z}{\gamma_m}\left[\int_0^{\infty}\left(2+\frac{\alpha}{(1-\alpha)}\right)\frac{e^{-A(\xi)t}}{A(\xi)} + \frac{1}{(A(\xi))^2}\right]\frac{\cos(y\xi)}{\xi^2+\gamma_m^2}d\xi. \tag{15}
$$

Figure 6 shows that by increasing the value of the differential parameter the velocity of the fluid decreases. Figures 7 and 8 show similar behavior of t and h, as observed in Figures 3 and 4, whereas in Figure 9, we calculated the value of h in order to validate our result in case of ramped type shear.

4.3. Case III (Oscillating Shear Stress)

Taking $f(t) = \sin(\omega t)H(t)$, in Equation (13) we get the velocity profile as:

$$
v(y,z,t) = \sum_{n=1}^{\infty}(-1)^{n+1}\frac{\tau_0}{\mu}\frac{\cos\gamma_m z}{\gamma_m}\left[\frac{2}{h}\sin(\omega t)\frac{e^{-y\,\gamma_m}}{\gamma_m} - \frac{4}{\pi h}\int_0^{\infty}\left\{\frac{1}{(A(\xi))^2+\omega^2}\right.\right.
$$
$$
\times\left(-A(\xi)e^{-A(\xi)t} + A(\xi)\cos(\omega t) + \omega\sin(\omega t) + \frac{\alpha}{1-\alpha}\right.
$$
$$
\left.\left.\times\left(\omega e^{-A(\xi)t}+\omega\cos(\omega t)+A(\xi)\sin(\omega t)\right)\right)\right\}\frac{\cos(y\xi)}{\xi^2+\gamma_m^2}d\xi\right]. \tag{16}
$$

Figures 10 and 11 show opposite behavior as compared to those of constant and ramped type shear for α and time t parameters, and Figure 12 shows the same behavior as the previous two cases for parameter h, i.e., distance between side walls.

5. Conclusions

This article is presented to obtain the exact solution in general form for the velocity of the fluid present over an infinite plate between two side walls using the definition of the Caputo–Fabrizio of fractional order differential operator, while the motion is produced due to shear stress as $\tau_0 f(t)$. The results are obtained by using Laplace and Fourier transformations for three different cases:

1.   Constant shear;
2.   Ramped type shear;
3.   Oscillating shear.

After the above discussion we have concluded the following results.

Firstly, keeping the condition on shear the response of the fractional fluid velocity is very quick, as compared with that of ordinary fluid velocity for all the cases. Secondly, in fractional fluid a small change in time parameter shows a clear difference for the profiles in all the cases listed above. Thirdly, for the above cases there are some values of h at which the motion of the fluid is unaffected by side walls.

Author Contributions: Cconceptualization, M.A. and S.U.H.; methodology, T.A. and S.I.; software, Z.A.K.; validation, I.K., and K.S.N.; formal analysis, M.A.; investigation, S.U.H.; resources, I.K.; writing—original draft preparation, T.A.A.; and writing—review and editing.

Acknowledgments: The authors acknowledge the anonymous referees for their useful suggestions.

References

1.   Podlubny, I. *Fractional Differential Equations: An Introduction to Fractional Derivatives, Fractional Differential Equations, to Methods of Their Solution and Some of Their Applications*; Elsevier: Amsterdam, The Netherlands, 1998; Volume 198.
2.   Torvik, P.J.; Bagley, R.L. On the appearance of the fractional derivative in the behavior of real materials. *J. Appl. Mech.* **1984**, *51*, 294–298. [CrossRef]
3.   Caputo, M. *Elasticità e dissipazione*; Zanichelli: Bologna, Italy, 1969.
4.   Suarez, L.; Shokooh, A. An eigenvector expansion method for the solution of motion containing fractional derivatives. *J. Appl. Mech.* **1997**, *64*, 629–635. [CrossRef]
5.   Michalski, M.W. *Derivatives of Noninteger Order and Their Applications*; Institute of Mathematics, Polish Academy of Sciences Warsaw: Warszawa, Poland, 1993.
6.   Gloeckle, W.G.; Nonnenmacher, T.F. Fractional integral operators and fox functions in the theory of viscoelasticity. *Macromolecules* **1991**, *24*, 6426–6434. [CrossRef]
7.   Ray, S.S.; Bera, R. An approximate solution of a nonlinear fractional differential equation by adomian decomposition method. *Appl. Math. Comput.* **2005**, *167*, 561–571.
8.   Babenko, Y. Non integer differential equation. In Proceedings of the 3rd International Conference on Intelligence in Networks, Bordeaux, France, 11–13 October 1994.
9.   Gaul, L.; Klein, P.; Kemple, S. Damping description involving fractional operators. *Mech. Syst. Signal Process.* **1991**, *5*, 81–88. [CrossRef]
10.   Ochmann, M.; Makarov, S. Representation of the absorption of nonlinear waves by fractional derivatives. *J. Acoust. Soc. Am.* **1993**, *94*, 3392–3399. [CrossRef]
11.   Kumar, S. A new fractional modeling arising in engineering sciences and its analytical approximate solution. *Alex. Eng. J.* **2013**, *52*, 813–819. [CrossRef]
12.   Mainardi, F. *Fractional Calculus and Waves in Linear Viscoelasticity: An Introduction to Mathematical Models*; World Scientific: Singapore, 2010.
13.   Podlubny, I.; Chechkin, A.; Skovranek, T.; Chen, Y.; Jara, B.M.V. Matrix approach to discrete fractional calculus ii: Partial fractional differential equations. *J. Comput. Phys.* **2009**, *228*, 3137–3153. [CrossRef]

14. Caputo, M.; Fabrizio, M. A new definition of fractional derivative without singular kernel. *Progr. Fract. Differ. Appl.* **2015**, *1*, 1–13.

15. Atangana, A. On the new fractional derivative and application to nonlinear fishers reaction–diffusion equation. *Appl. Math. Comput.* **2016**, *273*, 948–956.

16. Caputo, M.; Fabrizio, M. Applications of new time and spatial fractional derivatives with exponential kernels. *Progr. Fract. Differ. Appl.* **2016**, *2*, 1–11. [CrossRef]

17. Fetecau, C.; Vieru, D.; Fetecau, C.; Akhter, S. General solutions for magnetohydrodynamic natural convection flow with radiative heat transfer and slip condition over a moving plate. *Z. Naturforsch. A* **2013**, *68*, 659–667. [CrossRef]

18. Dokuyucu, M.A.; Celik, E.; Bulut, H.; Baskonus, H.M. Cancer treatment model with the caputo-fabrizio fractional derivative. *Eur. Phys. J. Plus* **2018**, *133*, 92. [CrossRef]

19. Riaz, M.; Zafar, A. Exact solutions for the blood flow through a circular tube under the influence of a magnetic field using fractional caputo-fabrizio derivatives. *Math. Model. Nat. Phenom.* **2018**, *13*, 8. [CrossRef]

20. Shah, N.A.; Imran, M.; Miraj, F. Exact solutions of time fractional free convection flows of viscous fluid over an isothermal vertical plate with caputo and caputo-fabrizio derivatives. *J. Prime Res. Math.* **2017**, *13*, 56–74.

21. Vieru, D.; Fetecau, C.; Fetecau, C. Time-fractional free convection flow near a vertical plate with newtonian heating and mass diffusion. *Therm. Sci.* **2015**, *19* (Suppl. 1), 85–98. [CrossRef]

22. Fetecau, C.; Vieru, D.; Fetecau, C. Effect of side walls on the motion of a viscous fluid induced by an infinite plate that applies an oscillating shear stress to the fluid. *Cent. Eur. J. Phys.* **2011**, *9*, 816–824. [CrossRef]

23. Haq, S.U.; Khan, M.A.; Shah, N.A. Analysis of magneto hydrodynamic flow of a fractional viscous fluid through a porous medium. *Chin. J. Phys.* **2018**, *56*, 261–269. [CrossRef]

24. Henry, B.I.; Langlands, T.A.; Straka, P. An introduction to fractional diffusion. In *Complex Physical, Biophysical and Econophysical Systems*; World Scientific: Singapore, 2010; pp. 37–89.

25. Hristov, J. Transient heat diffusion with a non-singular fading memory: From the cattaneo constitutive equation with jeffreys kernel to the caputofabrizio time-fractional derivative. *Therm. Sci.* **2016**, *20*, 757–762. [CrossRef]

26. Hristov, J. Derivatives with non-singular kernels from the caputo–fabrizio definition and beyond: Appraising analysis with emphasis on diffusion models. *Front. Fract. Calc.* **2017**, *1*, 270–342.

27. El-Lateif, A.M.A.; Abdel-Hameid, A.M. Comment on "solutions with special functions for time fractional free convection flow of brinkman-type fluid" by F. Ali et al. *Eur. Phys. J. Plus* **2017**, *132*, 407. [CrossRef]

28. Ahmed, N.; Shah, N.A.; Vieru, D. Natural convection with damped thermal flux in a vertical circular cylinder. *Chin. J. Phys.* **2018**, *56*, 630–644. [CrossRef]

29. Zafar, A.; Fetecau, C. Flow over an infinite plate of a viscous fluid with noninteger order derivative without singular kernel. *Alex. Eng. J.* **2016**, *55*, 2789–2796. [CrossRef]

30. Yang, P.; Lam, Y.C.; Zhu, K.Q. Constitutive equation with fractional derivatives for the generalized UCM model. *J. Non-Newton. Fluid Mech.* **2010**, *165*, 88–97. [CrossRef]

Integer and Non-Integer Order Study of the GO-W/GO-EG Nanofluids Flow by Means of Marangoni Convection

Taza Gul [1,2], Haris Anwar [1], Muhammad Altaf Khan [1], Ilyas Khan [3,*] and Poom Kumam [4,5,6,*]

[1] Department of mathematics, City University of Science and Information Technology, Peshawar 25000, Pakistan; tazagul@cusit.edu.pk (T.G.); harismathe@gmail.com (H.A.); makhan@cusit.edu.pk (M.A.K.)

[2] Department of Mathematics, Govt. Superior Science College Peshawar, Khyber Pakhtunkhwa, Peshawar 25000, Pakistan

[3] Faculty of Mathematics and Statistics, Ton Duc Thang University, Ho Chi Minh City 72915, Vietnam

[4] KMUTT-Fixed Point Research Laboratory, Room SCL 802 Fixed Point Laboratory, Science Laboratory Building, Department of Mathematics, Faculty of Science, King Mongkut's University of Technology Thonburi (KMUTT), 126 Pracha-Uthit Road, Bang Mod, Thrung Khru, Bangkok 10140, Thailand

[5] KMUTT-Fixed Point Theory and Applications Research Group, Theoretical and Computational Science Center (TaCS), Science Laboratory Building, Faculty of Science, King Mongkut's University of Technology Thonburi (KMUTT), 126 Pracha-Uthit Road, Bang Mod, Thrung Khru, Bangkok 10140, Thailand

[6] Department of Medical Research, China Medical University Hospital, China Medical University, Taichung 40402, Taiwan

* Correspondence: ilyaskhan@tdtu.edu.vn (I.K.); poom.kum@kmutt.ac.th (P.K.)

Abstract: Characteristically, most fluids are not linear in their natural deeds and therefore fractional order models are very appropriate to handle these kinds of marvels. In this article, we studied the base solvents of water and ethylene glycol for the stable dispersion of graphene oxide to prepare graphene oxide-water (GO-W) and graphene oxide-ethylene glycol (GO-EG) nanofluids. The stable dispersion of the graphene oxide in the water and ethylene glycol was taken from the experimental results. The combined efforts of the classical and fractional order models were imposed and compared under the effect of the Marangoni convection. The numerical method for the non-integer derivative that was used in this research is known as a predictor corrector technique of the Adams–Bashforth–Moulton method (Fractional Differential Equation-12) or shortly (FDE-12). The impact of the modeled parameters were analyzed and compared for both GO-W and GO-EG nanofluids. The diverse effects of the parameters were observed through a fractional model rather than the traditional approach. Furthermore, it was observed that GO-EG nanofluids are more efficient due to their high thermal properties compared with GO-W nanofluids.

Keywords: integer and non-integer order derivatives; GO-W/GO-EG nanofluids; Marangoni convection; FDE-12 numerical method

1. Introduction

Fractional order models are very useful in the study of nanofluids that contain small nanosized particles at the rate of small intervals rather than the traditional concept of integer order derivatives. A fractional order study has the credibility to explain the actual behavior of the physical parameters and is possible only in the case of the small intervals. The influences of the physical parameters in the classical models are limited and, in some cases, different from the fractional order models near the wall surface. Caputo [1] introduced the idea of fractional derivatives from the modified Darcy's law using the concept of unsteadiness. This idea was further modified by the researchers El Amin [2], Atangana and Alqahtani [3], and Alkahtani [4] by introducing varieties of new fractional derivatives and their applications. The fractional derivative concept can potentially be applied to the study of complicated control system problems. Yilun Shang [5] studied finite-time state consensus problems in continuous multi-agent systems with non-linear particles. Liu et al. [6] investigated the fixed-time event-triggered consensus control problem for multi-agent systems with non-linear uncertainties.

Advanced energy assets are the hot issue amid engineers and researchers as a response to rising energy demands. The base liquids have no sufficient thermal efficiency to fulfill the required demands of the industry. The small size of metal particles are used in common solvents to improve the thermal efficiency of the liquids. Water-, ethylene glycol-, and mineral oil-like convectional heat transfer fluids play an imperative role in many industrial and technological approaches such as heat generation, air-conditioning, chemical production, microelectronics, and transportation. The rate of change at small intervals has been examined by Atangana and Baleanu [7] to investigate the physical constraints of nanofluids for the heat transfer applications.

The physical aspects of the nanofluids and the role of the small sized nanoparticles in the enhancement of heat transfer applications using the traditional concept were introduced by Choi [8] to enhance the thermal efficiency of the nanofluids through nanoparticles.

The carbon family has the tendency to provide rapid cooling and fast thermal productivities. The experimental results demonstrated for carbon materials include the results of graphite nanoparticles, graphene oxides, and carbon nanotubes. Ellahi et al. [9] comprehensively discussed the effect of Carbon Nanotubes (CNT) nanofluid flow along a vertical cone with variable wall temperature. The results of both types of nanofluid can be obtained. Gul et al. [10] discussed effective Prandtl number model influences on $Al_2O_3 - H_2O$ and $Al_2O_3 - C_2H_6O_2$ nanofluids' spray along a stretching cylinder. Ellahi [11] worked on the effects of Magneto Hydrodynamic (MHD) and temperature-dependent viscosity on the flow of a non-Newtonian nanofluid in a pipe, using the analytical solution. Ellahi et al. [12] studied shiny film coating for multi-fluid flows of a rotating disk suspended with nanosized silver and gold particles. Khan et al. [13] worked on the Optimal Homotopy Analysis Method (OHAM) solution of Multi Walled Carbon Nanotubes and Single Walled Carbon Nanotubes (MWCNT/SWCNT) nanofluid thin film flow over a nonlinear extending disc.

Hummers and Offeman [14] developed a speedy and comparatively safe technique for the production of graphitic oxide from graphite in what is basically a crystalline substance of sulfuric acid H_2SO_4, potassium permanganate $KMnO_4$, and sodium nitrate $NaNO_3$.

The high thermal conductivity and characteristic lubricity of graphene make it a perfect claimant for the alteration of functional fluids. The solid particles, having an efficient thermal conductivity, are assorted to the base fluid to enhance the overall thermal conductivity of the fluid, as depicted in Maxwell [15]. Balandin et al. [16] examined the efficient thermal conductivity of single layer graphene in different solvents. Wei et al. [17] were pioneers in expressing the use of graphene oxide in ethylene glycol to enhance the thermal conductivity of ethylene glycol (EG). The graphene oxide nanosheets were set and isolated in EG and water at 5% capacity concentrations to enhance the thermal conductivity up to 60% compared with the base liquid EG.

Recently, Gul and Firdous [18] experimentally examined the stable dispersion of the graphene oxide in water and then analyzed the numerical study of the graphene oxide-water (GO-W) nanofluid between two rotating discs for the thermal applications.

Another type of convection which is used for temperature-dependent situations is called Marangoni convection. The existence of a spontaneous interface was first reported in 1855 by Thomson [19] and later represented in detail in 1865 by Marangoni [20] by spreading an oil droplet on a water surface, revealing that lower surface tension will spread on a liquid with higher surface tension.

In light of the previous meaningful discussion, the aim of this study was to examine the GO-W and graphene oxide and ethylene glycol (GO-EG) nanofluid flow under the effect of Marangoni convection using the classical and fractional order models. The comparison of the two types of nanofluids was conducted to investigate the impacts of the physical parameters. The physical and numerical outputs of the classical and fractional models were also compared and discussed. Sheikholeslami and Ganji [21] examined the Cu–H_2O nanofluid flow under the impact of Marangoni convection. The numerical approach to find the solution of a different type of problem was previously discussed [22–27]. The numerical scheme of Runge Kutta method of order 4 (RK-4) was used in their study to determine the impact of the physical parameters and numerical outputs.

The published work of Gul and Kiran [18] was extended by including the GO-EG nanofluid and a comparison of GO-EG and GO-W was made. Furthermore, integer and non-integer models h were compared under the effect of Marangoni convection. The fractional order differential equations were tackled numerically with the help of the Fractional Differential Equation-12 (FDE-12) technique [28–32]. A variety of numerical techniques are used to find the solutions of the classical models [33] and these techniques are further combined for the solutions of fractional order problems. Agarwal et al. [34] studied the neural network models using the GML synchronization and impulsive Caputo fractional differential equations. Morales-Delgado et al. [35] worked on the analytic solution for oxygen diffusion from capillaries to tissues involving external force effects using a fractional calculus approach. Khan et al. [36] researched the dynamics of the Zika virus with the Caputo fractional derivative. The physical configuration of the problem is shown in Figure 1.

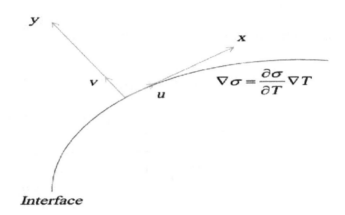

Figure 1. Geometry of the problem.

2. Problem Formulation

The two-dimensional Marangoni boundary layer flow of GO-W and GO-EG nanofluids is considered. The magnetic field is functional to the flow pattern in the transverse direction. The interface temperature is vigilant as a function of x. Assume that both base solvents (water and EG) contain GO nanoplatelets that are present in the thermally stable stage and no slippage. The flow under observation can be put into the following plan for GO nanofluids [14]:

$$\frac{\partial u}{\partial x} + \frac{\partial v}{\partial y} = 0, \tag{1}$$

$$u\left(\frac{\partial u}{\partial x}\right) + v\left(\frac{\partial v}{\partial y}\right) = v_{nf}\frac{\partial^2 u}{\partial y^2} - \frac{\sigma_{nf}B_0^2}{\rho_{nf}}u, \tag{2}$$

$$u\left(\frac{\partial T}{\partial x}\right) + v\left(\frac{\partial T}{\partial y}\right) = \frac{k_{nf}}{(\rho c_p)_{nf}}\left(\frac{\partial^2 T}{\partial y^2}\right), \tag{3}$$

where Equation (1) is the continuity equation, Equation (2) is the momentum equation, and Equation (3) is the energy equation. Exposed boundary conditions are expressed as:

$$v = 0,\ \frac{\mu_{nf}}{\mu_f}\left(\frac{\partial u}{\partial y}\right) = -\frac{\partial \sigma^{\otimes}}{\partial x} = \sigma_0\left(\gamma\frac{\partial T}{\partial x}\right),\ T = T_\infty + T_o X^2, X = \frac{x}{L},\ \text{at } y \to 0$$
$$u = 0,\ T = T_\infty,\ \text{at } y \to \infty. \tag{4}$$

The Marangoni conditions at the interface are revealed in Equation (4), taking the surface tension $\sigma^{\otimes} = \sigma[1 - \gamma(T - T_\infty)], \gamma = -\frac{1}{\sigma}\left(\frac{\partial \sigma^{\otimes}}{\partial T}\right)$, where γ stands for the surface tension temperature coefficient and σ represents the surface tension constant at the origin.

Also, u, v specify velocity components in the x-, y-directions. The interface and external flow of the temperature are represented by T, T_∞ respectively.

The effective $\rho_{nf}, \mu_{nf}, \sigma_{nf}, k_{nf}, (\rho c_p)_{nf}$ indicate the density, dynamic viscosity, electrical conductivity, thermal conductivity, and specific heat capacity of nanoplatelets, respectively, and are defined as:

$$\frac{(\rho c_p)_{nf}}{(\rho c_p)_f} = \left[1 - \phi + \frac{(\rho c_p)_s}{(\rho c_p)_f}\phi\right], \frac{\mu_{nf}}{\mu_f} = \frac{1}{(1-\phi)^{2.5}}, \frac{\rho_{nf}}{\rho_f} = \left[(1 - \phi) + \left(\frac{\rho_s}{\rho_f}\right)\phi\right],$$
$$\frac{\sigma_{nf}}{\sigma_f} = \left(1 + \frac{3(\sigma-1)\phi}{(\sigma+2)-(\sigma-1)\phi}\right). \tag{5}$$

where ϕ is the solid volume fraction and $\sigma_f, \rho_f, (\rho c_p)_f$ are the electrical conductivity, density, and specific heat capacity of the base fluids, respectively.

The similarity transformations are considered as [14]:

$$\eta = \frac{y}{L},\ \psi = v_f X f(\eta),\ u = \frac{\partial \psi}{\partial y},\ v = -\frac{\partial \psi}{\partial x},\ T = T_\infty + T_o X^2 \Theta(\eta). \tag{6}$$

Using the aforementioned assumption and condition, Equation (1) is verified identically, whereas Equations (2)–(4) are transformed in the following form:

$$\frac{\mu_{nf}}{\mu_f}\frac{\rho_{nf}}{\rho_f}\frac{\partial f^3(\eta)}{\partial \eta^3} + f(\eta)\frac{df^2(\eta)}{d\eta^2} - \left(\frac{df(\eta)}{d\eta}\right)^2 - M^2\frac{\sigma_{nf}}{\sigma_f}\left(\frac{df(\eta)}{d\eta}\right) = 0, \tag{7}$$

$$\left(\frac{k_{nf}}{k_f} + \frac{4}{3}\text{Rd}\right)\frac{d\Theta^2(\eta)}{d\eta^2} + \frac{(\rho c_p)_{nf}}{(\rho c_p)_f}\text{Pr}\left[f(\eta)\frac{d\Theta(\eta)}{d\eta} - 2\frac{df(\eta)}{d\eta}\Theta(\eta)\right] = 0. \tag{8}$$

$$f = 0,\ \frac{df^2}{d\eta^2} = -2(1 - \phi)^{2.5},\ \Theta = 1,\ \text{at } \eta = 0$$
$$\frac{df}{d\eta} = 0,\ \Theta = 0,\ \text{at } \eta \to \infty \tag{9}$$

where M, Pr, indicate the transformation of the magnetic parameter and Prandtl number, respectively, and are defined individually as:

$$M^2 = \frac{\sigma_f B_o^2 L^2}{\mu_o \rho_f},\ Pr = \frac{(\rho c_p)_f}{k_f}. \tag{10}$$

The local Nusselt number Nu_x is:

$$Nu = -2\frac{k_{nf}}{k_f}\left(\frac{\partial T}{\partial y}\right)_{y=0}. \tag{11}$$

3. Preliminaries on the Caputo Fractional Derivatives

The basic definition and properties related to non-integer or fractional derivatives derived by Caputo are as follows.

3.1. Definition 1

Let $b > 0$, $t > b$; $b, \alpha, t \in R$. The Caputo fractional derivative of order α of function $f \in C^n$ is given by:

$$_b^C D_t^\alpha f(t) = \frac{1}{\Gamma(n-\alpha)} \int_b^t \frac{f^{(n)}(\xi)}{(t-\xi)^{\alpha+1-n}} d\xi, \ n-1 < \alpha < n \in N. \tag{12}$$

3.2. Property 1

Let $f(t)$, $g(t) : [a, b] \to \mathfrak{R}$ be such that $_b^C D_t^\alpha f(t)$ and $_b^C D_t^\alpha g(t)$ exist almost everywhere and let c_1, $c_2 \in \mathfrak{R}$. Then $_b^C D_t^\alpha \{c_1 f(t) + c_2 g(t)\}$ exists almost everywhere and:

$$_b^C D_t^\alpha \{c_1 f(t) + c_2 g(t)\} = c_1 {}_b^C D_t^\alpha f(t) + c_2 {}_b^C D_t^\alpha g(t). \tag{13}$$

3.3. Property 2

The function $f(t) \equiv c$ is constant and therefore the fractional derivative is zero: $_b^C D_t^\alpha c = 0$. The general description of the fractional differential equation is assumed, including the Caputo concept:

$$_b^C D_t^\alpha x(t) = f(t, x(t)), \ \alpha \in (0, 1) \tag{14}$$

with the initial conditions $x_0 = x(t_0)$.

4. Solution Methodology

The variables were selected to alter Equations (7)–(9) into the system of the first order differential equations:

$$y_1 = \eta, \ y_2 = f, \ y_3 = f', \ y_4 = f'', \ y_5 = \Theta, \ y_6 = \Theta'. \tag{15}$$

The variables selected in Equation (15) were used for the classical (integer) system and Equations (7)–(9) are settled as:

$$y_1' = 1, \ y_2' = y_3, \ y_3' = y_4, \ y_4' = \left[\frac{\mu_{nf}\rho_{nf}}{\mu_f\rho_f}\right]^{-1}\left[-y_2 y_3 + (y_4)^2 + \frac{\sigma_{nf}}{\sigma_f}M^2 y_3\right], \ y_5' = y_6,$$
$$y_6' = \left[\frac{k_{nf}}{k_f} + \frac{4}{3}Rd\right]^{-1}\left[-\frac{(\rho c_p)_{nf}}{(\rho c_p)_f}\text{Pr}(y_2 y_6 - 2y_3 y_5)\right] \tag{16}$$

with initial conditions:

$$y_1 = 0, \ y_2 = 0, \ y_3 = u_1, \ y_4 = -2(1-\phi)^{2.5}, \ y_5 = 1, \ y_6 = u_2. \tag{17}$$

The first order ordinary differential equations system (15) is further transformed into the Caputo fractional order derivatives.

The FDE-12 technique was adopted for the fractional order differential equations. The final system and initial conditions are as follows:

$$\begin{pmatrix} D_\eta^\alpha y_1 \\ D_\eta^\alpha y_2 \\ D_\eta^\alpha y_3 \\ D_\eta^\alpha y_4 \\ D_\eta^\alpha y_5 \\ D_\eta^\alpha y_6 \end{pmatrix} = \begin{pmatrix} 1 \\ y_3 \\ y_4 \\ \left(\frac{\mu_{nf}\rho_{nf}}{\mu_f\rho_f}\right)^{-1}\left(-y_2 y_3 + (y_4)^2 + \frac{\sigma_{nf}}{\sigma_f}M^2 y_3\right) \\ y_6 \\ \left(\frac{k_{nf}}{k_f} + \frac{4}{3}Rd\right)^{-1}\left(-\frac{(\rho c_p)_{nf}}{(\rho c_p)_f}\text{Pr}(y_2 y_6 - 2y_3 y_5)\right) \end{pmatrix}, \begin{pmatrix} y_1 \\ y_2 \\ y_3 \\ y_4 \\ y_5 \\ y_6 \end{pmatrix} = \begin{pmatrix} 0 \\ 0 \\ u_1 \\ -2(1-\phi)^{2.5} \\ 1 \\ u_2 \end{pmatrix}. \tag{18}$$

5. Results and Discussions

The GO-W and GO-EG nanofluid flows under the effect of Marangoni convection were analyzed using the classical and fractional models for heat transfer applications. The impact of the physical parameters was obtained through the classical and fractional order models and compared. Moreover, the impact of the embedded parameters, comprising GO-W and GO-EG nanofluids, was compared, and it was observed that due to rich thermophysical properties the GO-EG nanofluid is a comparatively better heat transfer solvent.

In the following figures, an upward arrow shows an increasing effect while a downward arrow shows a decreasing effect.

The effect of the nanofluid volume fraction ϕ using the classical model versus the velocity profile $f(\eta)$ for the GO-W and GO-EG nanofluids is depicted in Figure 2. The rising values of ϕ lead to enhance the velocity field linearly in the classical model. Physically, the larger amount of nanoparticle volume fraction generates the friction force, and this force is more visible near the wall, reducing the flow motion. However, this impact is unclear in the classical model. The increase in the flow motion due to the rising values of ϕ indicates that the thermal efficiency of the nanofluid provides strength to the flow field. Moreover, this impact is comparatively high using the GO-EG nanofluids.

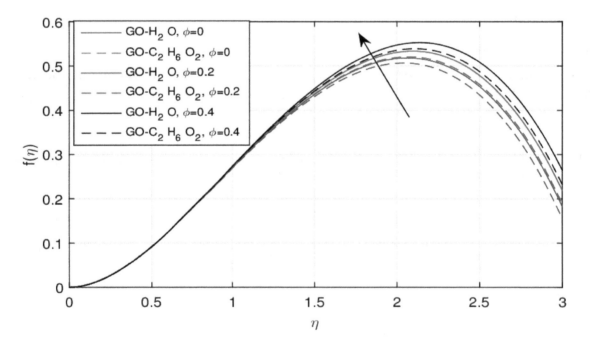

Figure 2. The impact of ϕ versus classical $f(\eta)$, when $M = 0.1$.

The effect of the nanofluid volume fraction ϕ using the fractional model for the same values of ϕ is shown in Figure 3. Near the wall surface the velocity field falls, because near the wall surface friction increases, which retards the velocity and increases after the critical point due to the reduction in friction. Physically, the larger nanoparticle volume fraction generates the friction force and this force is more visible near the wall, reducing the flow motion. This effect is clearer using the fractional model. The impact of the increasing values of ϕ versus the radial velocity field $f'(\eta)$ using the integer model is shown in Figure 4. The same effect as discussed above was observed. The larger amount of ϕ increases the value of $f'(\eta)$ in the integer order model. The impact of the increasing values of ϕ versus the radial velocity field $f'(\eta)$ using the fractional model is shown in Figure 5. The larger amount of ϕ reduces $f'(\eta)$ in the fractional order model near the wall surface and after the point of inflection the velocity enhances, as shown in Figure 5.

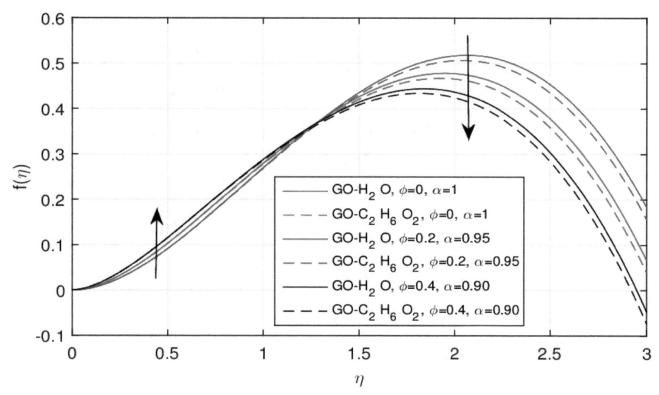

Figure 3. The impact of ϕ versus fractional $f(\eta)$, when $M = 0.1$.

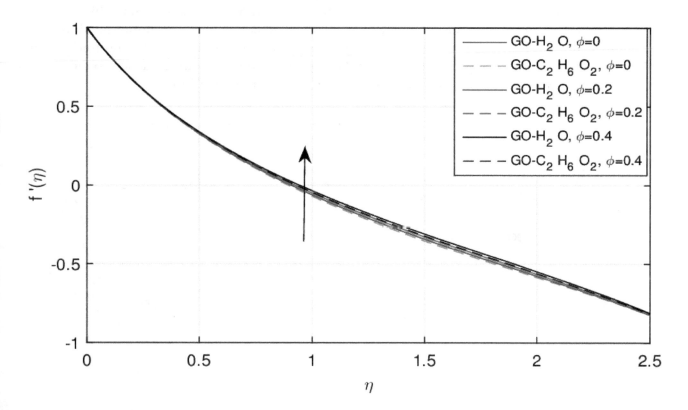

Figure 4. The impact of ϕ versus classical $f'(\eta)$, when $M = 0.1$.

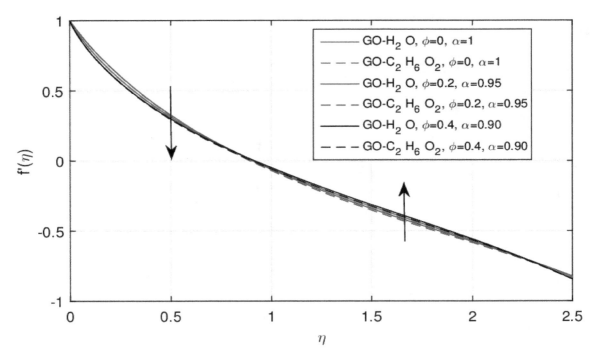

Figure 5. The impact of ϕ versus fractional $f'(\eta)$, when $M = 0.1$.

The influence of the larger values of the magnetic parameter M versus the temperature profile $\Theta(\eta)$ for the integer order and fractional order problems is shown in Figures 6 and 7, respectively. This is due to the Lorentz force, which results in resistance to the transport phenomena. This retarding force controls the GO-W and GO-EG nanofluid velocities, which is useful in numerous industrial and engineering applications such as heat transferring, industrial cooling, and nanofluid coolant. Mathematically, the magnetic parameter represents the ratio of the magnetic induction to the viscous force. Moreover, GO-EG was found to show more dominant results than GO-W.

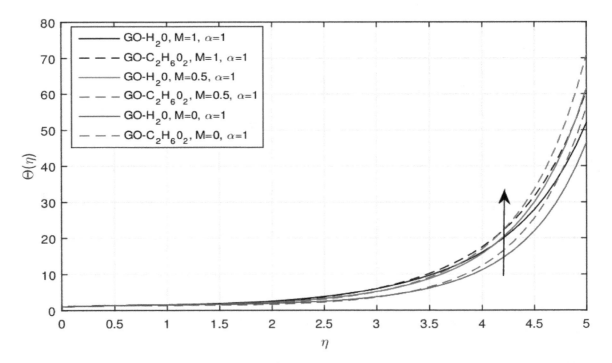

Figure 6. The impact of M versus classical $\Theta(\eta)$, when $\phi = 0.2, \mathrm{Pr} = 6.7$.

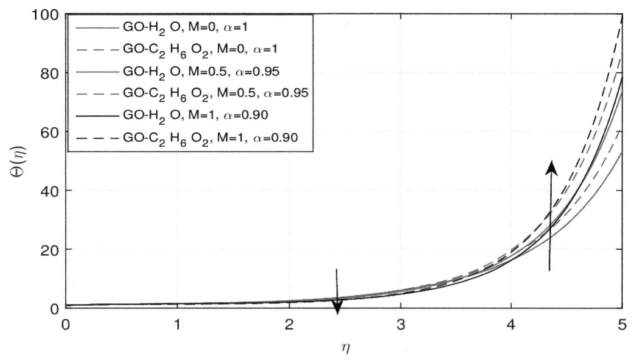

Figure 7. The impact of M versus fractional $\Theta(\eta)$, when $\phi = 0.2, \mathrm{Pr} = 6.7$.

The impact of ϕ using the classical model and the fractional order model versus the temperature profile $\Theta(\eta)$ for the GO-W and GO-EG nanofluids is depicted in Figures 8 and 9. The nanoparticle volume fraction ϕ is basically used as the heat transport agent parameter and its increase boosts up the temperature profile. In fact, the cohesive forces among the liquid molecules release with the increasing amount of ϕ and as a result the thermal boundary layer enhancement. This effect is comparatively efficient in GO-EG nanofluids due to their enriching thermophysical properties.

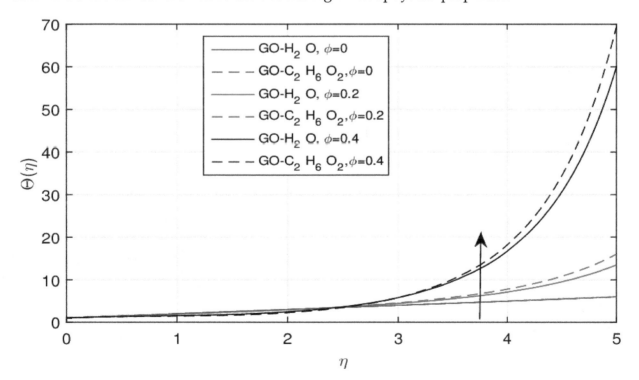

Figure 8. The impact of ϕ versus the integer order of $\Theta(\eta)$, when $M = 0.1, \mathrm{Pr} = 6.7$.

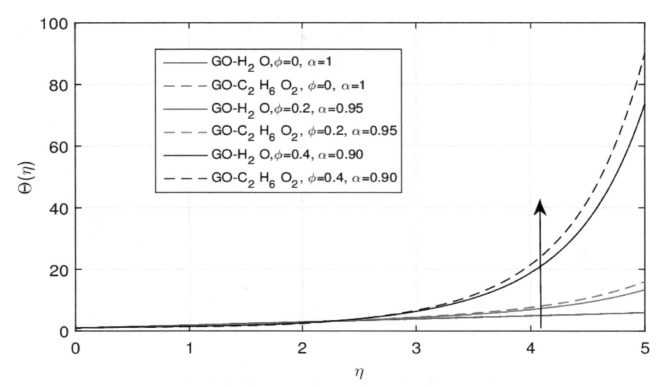

Figure 9. The impact of ϕ versus fractional $\Theta(\eta)$, when $M = 0.1, \mathrm{Pr} = 6.7$.

The impact of the magnetic parameter M versus the Nusselt number Nu of integer and fraction order problems is displayed in Figures 10 and 11, respectively. A higher value of the magnetic parameter enhances the temperature field and reduces the Nusselt number. This effect is slightly clearer using the fractional model for similar values of M, as shown in Figure 11. We noticed that in the cases of GO-EG and GO-W the temperature distribution is dominant and almost completely closed.

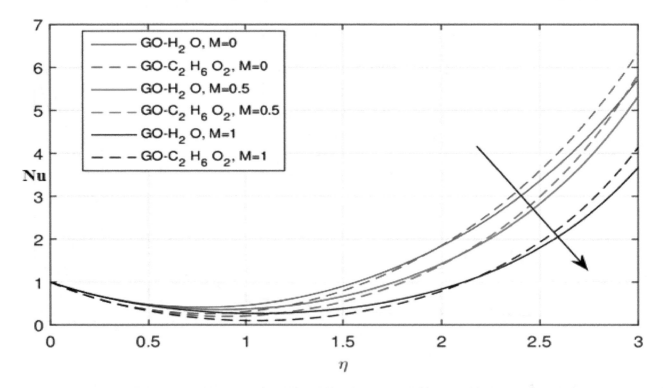

Figure 10. The impact of M versus classical Nu, when $\phi = 0.2, \mathrm{Pr} = 6.7$.

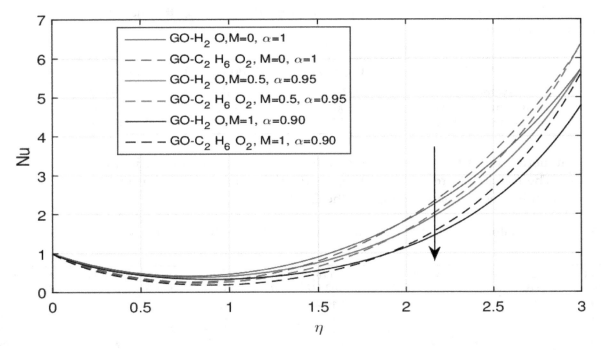

Figure 11. The impact of M versus fractional Nu, when $\phi = 0.2, \mathrm{Pr} = 6.7$.

The impact of the fractional order $\alpha = 1, 0.95, 0.90, 0.85$ versus Nu for both sorts of nanofluids is depicted in Figure 12. It was observed that the heat transfer and cooling efficiency of the GO-EG nanofluid is comparatively higher than the GO-W nanofluid. The Nusselt number increases near the wall surface and declines towards the free surface.

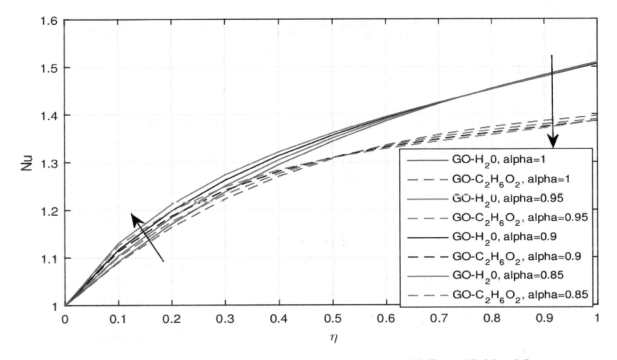

Figure 12. $\alpha = 1, 0.95, 0.90, 0.85$ versus Nu, when $\phi = 0.2, \mathrm{Pr} = 6.7, M = 0.5$.

The thermophysical properties of the two sorts of nanofluids (GO-W and GO-EG) were examined from the experimental results and are displayed in Table 1. These properties of the base fluids were initially calculated at 25 °C. The thermophysical properties were examined at a different temperature level from 25 °C to 40 °C.

Table 1. The experimental values (thermophysical properties) of water, ethylene glycol, and graphene oxide nanoparticles.

Model	ρ (kg/m^3)	C_p (kg^{-1}/k^{-1})	k (Wm^{-1}k^{-1})
Water (W)	997.1	4179	0.613
Graphene oxide (GO)	1800	717	5000
Ethylene glycol (EG)	1.115	0.58	0.1490

The numerical outputs for the heat transfer rate using the fractional order problem are displayed in Table 2. The fractional order $\alpha = 1, 0.95, 0.90, 0.85$ enhances the heat transfer rate in increasing intervals and this effect is relatively high in the GO-W nanofluid.

Table 2. $\alpha = 1, 0.95, 0.90, 0.85$ versus Nu, when $\phi = 0.2, \Pr = 6.7, M = 0.5$.

$\alpha = 1$ $\eta.$	$\Theta'(0)$ GO-W	$\Theta'(0)$ GO-EG	$\alpha = 0.95$ $\eta.$	$\Theta'(0)$ GO-W	$\Theta'(0)$ GO-EG	$\alpha = 0.9$ $\eta.$	$\Theta'(0)$ GO-W	$\Theta'(0)$ GO-EG	$\alpha = 0.85$ $\eta.$	$\Theta'(0)$ GO-W	$\Theta'(0)$ GO-EG
0.1	1.0921	1.0903	0.1	1.1039	1.1014	0.1	1.1167	1.1133	0.1	1.1304	1.1259
0.2	1.1708	1.1639	0.2	1.1845	1.1758	0.2	1.1983	1.1875	0.2	1.2121	1.1986
0.3	1.2380	1.2234	0.3	1.2504	1.2328	0.3	1.2624	1.2413	0.3	1.2736	1.2486
0.4	1.2953	1.2706	0.4	1.3051	1.2764	0.4	1.3141	1.2809	0.4	1.3221	1.2840
0.5	1.3442	1.3076	0.5	1.3509	1.3095	0.5	1.3567	1.3101	0.5	1.3617	1.3096
0.6	1.3862	1.3361	0.6	1.3899	1.3344	0.6	1.3928	1.3318	0.6	1.3952	1.3288
0.7	1.4225	1.3578	0.7	1.4235	1.3532	0.7	1.4242	1.3484	0.7	1.4250	1.3441
0.8	1.4544	1.3741	0.8	1.4534	1.3676	0.8	1.4527	1.3619	0.8	1.4527	1.3577
0.9	1.4832	1.3866	0.9	1.4809	1.3794	0.9	1.4796	1.3741	0.9	1.4800	1.3715
1.0	1.5098	1.3966	1.0	1.5072	1.3900	1.	1.5064	1.3866	1.0	1.5083	1.3872

6. Conclusions

The flow of the two types of nanofluids, GO-W and GO-EG, were analyzed for the augmentation of temperature. Numerical and theoretical analyses were carried out under the effect of Marangoni convection. The classical and fractional models were used to investigate the impact of the physical parameters for similar values of the constraint. It was observed that the outputs of the physical parameters over the velocity and temperature profiles in the classical model are limited, but in utilizing the fractional model the effect varies in each interval. The fractional order model specifies the outputs at the small number of intervals, leading to the accurate determination of the physical parameters, which is very necessary for industrial and engineering applications.

The main features of this study are as follows:

- The rising values of ϕ lead to the linear enhancement of the velocity field, which was observed more clearly in the non-integer case compared with the classical model.
- The increasing values of the magnetic parameter increase the temperature field and decrease the Nusselt number. This effect is somewhat better in the fractional case compared to the integer model.
- Due to the rising values of ϕ, the thermal boundary layer increases and this effect is somewhat better in the GO-EG nanofluid rather than the GO-W nanofluid.
- The cooling efficiency and heat transfer of the GO-EG nanofluid is far better than that of the GO-W nanofluid.
- With the Lorentz force, resistance arises in the transport phenomenon. This particular phenomenon controls the GO-W and GO-EG nanofluid velocities. Also, this effect is more visible in GO-EG than in GO-W.
- Due to the fractional order $\alpha = 1, 0.95, 0.90, 0.85$, the heat transfer rate enhances in growing increments and this effect is far better in the GO-W nanofluid compared with the GO-EG nanofluid.

7. Future Work

This mathematical model is extendable for future work considering gold nanoparticles, carbon nanotubes, porous media, variable viscosity, thermal radiation, and hall effects. The fractional ordered derivative scheme is also extendable using the Caputo–Fabrizio and Atangana–Baleanu operators.

Author Contributions: T.G. and H.A.; conceptualization, M.A.K. and T.G.; methodology, I.K.; software, I.K.; and P.K.; validation, T.G., M.A.K. and H.A.; formal analysis, I.K.; P.K.; investigation, T.G.; I.K.; writing—original draft preparation, T.G.; M.A.K. and P.K. I.K.; writing—review and editing.

Acknowledgments: This project was supported by the Theoretical and Computational Science (TaCS) Center under Computational and Applied Science for Smart Innovation Research Cluster (CLASSIC), Faculty of Science, KMUTT.

References

1. Caputo, M. Models of flux in porous media with memory. *Water Resour. Res.* **2000**, *36*, 693–705. [CrossRef]
2. El Amin, M.F.; Radwan, A.G.; Sun, S. Analytical solution for fractional derivative gas-flow equation in porous media. *Results Phys.* **2017**, *7*, 2432–2438. [CrossRef]
3. Atangana, A.; Alqahtani, R.T. Numerical approximation of the space-time Caputo-Fabrizio fractional derivative and application to groundwater pollution equation. *Adv. Differ. Equ.* **2016**, *2016*, 156–169. [CrossRef]
4. Alkahtani, B.S.T.; Koca, I.; Atangan, A. A novel approach of variable order derivative: Theory and Methods. *J. Nonlinear Sci. Appl.* **2016**, *9*, 4867–4876. [CrossRef]
5. Shang, Y. Finite-time consensus for multi-agent systems with fixed topologies. *Int. J. Syst. Sci.* **2012**, *43*, 499–506. [CrossRef]
6. Liu, J.; Yu, Y.; Wang, Q.; Sun, C. Fixed-time event-triggered consensus control for multi-agent systems with nonlinear uncertainties. *Neurocomputing* **2017**, *260*, 497–504. [CrossRef]
7. Atangana, A.; Baleanu, D. New fractional derivatives with non-local and non singular kernel: Theory and application to heat transfer model. *arXiv* **2016**, arXiv:1602.03408.
8. Choi, S.U.S. Enhancing thermal conductivity of fluids with nanoparticles, developments and applications of non-Newtonian flows. *FED-231IMD* **1995**, *66*, 99–105.
9. Ellahi, R.; Hassan, M.; Zeeshan, A. Study of Natural Convection MHD Nanofluid by Means of Single and Multi-Walled Carbon Nanotubes Suspended in a Salt-Water Solution. *IEEE Trans. Nanotechnol.* **2015**, *14*, 726–734. [CrossRef]
10. Gul, T.; Nasir, S.; Islam, S.; Shah, Z.; Khan, M.A. Effective prandtl number model influences on the Al_2O_3-H_2O and Al_2O_3-$C_2H_6O_2$ nanofluids spray along a stretching cylinder. *Arab. J. Sci. Eng.* **2019**, *2*, 1601–1616. [CrossRef]
11. Ellahi, R. The effects of MHD and temperature dependent viscosity on the flow of a non-Newtonian nanofluid in a pipe, Analytical solution. *Appl. Math. Model.* **2013**, *37*, 1451–1457. [CrossRef]
12. Ellahi, R.; Zeeshan, A.; Hussain, F.; Abbas, T. Study of Shiny Film Coating on Multi-Fluid Flows of a Rotating Disk Suspended with Nano-Sized Silver and Gold Particles: A Comparative Analysis. *Coatings* **2018**, *8*, 422. [CrossRef]
13. Taza, G.; Waris, K.; Muhammad, S.; Muhammad, A.K.; Ebenezer, B. MWCNTs/SWCNTs Nanofluid Thin Film Flow over a Nonlinear Extending Disc: OHAM solution. *J. Therm. Sci.* **2019**, *28*, 115–122. [CrossRef]
14. Hummers, W.S.; Offeman, R.E. Preparation of graphitic oxide. *J. Am. Chem. Soc.* **1958**, *80*, 1339. [CrossRef]
15. Balandin, A.A.; Ghosh, S.; Bao, W.; Calizo, I.; Teweldebrhan, D.; Miao, F.; Lau, C.N. Superior thermal conductivity of single-layer graphene. *Nano Lett.* **2008**, *8*, 902–907. [CrossRef]
16. Maxwell, J.C. *Treatise on Electricity and Magnetism*; Clarendon Press: Oxford, UK, 1873.
17. Wei, Y.; Huaqing, X.; Dan, B. Enhanced thermal conductivities of nanofluids containing graphene oxide nanosheets. *Nanotechnology* **2010**, *21*, 055705.

18. Gul, T.; Ferdous, K. The experimental study to examine the stable dispersion of the graphene nanoparticles and to look at the GO–H$_2$O nanofluid flow between two rotating disks. *Appl. Nanosci.* **2018**, *8*, 1711–1728. [CrossRef]

19. Thomson, J. On certain curious motions observable at the surface of wine and other alcoholic liquors. *Philos. Mag.* **1855**, *10*, 330–333. [CrossRef]

20. Marangoni, C. Ueber die Ausbreitung der Tropfeneiner Flussigkeit auf der Oberflache einer anderen. *Ann. Phys.* **1871**, *143*, 337–354. [CrossRef]

21. Sheikholeslami, M.; Ganji, D.D. Influence of magnetic field on CuOeH$_2$O nanofluid flow considering Marangoni boundary layer. *Int. J. Hydrog. Energy* **2017**, *42*, 2748–2755. [CrossRef]

22. Shirvan, K.M.; Ellahi, R.; Sheikholeslami, T.F.; Behzadmehr, A. Numerical investigation of heat and mass transfer flow under the influence of silicon carbide by means of plasmaenhanced chemical vapor deposition vertical reactor. *Neural Comput. Appl.* **2018**, *30*, 3721–3731. [CrossRef]

23. Barikbin, Z.; Ellahi, R.; Abbasbandy, S. The Ritz-Galerkin method for MHD Couette Fow of non-Newtonian fluid. *Int. J. Ind. Math.* **2014**, *6*, 235–243.

24. Hayat, T.; Saif, R.S.; Ellahi, R.; Muhammad, T.; Ahmad, B. Numerical study of boundary-layer flow due to a nonlinear curved stretching sheet with convective heat and mass conditions. *Results Phys.* **2017**, *7*, 2601–2606. [CrossRef]

25. Hayat, T.; Saif, R.S.; Ellahi, R.; Muhammad, T.; Ahmad, B. Numerical study for Darcy-Forchheimer flow due to a curved stretching surface with Cattaneo-Christov heat flux and homogeneous heterogeneous reactions. *Results Phys.* **2017**, *7*, 2886–2892. [CrossRef]

26. Javeed, S.; Baleanu, D.; Waheed, A.; Khan, M.S.; Affan, H. Analysis of Homotopy Perturbation Method for Solving Fractional Order Differential Equations. *Mathematics* **2019**, *7*, 40. [CrossRef]

27. Srivastava, H.M.; El-Sayed, A.M.A.; Gaafar, F.M. A Class of Nonlinear Boundary Value Problems for an Arbitrary Fractional-Order Differential Equation with the Riemann-Stieltjes Functional Integral and Infinite-Point Boundary Conditions. *Symmetry* **2018**, *10*, 508. [CrossRef]

28. Diethelm, K.; Freed, A.D. The Frac PECE subroutine for the numerical solution of differential equations of fractional order. In *Forschung und Wissenschaftliches Rechnen*; Heinzel, S., Plesser, T., Eds.; 1998 Gessellschaft fur Wissenschaftliche Datenverarbeitung: Gottingen, Germany, 1999; pp. 57–71.

29. Diethelm, K.; Ford, N.J.; Freed, A.D. Detailed error analysis for a fractional Adams method. *Numer. Algorithms* **2004**, *36*, 31–52. [CrossRef]

30. Saifullah Khan, M.A.; Farooq, M. A fractional model for the dynamics of TB virus. *Chaos Solitons Fractals* **2018**, *116*, 63–71.

31. Gul, T.; Khan, M.A.; Khan, A.; Shuaib, M. Fractional-order three-dimensional thin-film nanofluid flow on an inclined rotating disk. *Eur. Phys. J. Plus* **2018**, *133*, 500–5011. [CrossRef]

32. Gul, T.; Khan, M.A.; Noman, W.; Khan, I.; Alkanhal, T.A.; Tlili, I. Fractional Order Forced Convection Carbon Nanotubes Nanofluid Flow Passing Over a Thin Needle. *Symmetry* **2019**, *11*, 312. [CrossRef]

33. Ullah, S.; Khan, M.A.; Farooq, M.; Gul, T.; Hussai, F. A fractional order HBV model with hospitalization. *Discret. Contin. Dyn. Syst.* **2019**, 957–974. [CrossRef]

34. Agarwal, R.; Hristova, S.; O'Regan, D. Global Mittag-Leffler Synchronization for Neural Networks Modeled by Impulsive Caputo Fractional Differential Equations with Distributed Delays. *Symmetry* **2018**, *10*, 473. [CrossRef]

35. Morales-Delgado, V.F.; Gómez-Aguilar, J.F.; Saad, K.M.; Khan, M.A.; Agarwal, P. Analytic solution for oxygen diffusion from capillary to tissues involving external force effects: A fractional calculus approach. *Phys. A Stat. Mech. Its Appl.* **2019**, *523*, 48–65. [CrossRef]

36. Khan, M.A.; Ullah, S.; Farhan, M. The dynamics of Zika virus with Caputo fractional derivative. *AIMS Math.* **2019**, *4*, 134–146. [CrossRef]

9

Modified MHD Radiative Mixed Convective Nanofluid Flow Model with Consideration of the Impact of Freezing Temperature and Molecular Diameter

Umar Khan [1], Adnan Abbasi [2], Naveed Ahmed [3], Sayer Obaid Alharbi [4] Saima Noor [5], Ilyas Khan [6,*], Syed Tauseef Mohyud-Din [3] and Waqar A. Khan [7]

[1] Department of Mathematics and Statistics, Hazara University, Mansehra 21120, Pakistan; umar_jadoon4@yahoo.com

[2] Department of Mathematics, Mohi-ud-Din Islamic University Nerian Sharif, Azad Jammu & Kashmir 12080, Pakistan; adnan_abbasi89@yahoo.com

[3] Department of Mathematics, Faculty of Sciences, HITEC University Taxila Cantt, Punjab 47080, Pakistan; nidojan@gmail.com (N.A.); syedtauseefs@hotmail.com (S.T.M.-D.)

[4] Department of Mathematics, College of Science Al-Zulfi, Majmaah University, Al-Majmaah 11952, Saudi Arabia; so.alharbi@mu.edu.sa

[5] Department of Mathematics, COMSATS University Islamabad, Abbottabad 22010, Pakistan; saimanoor@ciit.net.pk

[6] Faculty of Mathematics and Statistics, Ton Duc Thang University, Ho Chi Minh City 72915, Vietnam

[7] Department of Mechanical Engineering, College of Engineering, Prince Mohammad Bin Fahd University, Al Khobar 31952, Saudi Arabia; wkhan1956@gmail.com

* Correspondence: ilyaskhan@tdtu.edu.vn

Abstract: Magnetohydrodynamics (MHD) deals with the analysis of electrically conducting fluids. The study of nanofluids by considering the influence of MHD phenomena is a topic of great interest from an industrial and technological point of view. Thus, the modified MHD mixed convective, nonlinear, radiative and dissipative problem was modelled over an arc-shaped geometry for Al_2O_3 + H_2O nanofluid at 310 K and the freezing temperature of 273.15 K. Firstly, the model was reduced into a coupled set of ordinary differential equations using similarity transformations. The impact of the freezing temperature and the molecular diameter were incorporated in the energy equation. Then, the Runge–Kutta scheme, along with the shooting technique, was adopted for the mathematical computations and code was written in Mathematica 10.0. Further, a comprehensive discussion of the flow characteristics is provided. The results for the dynamic viscosity, heat capacity and effective density of the nanoparticles were examined for various nanoparticle diameters and volume fractions.

Keywords: arched surface; nonlinear thermal radiation; molecular diameter; Al_2O_3 nanoparticles; streamlines; isotherms; RK scheme

1. Introduction

The liquids regularly used in heat transfer applications, such as water, propylene glycol, ethylene glycol, kerosene oil, engine oil and transformer oil, are extensively used in industry and in thermal power plants. Due to their reduced thermal conductivity, these liquids do not have effective heat transfer characteristics. However, for a great deal of industrial production, remarkable amounts of heat are required. The thermal conductivity of the solid materials, such as different metals and oxides, is very high in comparison with regular liquids. Thus, scientists and engineers hypothesized that the heat transfer in working fluids could be enhanced by mixing in the nanoparticles in above regular liquids. Finally, Choi [1] developed a colloidal composition that has effective heat transfer characteristics, as compared to regular fluids. Choi [1] unlocked a new innovative research area and researchers, scientists and engineers focused on the analysis of nanofluids. Before the development of nanofluids, heat transfer was a major problem from an industrial point of view, since considerable amounts of heat transfer were required for a great deal of technological and industrial production, and regular liquids failed to provide the desired amount of heat. Thermal conductivity plays a major role in the heat transfer rate of nanofluids. Thus, several theoretical models were proposed for thermal conductivity. The thermal conductivity model was developed by Maxwell in 1873 [2], who considered nanosized particles, and this can be considered the origin of the concept of the nanofluids.

Several theoretical models based on nanoparticle characteristics that take into account the effects of temperature and the shape and diameter of nanoparticles, as well as Brownian motion, have been presented. In 1935, Bruggemann [3] constructed a thermal conductivity correlation for spherical nanoparticles that was limited to high concentration patterns. The behavior of thermal conductivity was developed by Hamilton [4] in 1962, who explored the effects of nanoparticle shape. In 1996, Lu and Lin [5] proposed a model incorporating the effects of Brownian dynamics. The thermal conductivity model for the interaction between the nanoparticles and their surrounding liquid was developed by Koo and Kleinstreuer [6,7]. Xue [8] developed a thermal conductivity model for carbon nanotubes. In 2005, Prasher et al. [9] found a correlation by considering the influence of convection on the thermal conductivity of nanoparticles. In 2006, Li [10] made apparent the influence of temperature on thermal conductivity and outlined a correlation for Al_2O_3/H_2O and CuO/H_2O nanofluids. In 2011, Corcione [11] developed a model for $Al_2O_3 + H_2O$ nanofluids by incorporating the effects of freezing temperatures. By incorporating thermal conductivity models, the above researchers presented various models and described the heat transfer enhancement due to thermal conductivity. Some useful studies for nanofluids are described in [12–16].

Similar to nanoparticles, carbon nanotubes also have high thermal conductivity and unique mechanical and chemical properties. Carbon nanotubes are subcategorized as either single or multiple walled carbon nanotubes. The concept of colloidal suspension in relation to carbon nanotubes was presented by Iijima [17]. After the development of the thermal conductivity correlation for carbon nanotubes, many studies were presented outlining thermal enhancement due to suspended carbon nanotubes. Recently, Ahmed et al. [18] explored the influence of thermal radiation and viscous dissipation on the flow of water suspended by carbon nanotubes. The effect of thermophysical characteristics of the nanotubes on the heat transfer enhancement water over a curved surface and non-parallel walls was described in [19,20], respectively.

Recently, the flow over an arc-shaped geometry has become a point of interest. Reddy et al. [21] recently modified the curve-shaped flow model for nonlinear radiative heat flux. They also examined the impact of the cross-diffusion phenomenon on heat and mass transfer. Another useful mechanism to enhance the fluid temperature is ohmic heating, which produces extra heat in the conductor, with electrons supplying energy to the atoms of the conductor through collisions. In 2017, Hayat et al. [22] examined the effects of resistive heating on the curve surface flow.

A literature review revealed that there have not yet been any studies of the impact of freezing temperatures and the diameter of nanoparticles on the flow of incompressible fluids due to the effects of nonlinear radiative heat flux, viscous dissipation, mixed convection and Lorentz forces. This study is presented to cover this significant gap. The nanofluids $Al_2O_3 + H_2O$ were used to study the characteristics of the flow and other effective thermophysical properties, such as effective density, heat capacity and thermal conductivity. The results for shear stress and local heat transfer are also described and discussed comprehensively. Finally, major findings of the study is presented.

2. Model Formulation

We considered the laminar time independent and the incompressible flow of the $Al_2O_3 + H_2O$ nanofluid by taking into account the influence of a nonlinear, radiative heat flux and the imposed variable magnetic field over an arc geometry situated in the curvilinear frame r and s. Further, the r-axis was perpendicular, and the arc was placed in the direction of s. The velocity and magnetic field were functions of s and mathematically described as below:

$$\hat{u}_w(s) = \frac{b}{s^{-m}}, \hat{B}(s) = \frac{B_0}{\left(s^{0.5(m-1)}\right)^{-1}},$$

It was assumed that the induced magnetic field was inconsequential, and therefore was not taken into consideration. The temperatures at the arched and the free surface were $\hat{T}_w$ and $\hat{T}_\infty$, respectively. The value of $m > 1$ represented the flow over an arc shaped, which was nonlinearly stretched. $m = 1$ was for the flow of a linearly stretching geometry. Figure 1 presents the flow description in a curvilinear frame. The flow chart of the study presented in Figure 2.

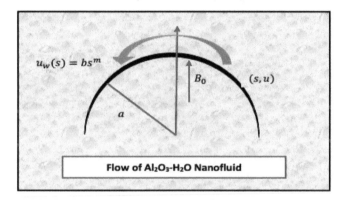

Figure 1. Flow description in a curvilinear frame.

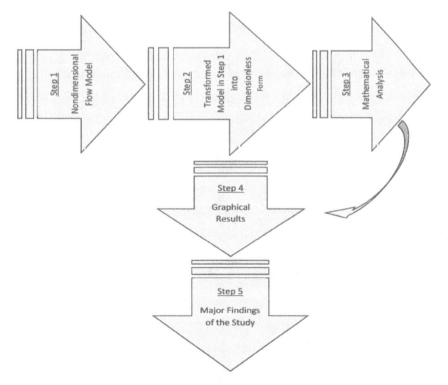

Figure 2. Flowchart of the study.

The $Al_2O_3 + H_2O$ nanofluid flow, incorporating the phenomena of Lorentz forces, viscous dissipation and nonlinear radiative heat flux, is described by the following system [21]:

$$\frac{\partial}{\partial r}(\hat{v}(r+a)) = 0, \tag{1}$$

$$\hat{u}^2 = \frac{\partial p}{\partial r}(r+a)(\rho_{nf})^{-1}, \tag{2}$$

$$\rho_{nf}\left(\hat{v}\frac{\partial \hat{u}}{\partial r} + \left(\frac{a}{r+a}\right)\hat{u}\frac{\partial \hat{u}}{\partial s} + \hat{u}\frac{\hat{v}}{(r+a)}\right) = -\frac{a}{(r+a)}\frac{\partial p}{\partial s} + \mu_{nf}\left(\frac{\partial^2 \hat{u}}{\partial r^2} + \left(\frac{1}{r+a}\right)\frac{\partial \hat{u}}{\partial r} - \left(\frac{1}{r+a}\right)^2\hat{u}\right) - \sigma_{nf}B_0^2\hat{u} + g(\rho\beta)_{nf}(T-T_\infty), \tag{3}$$

$$\frac{a}{(r+a)}\hat{u}\frac{\partial \hat{T}}{\partial s} + \hat{v}\frac{\partial \hat{T}}{\partial r} = \frac{k_{nf}}{(\rho C_p)_{nf}}\left(\frac{1}{(r+a)}\frac{\partial \hat{T}}{\partial r} + \frac{\partial^2 \hat{T}}{\partial r^2}\right) + \frac{\mu_{nf}}{(\rho C_p)_{nf}}\left(\frac{\partial \hat{u}}{\partial r} - \frac{\hat{u}}{(r+a)}\right)^2 - \frac{1}{(\rho C_p)_{nf}(r+a)}\frac{\partial}{\partial r}(q_R(r+a)), \tag{4}$$

with boundary conditions at the curve and far from the curve of:
At $r = 0$:

$$\left.\begin{array}{c}\hat{u} = bs^m \\ \hat{v} = 0 \\ \hat{T} = T_w\end{array}\right\}. \tag{5}$$

At $r \to \infty$:

$$\left.\begin{array}{c}\hat{u} \to 0 \\ \frac{\partial \hat{u}}{\partial r} \to 0 \\ \hat{T} \to T_\infty\end{array}\right\}. \tag{6}$$

The expression for nonlinear radiative heat flux is described as:

$$q_R = -4\frac{\hat{\sigma}}{3\hat{k}}\frac{\partial}{\partial r}(\hat{T}^4) = -\frac{16}{3}\frac{\hat{\sigma}}{\hat{k}}\hat{T}^3\frac{\partial \hat{T}}{\partial r}, \tag{7}$$

Equations (1)–(4) are the conservation of mass, momentum and energy, respectively. Further, the Stefan–Boltzmann law and adsorption coefficient are $\hat{\sigma}$ and $\hat{k}$, respectively. To enhance the thermal and physical characteristics of the model, the following effective models for thermophysical characteristics were used [23]:

$$\rho_{nf} = \left[(1-\phi) + \frac{\phi\rho_p}{\rho_f}\right]\rho_f, \tag{8}$$

$$(\rho C_p)_{nf} = \left[(1-\phi) + \frac{\phi(\rho C_p)_p}{(\rho C_p)_f}\right](\rho C_p)_f, \tag{9}$$

$$\mu_{nf} = \mu_f\left(1 - 34.87\left(\frac{d_{particle}}{d_{fluid}}\right)^{-0.3}\phi^{1.03}\right)^{-1}. \tag{10}$$

$$k_{nf} = k_f\left(1 + 4.4Re_b^{0.4}Pr^{0.66}\left(\frac{T}{T_{freezing}}\right)^{10}\left(\frac{k_p}{k_f}\right)^{0.03}\phi^{0.66}\right), \tag{11}$$

$$(\rho\beta)_{nf} = \left[(1-\phi) + \frac{\phi(\rho\beta)_p}{(\rho\beta)_f}\right](\rho\beta)_f, \tag{12}$$

$$\sigma_{nf} = \sigma_f\left[1 + \frac{3(\Theta-1)\phi}{(\Theta+2) - (\Theta-1)\phi}\right], \text{ where, } \Theta = \frac{\sigma_p}{\sigma_f}. \tag{13}$$

In Equation (11), Re_b shows the Reynolds number due to Brownian motion, and is described in the following pattern:

$$Re_b(\mu_f) = d_p \rho_f u_b, \tag{14}$$

In Equation (14), the velocity of Brownian motion is calculated by the formula:

$$u_b = 2Tk_b(\pi d_p^2 \mu_f), \tag{15}$$

Here, k_b is the Stefan–Boltzmann coefficient and its value is 1.380648×10^{-23} (JK^{-1}). d_p represents the molecular diameter which is calculated by the expression [24]:

$$d_f = 6M^*(N^* \rho_f \pi)^{-1}, \tag{16}$$

The molecular weight of the regular liquid and Avogadro number are denoted by M^* and N^*, respectively. Further, the value of d_f is calculated as:

$$d_f = \left(\frac{6 \times 0.01801528}{998.62 \times (6.022 \times 10^{23}) \times \pi}\right)^{\frac{1}{3}} = 3.85 \times 10^{-10} \text{ nm}, \tag{17}$$

Table 1 shows the data for thermal conductivity, effective density, thermal expansion coefficient, and effective dynamic viscosity [23]:

Table 1. Thermal and physical characteristics of the fluid phase and nanoparticles at $T = 310$ K [23].

Properties	d_p (nm)	ρ (kg/m³)	β (1/k)	c_p (J/Kg K)	μ_f (kg/ms)	k (W/mk)	σ (S/m)
H_2O	0.385	993	36.2×10^5	4178	695×10^6	0.628	0.005
Al_2O_3	33	3970	0.85×10^5	765	-	40	0.05×10^6

The similarity transformations are described in the following set of equations [21]:

$$\left.\begin{array}{c} \eta = \sqrt{\frac{b}{v_{bf}}} s^{0.5(m-1)} r \\ \hat{u} = bs^m L' \\ p = \rho_{bf} b^2 s^{2m} P' \\ \hat{v} = -\frac{a}{(r+a)} \sqrt{bv_{bf}} s^{0.5(m-1)} \{0.5(m+1)L + 0.5(m-1)\eta L'\} \\ N = \frac{T-T_\infty}{T_w - T_\infty} \end{array}\right\} \tag{18}$$

To analyze the phenomena of nonlinear radiative heat flux, the following expression was used:

$$\hat{T} = T_\infty(1 + (\beta_w - 1)N) \tag{19}$$

The ratio of wall and free surface temperature was denoted by β_w.

The following model was attained after incorporating the similarity transformations and partial derivatives in Equations (1)–(4):

$$\frac{1}{\left(1 - 34.87\left(\frac{d_p}{d_f}\right)^{-0.3} \phi^{1.03}\right)} \left[(\eta + K)^3 L'''' + 2(\eta + K)^2 L''' - (\eta + K)L'' + L'\right] -$$

$$\left[1 + \frac{3(\Theta - 1)\phi}{(\Theta + 2) - (\Theta - 1)\phi}\right] M\left[(\eta + K)^3 L'' + (\eta + K)^2 L'\right] + \left((1 - \phi) + \frac{\phi(\rho\beta)_p}{(\rho\beta)_f}\right)$$

$$\alpha\left((\eta + K)^3 N' + (\eta + K)^2 L\right) +$$

$$\left\{(1 - \phi) + \frac{\phi\rho_p}{\rho_f}\right\} K \left[\begin{array}{c} 0.5(1 - 3m)(\eta + K)(L')^2 + 0.5(m + 1)(\eta + K)LL'' \\ -0.5(m + 1)LL' + 0.5(m + 1)(\eta + K)^2 LL''' + \\ (\eta + K)^2 0.5(1 - 3m)L'L'' \end{array}\right] = 0, \tag{20}$$

$$Rd\Big[3(N')^3(1 + (\beta_w - 1)N)^2(\beta_w - 1)(\eta + K) + (1 + (\beta_w - 1)N)^3((\eta + K)N'' + N')\Big] +$$
$$\left[1 + 4.4Re_b^{0.4}\Pr^{0.66}\left(\frac{T}{T_{freezing}}\right)^{10}\left(\frac{k_p}{k_f}\right)^{0.03}\phi^{0.66}\right]((\eta + K)N'' + N') +$$
$$\frac{PrEc}{\left(1 - 34.87\left(\frac{d_p}{d_f}\right)^{-0.3}\phi^{1.03}\right)(\eta + K)}((\eta + K)L'' - L')^2 = 0. \tag{21}$$

In Equations (20) and (21), L and N are the functions of η. After solving, L and N provide the velocity and temperature distributions, respectively.

$$L(\eta) = 0, \ L'(\eta) = 1, \ N(\eta) = 1, \tag{22}$$

At $\eta \to \infty$:

$$L'(\eta) \to 0, L''(\eta) \to 0, N(\eta) \to 0, \tag{23}$$

Dimensionless quantities were described by the following formulas:

$$K = a\ \sqrt{\frac{b}{v_{bf}}}, Ec = \frac{b^2 s^{2m}}{(C_p)_f(T_w - T_\infty)}, Rd = 16\hat{\sigma}\frac{T_\infty^3}{3\hat{k}k_f}, M = \frac{\sigma_f B_0^2}{\rho_f b}, \Pr = \frac{(c_p)_f \mu_f}{k_f},$$
$$\alpha = Gr_s\left(Re_s^2\right)^{-1}, Re_s = bs^2\left(v_f\right)^{-1}, Gr_s = s^3(T_w - T_\infty)g\beta\left(v_f\right)^{-2}. \tag{24}$$

Moreover, the dimensional formula for skin friction and local Nusselt number were defined as:

$$C_f = \tau_{rs}\left(\rho_f\ \hat{u}_w^2\right)^{-1}, Nu_s = sq_w\left(k_f(T_w - T_\infty)\right)^{-1}, \tag{25}$$

where,

$$\tau_{rs} = \left(\frac{\partial \hat{u}}{\partial r} - \frac{\hat{u}}{(a + r)}\right) \downarrow r = 0, q_w = -k_{nf}\frac{\partial \hat{T}}{\partial r} \downarrow_{r=0}, \tag{26}$$

Drawing the values from Equation (26) into Equation (25), the following dimensionless formulas were obtained:

$$C_f(Re_s)^{\frac{1}{2}} = \frac{1}{\left(1 - 34.87\left(\frac{d_p}{d_f}\right)^{-0.3}\phi^{1.03}\right)}\left(L''(0) - L'(0)K^{-1}\right), \tag{27}$$

$$Nu_s(Re_s)^{-\frac{1}{2}} = -\left(\left(1 + 4.4Re_b^{0.4}\Pr^{0.66}\left(\frac{T}{T_{freezing}}\right)^{10}\left(\frac{k_p}{k_f}\right)^{0.03}\phi^{0.66}\right) + Rd\beta_w^3\right)N'(0). \tag{28}$$

3. Mathematical Analysis

Highly nonlinear and coupled systems of differential equations usually possess no closed-form solution. Our flow model for $Al_2O_3 + H_2O$ is a highly nonlinear fourth-order model defined at a semi-infinite domain. Therefore, the model was tackled numerically using the Runge-Kutta scheme [13,16,25], as the RK scheme is used for the first order initial value problem (IVP). First the following substitutions were made and the model was reduced into first order IVP.

$$h_1 = L, h_2 = L', h_3 = L'', h_4 = L''', h_5 = N, h_6 = N'. \tag{29}$$

After the successful transformation into the first order IVP, Mathematica 10.0 was used and the system was solved successfully.

4. Graphical Results and Discussion

This section emphasizes the flow and thermophysical characteristics of the fluid phase and the nanoparticles of Al_2O_3. The values for the thermophysical characteristics were calculated at 310 K [23]. The results for the shear stress and heat transfer rate are elaborated using bar charts and are discussed comprehensively.

4.1. Velocity and Temperature Distribution

Magnetic field phenomena are of a great significance from an industrial point of view. Many industrial productions contain impurities that need to be removed. However, the magnetic parameter opposes the fluid motion, and the impurities remain at the bottom and the nanofluid velocity drops. The influence of Lorentz forces on the velocity distribution of $Al_2O_3 + H_2O$ nanofluids is elaborated in Figure 3a. It was shown that the applied magnetic field opposed the nanofluid motion, and the velocity of the $Al_2O_3 + H_2O$ nanofluid dropped. The velocity declined more slowly for a weaker magnetic field, and a rapid decrement in the nanofluid velocity was observed for a stronger magnetic field. Near the arched surface, variations in the velocity $(L'(\eta))$ were almost negligible. This behavior of the velocity distribution was due to the friction between the surface and the nanolayer of $Al_2O_3 + H_2O$. In the successive nanolayers, the velocity field was altered significantly. These influences became negligible far from the curve and showed an asymptotic pattern of velocity distribution at the free surface. Figure 3b shows the velocity distribution of the parameter m. The velocity of the $Al_2O_3 + H_2O$ nanofluid dropped rapidly for m in comparison with M. As the values for parameter m became larger, the velocity decreased promptly.

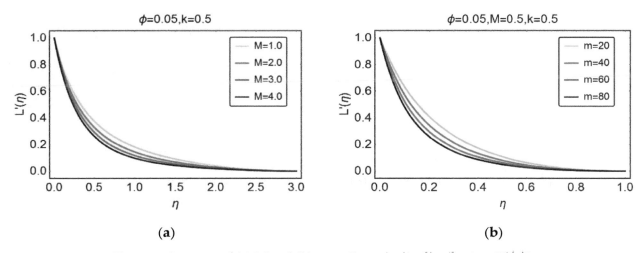

Figure 3. Impacts of (**a**) M and (**b**) m on the velocity distribution $(L'(\eta))$.

The effects of surface curvature on the velocity $(L'(\eta))$ are elucidated in Figure 4a. Altering the surface curvature caused the velocity to increase. For a smaller curvature, the velocity increased slowly and then vanished asymptotically far away from the surface. Similar behavior of the velocity of the $Al_2O_3 + H_2O$ nanofluid is depicted in Figure 4b. For α, a prominent behavior of the velocity was noticed in $0.5 \leq \eta \leq 1.5$. Besides this, the velocity $(L'(\eta))$ was almost inconsequential.

The temperature distribution $(N(\eta))$ for the radiation parameter (Rd) and Eckert number is highlighted in Figure 5a,b, respectively. In Figure 5a, we can see rapid drops in the temperature that were investigated by altering the radiation parameter (Rd). The temperature $(N(\eta))$ dropped rapidly for a stronger radiation parameter, however at the free surface this was almost negligible and vanished asymptotically. The Eckert number, which appeared due to viscous dissipation, played a vibrant role in the heat transfer enhancement. These effects are elucidated in Figure 5b. It was obvious that the temperature of $Al_2O_3 + H_2O$ nanofluid grew for the more dissipative nanofluid. For the larger Ec, the temperature distribution rose rapidly.

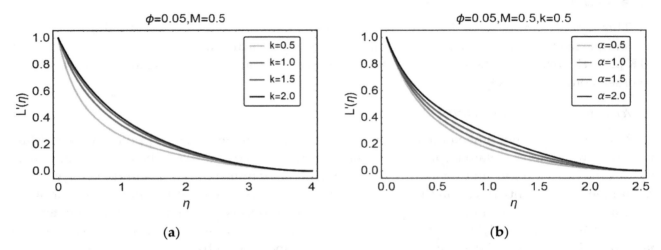

Figure 4. Impacts of (**a**) K and (**b**) α on the velocity distribution ($L'(\eta)$).

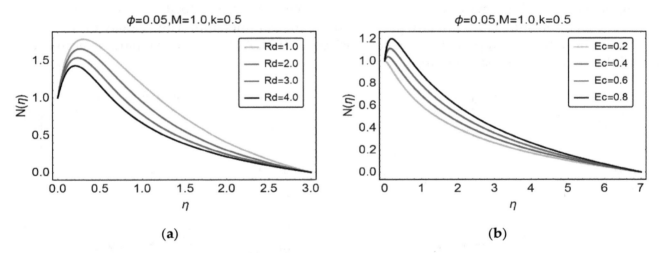

Figure 5. Impacts of (**a**) Rd and (**b**) Ec on the temperature distribution ($N(\eta)$).

4.2. Streamlines and Isotherms

This subsection is devoted to analyzing the behavior of the streamlines and isotherm patterns by altering different pertinent flow parameters. Figure 6 presents the streamline pattern by a varying magnetic parameter (M). For a smaller magnetic parameter, the streamlines were more curved near the surface and for a stronger M, the streamlines assumed a less curved shape. At the free stream, these streamlines became straight, since, as already mentioned for the parameter $m > 1$, it showed that it was a nonlinear stretching curved surface. The streamline pattern versus m is elaborated in Figure 7. It was noticed that by decreasing the parameter m, the streamlines assumed a more curved pattern. The curvature parameter showed a fascinating pattern for the streamlines, in comparison with M and m. These alterations are illustrated in Figure 8. Figure 9 depicts the flow pattern for varying α. For a higher α, the streamlines shrank and became almost straight at the top. Figure 10 elaborates the isotherm pattern for the radiation parameter (Rd). When there was more radiative nanofluid, the isotherms increased, and vice versa. Further, a 3D scenario of the isotherms is depicted in Figure 11.

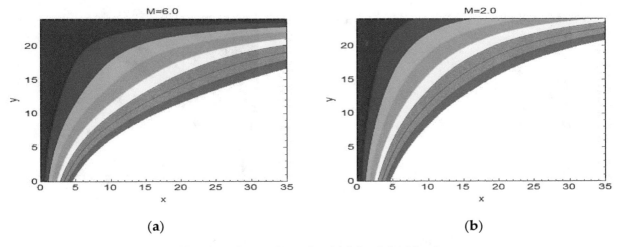

Figure 6. Streamlines for (**a**) $M = 6$ (**b**) $M = 2$.

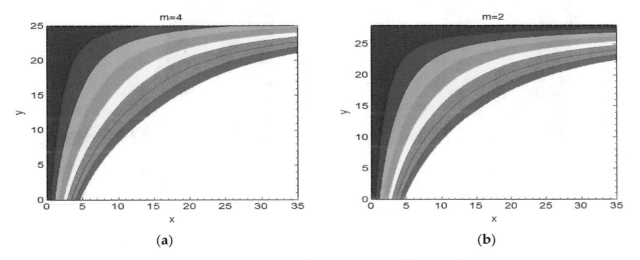

Figure 7. Streamlines for (**a**) $m = 4$ and (**b**) $m = 2$.

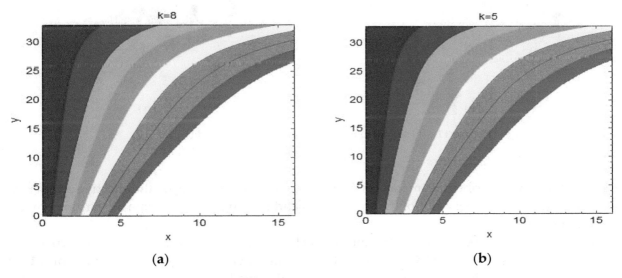

Figure 8. Streamlines for (**a**) $k = 8$ and (**b**) $k = 5$.

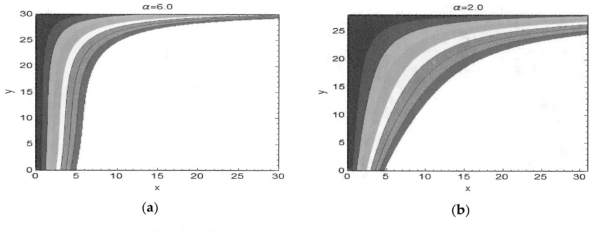

Figure 9. Streamlines for (**a**) $\alpha = 6$ and (**b**) $\alpha = 2$.

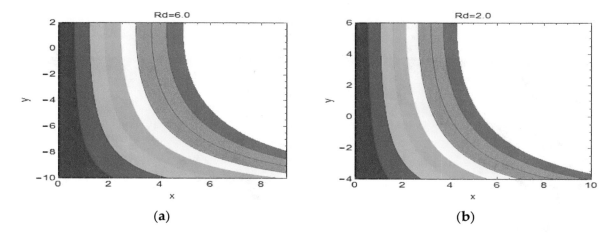

Figure 10. Isotherms for (**a**) $Rd = 6$ and (**b**) $Rd = 2$.

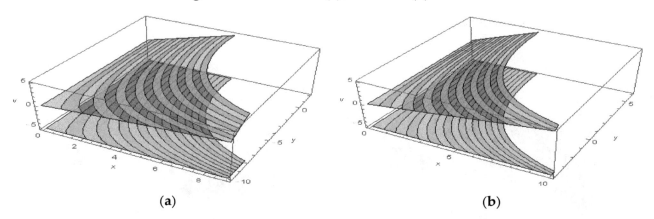

Figure 11. 3D scenario of Figure 10; (**a**) $Rd = 6$ and (**b**) $Rd = 2$.

4.3. Thermophysical Characteristics

This subsection describes the impacts of the volume fraction factor (ϕ) on the effective characteristics of the nanofluids, and the behavior of the shear stress and local heat transfer rate by varying embedded flow parameters.

Figure 12 describes the effects of the volume fraction (ϕ,) and diameter of the nanoparticles on the effective dynamic viscosity of $Al_2O_3 + H_2O$. The volume fraction of Al_2O_3 showed a vibrant role in enhancing the dynamic viscosity of the nanofluid. The observed high dynamic viscosity corresponded to a greater volume fraction. On the other hand, the nanoparticle diameter (d_p) induced inverse variations in the dynamic viscosity. Increasing the diameter of the nanoparticles caused the dynamic viscosity to drop. This means that nanoparticles with a smaller diameter are important to enhance the dynamic viscosity of nanofluids.

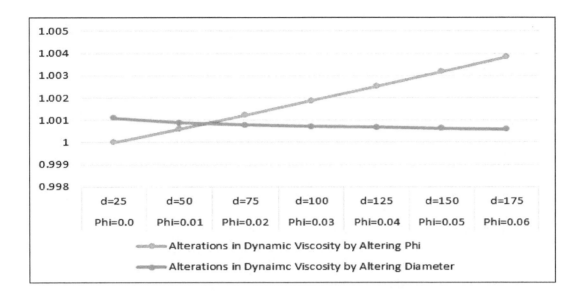

Figure 12.　The impact of the volume fraction (ϕ) and the nanoparticle diameter (d_p) on the dynamic viscosity.

Figure 13 highlights the effective density (ρ_{nf}) and heat capacity ($(\rho c_p)_{nf}$) of the nanofluid versus ϕ. ϕ and ρ_{nf} were in direct proportion to each other, and the effective heat capacity dropped when ϕ increased. Therefore, smaller values of ϕ enhanced the effective heat capacity. Due to the high volume fraction, the colloidal suspension $Al_2O_3 + H_2O$ became denser, which enhanced the effective density (ρ_{nf}). Figure 14 highlights that the volume fraction (ϕ) and the effective electrical conductivity were in inverse proportion to each other.

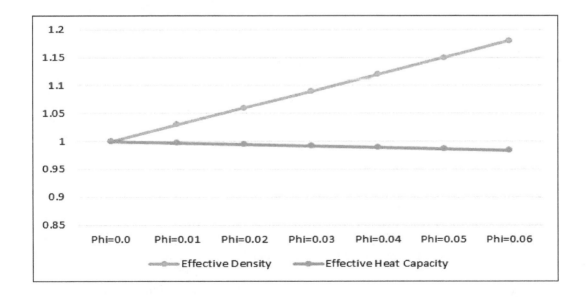

Figure 13. The impact of ϕ on the effective density and heat capacity.

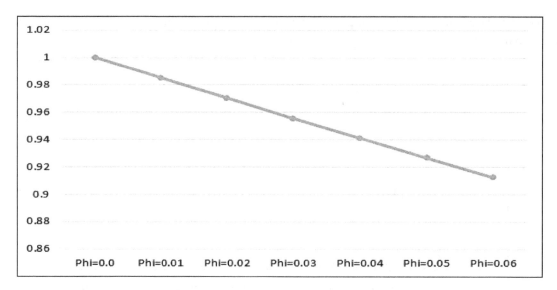

Figure 14. The impact of ϕ on the effective electrical conductivity.

4.4. Skin Fraction and Heat Transfer Rate

The shear stress and the local rate of heat transfer are very interesting and have attained great significance from an industrial point of view. The radiation parameter, Eckert number, and the curvature parameter all play a significant role in the shear stress and local Nusselt number. Figure 15 describes the heat transfer behavior for *Rd, Ec, K* and the volume fraction of the nanoparticles. It was noted that the more radiative nanofluids favored the heat transfer. On the other hand, a smaller amount of the heat transfer was noticed for a higher Eckert number. Therefore, the less dissipative fluids similarly favored the heat transfer. For a more curved surface, a large curvature worked against the heat transfer. At greater volume concentrations the heat transfer rate grew slowly.

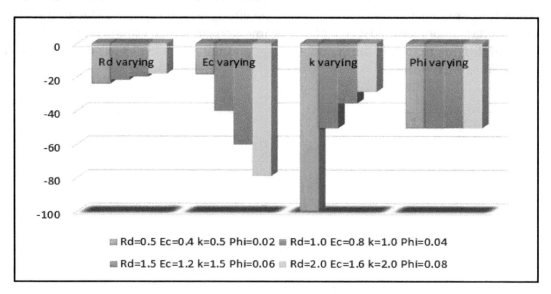

Figure 15. The impact of various flow parameters on the local heat transfer.

Figure 16 elucidates the shear stress behavior for the mixed convection parameter (α), magnetic parameter (M), curvature of the surface, and m. For a more convective fluid, the heat transfer increased rapidly at the surface. The magnetic parameter *(M)* highlighted the reverse behavior of the shear stress. With an increase of the parameter M, the shear stress dropped quickly, and an almost negligible influence of m on the shear stresses was also observed. Significant alterations were pointed out for surface curvature, as the curve was along the circle loop of radius a. Therefore, for a smaller radius, shear stress declined promptly. Increasing the radius of the loop caused the surface curvature to become larger, which favored the shear stress.

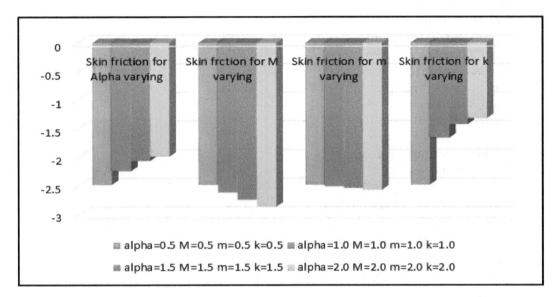

Figure 16. The impact of various parameters on the shear stress.

5. Conclusions

The mixed convective laminar flow of water, composed by Al_2O_3 nanofluids in the presence of Lorentz forces and nonlinear radiative heat flux, was examined over an arc-shaped geometry. To enhance the heat transfer rate, a thermal conductivity model that considered the impact of freezing temperature and molecular diameter was used. It was found that the nanofluid velocity ($L'(\eta)$) dropped for a stronger magnetic parameter (M), which is very significant from an industrial point of view. Further, the velocity of the nanofluid increased due to the mixed convection and larger curvature of the surface. The temperature $N(\eta)$ intensified for more dissipative fluid, and an inverse relationship between the temperature and the thermal radiation parameter was found. The dynamic viscosity of the nanofluid increased with the volume fraction, and the diameter of the nanoparticles showed reverse alterations for dynamic viscosity. The nanofluid became denser for a high volume fraction, and the electrical conductivity dropped. It was found that the heat transfer reduced the surface of a smaller curvature and intensified for larger curvatures. A more convective fluid and a larger curvature better opposed the shear stress. Finally, when considering the influence of the freezing temperature and the molecular diameter, the nanofluid flow model is very useful for heat transfer in comparison with existing studies.

Author Contributions: Conceptualization, U.K. and A.A.; methodology, Adnan, N.A.; software, S.N.; validation, writing—review and editing, I.K. and W.A.K.; supervision, S.T.M.-D; Funding acquisition, S.O.A. All the authors contributed to the manuscript equally.

Nomenclature

u	Component of the velocity
a	Radius
T	Temperature
T_∞	Temperature far from the surface
k_f	Thermal conductivity of the host fluid
k_s	Thermal conductivity of the nanoparticles
k_{nf}	Effective thermal conductivity of the nanofluid
$(C_p)_f$	Heat capacity of the host fluid
σ_s	Electrical conductivity of the nanoparticles

σ_{nf}	Electrical conductivity of the nanofluid
$\mu)_f$	Dynamic viscosity of the fluid
M^*	Molecular weight
d_f	Molecular diameter
β_w	Temperature ratio parameter
Pr	Prandtl number
M	Hartmann number
$L(\eta)$	Dimensionless velocity
Nu	Nusselt number
v	Component of the velocity
p	Pressure
T_w	Temperature at the surface
ρ_f	Density of the host fluid
ρ_s	Density of the nanoparticles
ρ_{nf}	Effective density of the nanofluid
$(C_p)_s$	Heat capacity of the nanoparticles
$(C_p)_{nf}$	Heat capacity of the nanofluid
σ_f	Electrical conductivity of the host fluid
μ_{nf}	Effective dynamic viscosity
ϕ	Volume fraction of the nanoparticles
N^*	Avogadro number
k_b	Stefan Boltzmann constant
Rd	Radiation parameter
Ec	Eckert number
K	Curvature parameter
$N(\eta)$	Dimensionless temperature
C_f	Skin fraction coefficient

References

1. Choi, S. Enhancing thermal conductivity of fluids with nanoparticles in developments and applications of non-newtonians flows. *ASME J. Heat Transf.* **1995**, *66*, 99–105.
2. Clerk, M.J. *Treatise on Electricity and Magnetism*; Oxford University Press: Oxford, UK, 1873.
3. Bruggeman, D.A.G. Berechnung verschiedener physikalischer konstanten von heterogenen substanzen, I—dielektrizitatskonstanten und leitfahigkeiten der mischkorper aus isotropen substanzen. *Ann. Phys. Leipz.* **1935**, *24*, 636–679. [CrossRef]
4. Hamilton, H.L.; Crosser, O.K. Thermal conductivity of heterogeneous two-component systems. *Ind. Eng. Chem. Fundam.* **1962**, *1*, 187–191. [CrossRef]
5. Lu, S.; Lin, H. Effective conductivity of composites containing aligned spherical inclusions of finite conductivity. *J. Appl. Phys.* **1996**, *79*, 6761–6769. [CrossRef]
6. Koo, J.; Kleinstreuer, C. A new thermal conductivity model for nanofluids. *J. Nanopart. Res.* **2004**, *6*, 577–588. [CrossRef]
7. Koo, J.; Kleinstreuer, C. Laminar nanofluid flow in micro-heat sinks. *Int. J. Heat Mass Transf.* **2005**, *48*, 2652–2661. [CrossRef]
8. Xue, Q.Z. Model for thermal conductivity of carbon nanotube-based composites. *Phys. B Phys. Condens. Matter.* **2005**, *368*, 302–307. [CrossRef]
9. Prasher, R.; Bhattacharya, P.; Phelan, P.E. Thermal conductivity of nanoscale colloidal solutions (nanofluids). *Phys. Rev. Lett.* **2005**, *92*, 25901. [CrossRef]
10. Li, C.H.; Peterson, G.P. Experimental investigation of temperature and volume fraction variations on the effective thermal conductivity of nanoparticle suspensions (nanofluids). *J. Appl. Phys.* **2006**, *99*. [CrossRef]
11. Corcione, M. Rayleigh–Be´nard convection heat transfer in nanoparticle suspensions. *Int. J. Heat Fluid Flow* **2011**, *32*, 65–77. [CrossRef]
12. Sheikholeslami, M.; Li, Z.; Shamlooei, M. Nanofluid MHD natural convection through a porous complex shaped cavity considering thermal radiation. *Phys. Lett. A* **2018**, *382*, 1615–1632. [CrossRef]

13. Ahmed, N.; Khan, A.U.; Mohyud-Din, S.T. Influence of an effective prandtl number model on squeezed flow of $\gamma Al\_2 O\_3$-$H\_2 O$ and $\gamma Al\_2 O\_3$-$C\_2 H\_6 O\_2$ nanofluids. *J. Mol. Liq.* **2017**, *238*, 447–454. [CrossRef]

14. Sheikholeslami, M.; Zia, Q.M.Z.; Ellahi, R. Influence of induced magnetic field on free convection of nanofluid considering Koo-Kleinstreuer-Li (KKL) correlation. *Appl. Sci.* **2016**, *6*, 324. [CrossRef]

15. Asadullah, A.M.; Khan, U.; Naveed, A.; Mohyud-Din, S.T. Analytical and numerical investigation of thermal radiation effects on flow of viscous incompressible fluid with stretchable convergent/divergent channels. *J. Mol. Liq.* **2016**, *224*, 768–775.

16. Khan, U.; Naveed, A.A.; Mohyud-Din, S.T. 3D squeezed flow of $\gamma Al\_2 O\_3$-$H\_2 O$ and $\gamma Al\_2 O\_3$-$C\_2 H\_6 O\_2$ nanofluids: A numerical study. *Int. J. Hydrog. Energy* **2017**, *42*, 24620–24633. [CrossRef]

17. Iijima, S. Helical microtubules of graphitic carbon. *Nature* **1991**, *354*, 56–58. [CrossRef]

18. Naveed, A.A.; Khan, U.; Mohyud-Din, S.T. Influence of thermal radiation and viscous dissipation on squeezed flow of water between two riga plates saturated with carbon nanotubes. *Colloids Surf. A Physciochem. Eng. Asp.* **2017**, *522*, 389–398.

19. Saba, F.; Naveed, A.; Hussain, S.; Khan, U.; Mohyud-Din, S.T.; Darus, M. Thermal analysis of nanofluid flow over a curved stretching surface suspended by carbon nanotubes with internal heat generation. *Appl. Sci.* **2018**, *8*, 395. [CrossRef]

20. Khan, U.; Naveed, A.; Mohyud-Din, S.T. Heat transfer effects on carbon nanotubes suspended nanofluid flow in a channel with non-parallel walls under the effect of velocity slip boundary condition: A numerical study. *Neural Comput. Appl.* **2017**, *28*, 37–46. [CrossRef]

21. Reddy, J.V.R.; Sugunamma, V.; Sandeep, N. Dual solutions for nanofluid flow past a curved surface with nonlinear radiation, soret and dufour effects. *J. Phys. Conf. Ser.* **2018**, *1000*, 12152. [CrossRef]

22. Hayat, T.; Qayyum, S.; Imtiaz, M.; Alsaedi, A. Double stratification in flow by curved stretching sheet with thermal radiation and joule heating. *J. Therm. Sci. Eng. Appl.* **2017**, *10*. [CrossRef]

23. Alsabery, A.I.; Sheremet, M.A.; Chamkha, A.J.; Hashim, I. MHD convective heat transfer in a discretely heated square cavity with conductive inner block using two-phase nanofuid model. *Sci. Rep.* **2018**, *8*, 1–23. [CrossRef]

24. Corcione, M. Empirical correlating equations for predicting the efective thermal conductivity and dynamic viscosity of nanofuids. *Energy Convers. Manag.* **2011**, *52*, 789–793. [CrossRef]

25. Naveed, A.A.; Khan, U.; Mohyud-Din, S.T. Unsteady radiative flow of chemically reacting fluid over a convectively heated stretchable surface with cross-diffusion gradients. *Int. J. Therm. Sci.* **2017**, *121*, 182–191.

Peristaltic Blood Flow of Couple Stress Fluid Suspended with Nanoparticles under the Influence of Chemical Reaction and Activation Energy

Rahmat Ellahi [1,2,*], Ahmed Zeeshan [2], Farooq Hussain [2,3] and A. Asadollahi [4]

[1] Center for Modeling & Computer Simulation, Research Institute, King Fahd University of Petroleum & Minerals, Dhahran 31261, Saudi Arabia

[2] Department of Mathematics & Statistics, FBAS, IIUI, Islamabad 44000, Pakistan; ahmad.zeeshan@iiu.edu.pk (A.Z.); farooq.hussain@buitms.edu.pk (F.H.)

[3] Department of Mathematics, (FABS), BUITMS, Quetta 87300, Pakistan

[4] Department of Mechanical Engineering & Energy Processes, Southern Illinois University, Carbondale, IL 62901, USA; arash.asadollahi@siu.edu

* Correspondence: rellahi@alumni.ucr.edu

Abstract: The present study gives a remedy for the malign tissues, cells, or clogged arteries of the heart by means of permeating a slim tube (i.e., catheter) in the body. The tiny size gold particles drift in free space of catheters having flexible walls with couple stress fluid. To improve the efficiency of curing and speed up the process, activation energy has been added to the process. The modified Arrhenius function and Buongiorno model, respectively, moderate the inclusion of activation energy and nanoparticles of gold. The effects of chemical reaction and activation energy on peristaltic transport of nanofluids are also taken into account. It is found that the golden particles encapsulate large molecules to transport essential drugs efficiently to the effected part of the organ.

Keywords: chemical reaction; activation energy; peristalsis; couple stress fluid; nanoparticle; Keller-box method

1. Introduction

In any living organism peristaltic motion is mainly caused by the contraction and expansion of some flexible organs. This applies a pressure force to drive fluids, for example, blood in veins, urine to bladder, and transport of medicines to desired locations are a few common biological examples. The rapid developments in nano-science have noticeably revolutionized almost every field of life, particularly in medical sciences. The advent of nano-technology in medicines has brought miraculous changes by reshaping the primitive methods of treatment. Nowadays, in developed countries operations are preferably performed without involving any prunes and cuts, which was once thought to be very complex and menacing for cancer treatment, brain tumors, lithotripsy, etc. Regardless of many other uses of nanofluids in industrial and practical settings, the primary objective of nanoparticles is the enhancement of heat transfer [1]. It is mainly due to their high conductivity. In addition to the size and type of nanoparticles, other factors, such as temperature, volume fraction, and thermal conductivity are also very important to maximize the thermal conductivity. In pursuit of attaining such enhancement in the system, with the passage of time many useful models based on the physical properties of the matter have been developed. On the said topic, scholars have made full use of these models in their analyses, experiments, and conditions, which have been discussed here very briefly. For instance, the investigation of Tripathi and Beg [2] explains the application of peristaltic micropumps and novel drug delivery systems in pharmacological engineering. They formulated their study with the help of the Buongiornio nanofluid model and treated blood as Newtonian fluid. El-dabe et al. [3] have explained the significant contribution of nanofluids in peristalsis. They produced their results for flexible wall properties, lubrication, MHD, and porosity. As generally it is believed that the motion of blood is likely to be non-Newtonian, therefore, Swarnalathamma and Krishana [4] have studied the physiological flow of the blood in the micro circulatory system by taking account of the particle size effect. They considered couple stress fluid for the given peristaltic analysis, which is further affected by magnetic fields. The effects of channel inclination are studied by Shit and Roy [5]. The couple stress fluid influenced by constant application of magnetic fields is used as the base fluid. Jamalabad et al. [6] reported the effects of biomagnetic blood flow through a stenosis artery by means of non-Newtonian flow of a Carreau-Yasuda fluid model. They carried out a numerical simulation of an unsteady blood flow problem. Hosseini et al. [7] have presented the thermal conductivity of a nanofluid model. To perform this investigation, the nanofluid model is considered as the function of thermal conductivity of nanoparticles, base fluid, and interfacial shell properties by considering temperature as the most effective of parameters involved in the study. The most noteworthy contributions on the matter can be seen in the list of references [8–19]. Furthermore, activation energy has a key role in industries, in particular, effectively aggravating slow chemical reactions in chemistry laboratories to improve the efficiency of various mechanisms by adding activation energy to respective physical and mechanical processes. Mustafa et al. [20] have proposed a chemical and activation energy MHD-effected mix convection flow of nanofluids. In this study the flow over the vertical sheet expands due to high temperature and causes the fluid motion is analyzed numerically. A few of the latest works related to this present work have been listed in [21–28].

In view of the existing literature, one can feel the application of nanotechnology in medical science opens a new dimension for researchers to turn their attention towards the effective role of chemical reaction and activation energy [29–31], since nanoparticles help in treating different diseases by means of the peristaltic movement of blood. Such biological transport of blood helps to deliver drugs or medicine effectively to the damaged tissue or organ. As a matter of fact, this effort is devoted to inspecting the simultaneous effects of chemical reaction and activation energy for the peristaltic flow of couple stress nanofluids in a single model, which is yet not available in literature, and could have dual applications in expediting the treatment process.

2. Formulism

The inner tube is of a rigid configuration, while the outer tube is flexible in nature as shown in Figure 1. The sinusoidal waves travel with a constant speed through its walls, due to the stress caused by an unsteady movement of heated nanofluid through the space between both tubes. The general form of equations governing the two-dimensional flows are given as:

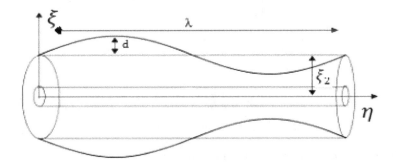

Figure 1. Configuration of coaxial tubes.

Conservation of mass
$$\nabla.\vec{V} = 0. \tag{1}$$

Conservation of momentum

$$\rho_f \left(\vec{V}.\vec{\nabla} \right) \vec{V} = -\vec{\nabla} P^* + \mu \nabla^2 \vec{V} + \left[\varphi \rho_p + (1-\varphi)\{\rho_f(1-\beta_T(v-v_w))\} \right] g - \gamma_1 \nabla^4 \vec{V}. \tag{2}$$

Thermal energy

$$(\rho c)_f \left(\vec{V}.\vec{\nabla} \right) v = k\nabla^2 v + (\rho c)_p \left[D_b \vec{\nabla} \varphi . \vec{\nabla} v + \frac{D_T}{v_w} \vec{\nabla} v . \vec{\nabla} v \right]. \tag{3}$$

Concentration of nanoparticles

$$\rho_p \left(\vec{V}.\vec{\nabla} \right) \varphi = D_b \nabla^2 \varphi + \frac{D_T}{v_w} \nabla^2 v - k_r^2 \left(\frac{v}{v_w} \right)^n (\varphi - \varphi_w) \exp\left(\frac{-E_a}{kv} \right). \tag{4}$$

One can easily identify that the last term in the momentum equation describes the velocity of couple stress fluid involving a constant associated with the couple stress fluid γ_1. The last term in Equation (4) on the right side is known as "Arrhenius term", which shows the effects of chemical reaction and activation energy incorporated to a nanofluid. The radial and axial velocity components of nanofluids are respectively defined by $u(\xi, \eta)$ and $w(\xi, \eta)$ in two concentric tubes, such that there is no rotation about their axes. A peristaltic flow of a heated couple stress fluid carrying the gold nanoparticles (GNPs) through these coaxial tubes due to the contraction and expansion of flexible walls of the outer tube is assumed. If the two-dimensional peristaltic motion of the concerned nanofluid is denoted by $[u(\xi, \eta) \ 0 \ w(\xi, \eta)]$, then the Equations (1)–(4) in the component's form will take the following form:

$$\frac{u}{\xi} + \frac{\partial u}{\partial \xi} + \frac{\partial w}{\partial \eta} = 0, \tag{5}$$

$$\begin{aligned} \rho_f \left(\frac{\partial u}{\partial t} + u\frac{\partial u}{\partial \xi} + w\frac{\partial u}{\partial \eta} \right) \\ = -\gamma_1 \left[\frac{\partial^4 u}{\partial \xi^4} + \frac{\partial^4 u}{\partial \eta^4} + \frac{\partial^4 u}{\partial \xi^2 \partial \eta^2} + \frac{\partial^4 u}{\partial \eta^2 \partial \xi^2} + 2\frac{\partial^3 u}{\xi \partial \xi^3} + \frac{\partial^3 u}{\xi \partial \xi \partial \eta^2} + \frac{\partial^3 u}{\xi \partial \eta^2 \partial \xi} \right. \\ \left. + \frac{\partial^2 u}{\xi^2 \partial \xi^2} + \frac{\partial u}{\xi^3 \partial \xi} \right] - \frac{\partial P}{\partial \xi} + \mu \left(\frac{\partial^2 u}{\partial \xi^2} + \frac{\partial u}{\xi \partial \xi} + \frac{\partial^2 u}{\partial \eta^2} \right), \end{aligned} \tag{6}$$

$$\begin{aligned} \rho_f \left(\frac{\partial w}{\partial t} + u\frac{\partial w}{\partial \xi} + w\frac{\partial w}{\partial \eta} \right) \\ = -\gamma_1 \left[\frac{\partial^4 w}{\partial \xi^4} + \frac{\partial^4 w}{\partial \eta^4} + \frac{\partial^4 w}{\partial \xi^2 \partial \eta^2} + \frac{\partial^4 w}{\partial \eta^2 \partial \xi^2} + \frac{\partial^3 w}{\xi \partial \xi^3} + \frac{\partial^3 w}{\xi \partial \xi \partial \eta^2} + \frac{\partial^3 w}{\xi \partial \eta^2 \partial \xi} \right. \\ \left. - \frac{\partial^2 w}{\xi^2 \partial \xi^2} + \frac{\partial w}{\xi^3 \partial \xi} \right] + \left[\varphi \rho_p + (1-\varphi)\{\rho_f(1-\beta_T(v-v_w))\} \right] g - \frac{\partial P}{\partial \eta} \\ + \mu \left(\frac{\partial^2 w}{\partial \xi^2} + \frac{1}{\xi}\frac{\partial w}{\partial \xi} + \frac{\partial^2 w}{\partial \eta^2} \right), \end{aligned} \tag{7}$$

$$\begin{aligned} (\rho c)_f \left(\frac{\partial v}{\partial t} + u\frac{\partial v}{\partial \xi} + w\frac{\partial v}{\partial \eta} \right) \\ = (\rho c)_p \left[D_b \left\{ \frac{\partial \varphi}{\partial \xi}\frac{\partial v}{\partial \xi} + \frac{\partial \varphi}{\partial \eta}\frac{\partial v}{\partial \eta} \right\} + \frac{D_T}{v_w} \left\{ \left(\frac{\partial v}{\partial \xi} \right)^2 + \left(\frac{\partial \varphi}{\partial \eta} \right)^2 \right\} \right] \\ + k \left(\frac{\partial^2 v}{\partial \xi^2} + \frac{1}{\xi}\frac{\partial v}{\partial \xi} + \frac{\partial^2 v}{\partial \eta^2} \right), \end{aligned} \tag{8}$$

$$\begin{aligned} \left(\frac{\partial \varphi}{\partial t} + u\frac{\partial \varphi}{\partial \xi} + w\frac{\partial \varphi}{\partial \eta} \right) \\ = D_b \left(\frac{\partial^2 \varphi}{\partial \xi^2} + \frac{1}{\xi}\frac{\partial \varphi}{\partial \xi} + \frac{\partial^2 \varphi}{\partial \eta^2} \right) + \frac{D_T}{v_w} \left(\frac{\partial^2 v}{\partial \xi^2} + \frac{1}{\xi}\frac{\partial v}{\partial \xi} + \frac{\partial^2 v}{\partial \eta^2} \right) \\ - k_r^2 \left(\frac{v}{v_w} \right)^n (\varphi - \varphi_w) \exp\left(\frac{-E_a}{kv} \right). \end{aligned} \tag{9}$$

The corresponding boundary at the extreme wall.
At the rigid wall:

$$\left. \begin{aligned} &(i). \ w(\xi) = 0, \\ &(ii). \ v(\xi) = v_m, \\ &(iii). \ \varphi(\xi) = \varphi_m. \end{aligned} \right\} ; \text{ When } \xi = \xi_1. \tag{10}$$

At the flexible wall:

$$\left.\begin{array}{l}(i).\ w(\xi) = 0, \\ (ii).\ v(\xi) = v_w, \\ (iii).\ \varphi(\xi) = \varphi_w.\end{array}\right\}; \text{When } \xi = \xi_2. \tag{11}$$

As the unsteady peristaltic flow of nanofluids in the laboratory frame (ξ, η) is considered, thus a wave frame (ξ^*, η^*), which moves corresponding to the wave that travels on the flexible and parallel walls of the outer tube, is taken into account. Let "c" be the constant velocity of the wave frame, such that:

$$\xi^* = \xi; \quad \eta^* = \eta - ct; \quad u^* = u; \quad w^* = w - c;$$
$$v^*(\xi^*, \eta^*) = v(\xi, \eta, t); \quad \varphi^*(\xi^*, \eta^*) = \varphi(\xi, \eta, t). \tag{12}$$

In view of the transformation given in Equation (10), the governing Equations (6)–(9) in wave frame can be written as:

$$\rho_f\left[\left(\frac{-\delta c^2}{\lambda}\right)\frac{\partial \bar{u}}{\partial \eta} + \left(\frac{\delta^2 c^2}{\xi_2}\right)\bar{u}\frac{\partial \bar{u}}{\partial \bar{\xi}} + \left(\frac{\delta c^2}{\lambda}\right)\bar{w}\frac{\partial \bar{u}}{\partial \eta}\right]$$
$$= -\gamma_1\left[\left(\frac{\delta c}{\xi_2^4}\right)\frac{\partial^4 \bar{u}}{\partial \bar{\xi}^4} + \left(\frac{\delta c}{\lambda^4}\right)\frac{\partial^4 \bar{u}}{\partial \eta^4} + \left(\frac{\delta c}{\lambda^2\xi_2^2}\right)\frac{\partial^4 \bar{u}}{\partial \bar{\xi}^2\partial \eta^2} + \left(\frac{\delta c}{\lambda^2\xi_2^2}\right)\frac{\partial^4 \bar{u}}{\partial \eta^2\partial \bar{\xi}^2}\right.$$
$$+ \left(\frac{2\delta c}{\xi_2^4}\right)\frac{\partial^3 \bar{u}}{\bar{\xi}\partial \bar{\xi}^3} + \left(\frac{\delta c}{\lambda^2\xi_2^2}\right)\frac{\partial^3 \bar{u}}{\bar{\xi}\partial \bar{\xi}\partial \eta^2} + \left(\frac{\delta c}{\lambda^2\xi_2^2}\right)\frac{\partial^3 \bar{u}}{\bar{\xi}\partial \eta^2\partial \bar{\xi}} + \left(\frac{\delta c}{\xi_2^4}\right)\frac{\partial^2 \bar{u}}{\bar{\xi}^2\partial \bar{\xi}^2}$$
$$\left.+ \left(\frac{\delta c}{\xi_2^4}\right)\frac{\partial \bar{u}}{\bar{\xi}^3\partial \bar{\xi}}\right] - \left(\frac{\mu\lambda c}{\xi_2^3}\right)\frac{\partial \bar{P}}{\partial \bar{\xi}} + \mu\left[\left(\frac{\delta c}{\xi_2^2}\right)\frac{\partial^2 \bar{u}}{\partial \bar{\xi}^2} + \left(\frac{\delta c}{\xi_2^2}\right)\frac{\partial \bar{u}}{\bar{\xi}\partial \bar{\xi}} + \left(\frac{\delta c}{\lambda^2}\right)\frac{\partial^2 \bar{u}}{\partial \eta^2}\right], \tag{13}$$

$$\rho_f\left[-c\frac{\partial(w^*+c)}{\partial \eta^*} + u^*\frac{\partial(w^*+c)}{\partial \xi^*} + (w^*+c)\frac{\partial(w^*+c)}{\partial \eta^*}\right]$$
$$= -\gamma_1\left[\frac{\partial^4(w^*+c)}{\partial \xi^{*4}} + \frac{\partial^4(w^*+c)}{\partial \eta^{*4}} + \frac{\partial^4(w^*+c)}{\partial \xi^{*2}\partial \eta^{*2}} + \frac{\partial^4(w^*+c)}{\partial \eta^{*2}\partial \xi^{*2}}\right.$$
$$+ \frac{2\partial^3(w^*+c)}{\xi^*\partial \xi^{*3}} + \frac{\partial^3(w^*+c)}{\xi^*\partial \xi^*\partial \eta^{*2}} + \frac{\partial^3(w^*+c)}{\xi^*\partial \eta^{*2}\partial \xi^*} + \frac{\partial^2(w^*+c)}{\xi^{*2}\partial \xi^{*2}}$$
$$\left.+ \frac{\partial(w^*+c)}{\xi^{*3}\partial \xi^*}\right] + \left[\varphi^*\rho_p + (1-\varphi^*)\left\{\rho_f(1-\beta_T(v^*-v_w))\right\}\right]g - \frac{\partial P^*}{\partial \xi^*}$$
$$+ \mu\left[\frac{\partial^2(w^*+c)}{\partial \xi^{*2}} + \frac{\partial(w^*+c)}{\xi^*\partial \xi^*} + \frac{\partial^2(w^*+c)}{\partial \eta^{*2}}\right]. \tag{14}$$

$$u^*\frac{\partial v^*}{\partial \xi^*} - c\frac{\partial v^*}{\partial \eta^*} + (w^*+c)\frac{\partial v^*}{\partial \eta^*} = \frac{k}{(\rho c)_f}\left(\frac{\partial^2 v^*}{\partial \xi^{*2}} + \frac{1}{\xi^*}\frac{\partial v^*}{\partial \xi^*} + \frac{\partial^2 v^*}{\partial \eta^{*2}}\right) + \frac{(\rho c)_p}{(\rho c)_f}$$
$$\left[D_b(\psi_m - \psi_w)\left\{\frac{\partial \varphi^*}{\partial \xi^*}\frac{\partial u^*}{\partial \xi^*} + \frac{\partial \varphi^*}{\partial \eta^*}\frac{\partial v^*}{\partial \eta^*}\right\} + \frac{D_T(v_m-v_w)}{v_w}\left\{\left(\frac{\partial v^*}{\partial \xi^*}\right)^2 + \left(\frac{\partial \varphi^*}{\partial \eta^*}\right)^2\right\}\right]. \tag{15}$$

$$\left[u^*\frac{\partial \varphi^*}{\partial \xi^*} + w^*\frac{\partial \varphi^*}{\partial \eta^*}\right] = \frac{D_T}{v_w}\left(\frac{\partial^2 v^*}{\partial \xi^{*2}} + \frac{1}{\xi^*}\frac{\partial v^*}{\partial \xi^*} + \frac{\partial^2 v^*}{\partial \eta^{*2}}\right) - k_r^2\left(\frac{v^*}{v_w}\right)^n(\varphi^* - \varphi_w)\exp\left(\frac{-E_a}{kv^*}\right)$$
$$+ D_b\left(\frac{\partial^2 \varphi^*}{\partial \xi^{*2}} + \frac{1}{\xi^*}\frac{\partial \varphi^*}{\partial \xi^*} + \frac{\partial^2 \varphi^*}{\partial \eta^{*2}}\right). \tag{16}$$

3. Results

Dealing with an unsteady peristaltic transport of couple stress fluid suspended with heated golden nano-sized particles ends up with a system of ordinary differential equations. These differential equations were mutually intermingled with each other, involving a nonlinear composition. Therefore, for such complex geometry, an exact solution was not possible. This means one has to turn to a numerical scheme suitable for tackling the said issue. In order to achieve the desired goal, first we have to make the entire system a non-dimensional form, by using the following transformations along with Oberbeck-Boussinesq approximation and long wave length assumption:

$$\frac{\xi^*}{\xi_2} = \bar{\xi}; \quad \frac{\eta^*}{\lambda} = \bar{\eta}; \quad \frac{u^*}{c\,\delta} = \bar{u}; \quad \frac{w^*}{c} = \bar{w}; \quad \frac{\xi_2}{\lambda} = \delta; \quad \frac{\xi_2^2\,P^*}{c\,\lambda\,\mu} = \bar{P}; \quad \frac{(\rho c)_p}{(\rho c)_f} = \tau;$$

$$\frac{k}{(\rho c)_f} = \alpha; \quad \sqrt{\frac{\mu}{\gamma_1}}\,\xi_2 = \gamma; \quad 1 + \bar{\epsilon}\,\cos(2\pi\bar{\eta}) = R_2; \quad E^* = \frac{E_a}{k\,v_w};$$

$$\frac{\xi_2^2\,(v_m - v_w)\,(1 - \varphi_w) g\,\rho_f\,\beta_T}{c\,\mu} = G_r; \quad D_b(\varphi_m - \varphi_w) = N_b; \quad A^* = \frac{k_r^2}{D_b}; \tag{17}$$

$$\frac{v^* - v_w}{v_m - v_w} = \bar{v}; \quad \frac{\varphi^* - \varphi_w}{\varphi_m - \varphi_w} = \bar{\varphi}; \quad \frac{D_T\,(v_m - v_w)}{v_w} = N_t; \quad \frac{d}{\xi_2} = \bar{\epsilon};$$

$$\frac{\xi_2^2\,(\varphi_m - \varphi_w)\,(\rho_p - \rho_f) g}{c\,\mu} = B_r; \quad \frac{\xi_1}{\xi_2} = R_1; \quad \beta^* = \frac{(v_m - v_w)}{v_w}.$$

Equations (13)–(16) in dimensionless form can be obtained as:

$$\frac{d\bar{P}}{d\bar{\eta}} = \frac{d^2\bar{w}}{d\bar{\xi}^2} + \frac{1}{\bar{\xi}}\frac{d\bar{w}}{d\bar{\xi}} - \frac{1}{\gamma^2}\left[\frac{d^4\bar{w}}{d\bar{\xi}^4} + \frac{2}{\bar{\xi}}\frac{d^3\bar{w}}{d\bar{\xi}^3} + \frac{d^2\bar{w}}{\bar{\xi}^2 d\bar{\xi}^2} + \frac{d\bar{w}}{\bar{\xi}^3 d\bar{\xi}}\right] + B_r\bar{\varphi} + G_r\,\bar{v}, \tag{18}$$

$$\alpha\left(\frac{\partial^2\bar{v}}{\partial\bar{\xi}^2} + \frac{1}{\bar{\xi}}\frac{\partial\bar{v}}{\partial\bar{\xi}}\right) + \tau\left\{N_b\left(\frac{\partial\bar{\varphi}}{\partial\bar{\xi}}\right)\left(\frac{\partial\bar{v}}{\partial\bar{\xi}}\right) + N_t\left(\frac{\partial\bar{v}}{\partial\bar{\xi}}\right)^2\right\} = 0, \tag{19}$$

$$N_b\left(\frac{\partial^2\bar{\varphi}}{\partial\bar{\xi}^2} + \frac{1}{\bar{\xi}}\frac{\partial\bar{\varphi}}{\partial\bar{\xi}}\right) + N_t\left(\frac{\partial^2\bar{v}}{\partial\bar{\xi}^2} + \frac{1}{\bar{\xi}}\frac{\partial\bar{v}}{\partial\bar{\xi}}\right) - \{A^*(\beta^*\bar{v} + 1)^n\,N_b\}\bar{\varphi}\,\exp\left(\frac{-E^*}{\beta^*\bar{v} + 1}\right) = 0. \tag{20}$$

Also, the corresponding boundary conditions in dimensionless form are as follows.
At the rigid wall:

$$\left.\begin{array}{l}(i).\ \bar{w}(\bar{\xi}) = -1, \\ (ii).\ \bar{v}(\bar{\xi}) = 1, \\ (iii).\ \bar{\varphi}(\bar{\xi}) = 1.\end{array}\right\}; \text{ When } \bar{\xi} = R_1. \tag{21}$$

At the flexible wall:

$$\left.\begin{array}{l}(i).\ \bar{w}(\bar{\xi}) = -1, \\ (ii).\ \bar{v}(\bar{\xi}) = 0, \\ (iii).\ \bar{\varphi}(\bar{\xi}) = 0.\end{array}\right\}; \text{ When } \bar{\xi} = R_2. \tag{22}$$

Finally, to obtain reliable solutions of Equations (18)–(20) subject to corresponding boundary conditions given in Equations (21) and (22), the most efficient numerical approach, Keller-box scheme, [32] is utilized. This method is much faster and more flexible to use as compared to other methods. It has been extensively used and tested on boundary layer flows. By means of said method, the solution can be attained by using four steps: (i) First reduce the system of equations to a first order system; (ii) then write the difference equations by means of central differences; (iii) now linearize the resulting nonlinear equation by Newton's method, if needed; and (iv) finally the block-tridiagonal-elimination technique is used to solve the linear system.

4. Discussion

This graphical section is relevant to the effectively contributing parameters, which influence axial velocity of couple stress fluid, temperature of nanofluid, and concentration of nano sized Hafnium particles, respectively. The involved parameters have a greater impression on the flow, namely, couple stress parameter γ, Brownian motion N_b and thermophoresis parameters N_t, thermophoresis diffusion G_r, and Brownian parameter B_r emerging due to the presence of heat and metallic particles. Moreover, a modified Arrhenius mathematical term yields some additional parameters, such as reaction rate A^*, activation energy E^*, temperature difference parameter β^*, and the fitted rate constant n, assuming the contribution of peristaltic pressure to be constant. To make this more systematic, the main portion is further divided into four subsections.

4.1. Axial Velocity

Axial velocity is spotted in Figures 2–4 for couple stress parameter, Brownian diffusion constant, and Grashof number. Axial velocity, as shown in Figure 2, accelerates in response to an increases in couple stress parameter. This is mainly due to the decrease in friction, which arises from the particle (i.e., base-fluid particles) additives, which constitute a size-dependent effect in couple stress fluids. In addition to the preceding remark, the rotational field of fluid particles is minimal as well. The peristaltic motion of outer walls of the tube also contributes by rapidly pushing the fluid in the axial direction, as B_r gets numerically variated in Figure 3. Figure 4 displays a different picture of the velocity of the fluid for the case of thermophoresis diffusion constant. The diagram basically describes the influence of buoyancy in terms of Grashof number G_r. As one can see from Equation (17), the buoyancy effects are mainly due to gravity and temperature difference. Therefore, increase in G_r attenuates the fluid's momentum by aggravating buoyant force. This brings a vivid decline in the velocity of the fluid. Furthermore, the relation defining Gr suggests that if $Gr > 0$, then this physically describes the heating of the nanofluid, while a reverse case can be expected for $Gr < 0$.

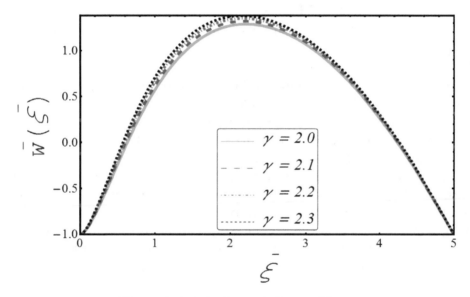

Figure 2. Axial velocity influenced by γ.

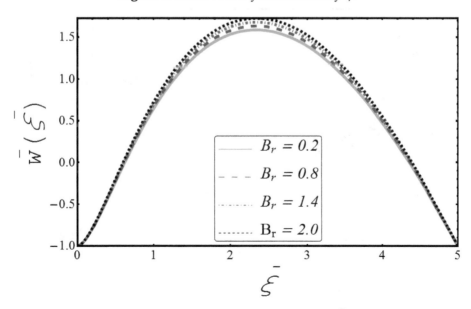

Figure 3. Axial velocity influenced by B_r.

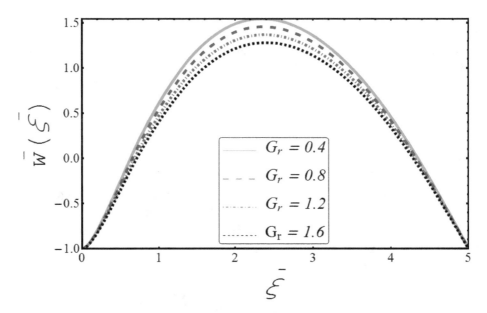

Figure 4. Axial velocity influenced by G_r.

4.2. Thermal Distribution

The temperature distribution of the nanofluid in the presence of additional chemical reaction and activation energy are portrayed in Figures 5 and 6. The variation of the Brownian motion parameter has noticeable effects on the nanofluid temperature, as the Brownian motion is generated due the collision of nanoparticles, driving the particles to a random motion. The collision of the particles, whether mutual or with the fluid molecules, is enhanced by the inward contraction of the flexible walls. Due to this factor Brownian motion parameter, N_b accumulates some additional thermal energy in the fluid, as shown in Figure 5. The nanoparticles were further thermally charged by the increase in N_t. It is important to keep in mind that the thermophoresis forces become stronger in the response of larger values of N_t, which finally result in higher temperature, as seen in Figure 6. Sometimes, such variations are credited to the thermal boundary layer thickness as well. Obviously, this increase in fluid temperature is due to increase in the random motion of nanoparticles when the above-mentioned parameters are increased.

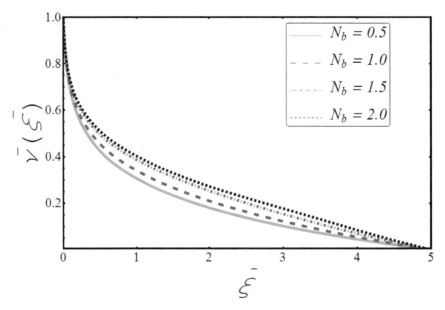

Figure 5. Temperature influenced by N_b.

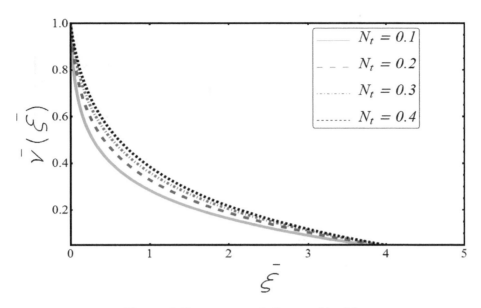

Figure 6. Temperature influenced by N_t.

4.3. Nanoparticle Concentration Profile

The concentration of golden particles is observed in Figures 7 and 8, when the Brownian motion parameter and thermophoresis parameter, respectively, are given higher numeric values. The random motion of the nanoparticles is seen to be faster in response to increase in the values of said parameters, which makes diffusion of nanoparticles rapid and fast. Therefore, rising curves show an increase in the concentration of nanoparticles. Moreover, this contribution of Brownian motion identifies the quick movement of hotter gold particle, from the region of higher temperature to lower temperature. The thermophoresis forces also bring positive effects on the golden particles by making the concentration strong against the higher numerical variation in N_t, as is noticeable in Figure 8. With the same trend of influence, an onward surge of activation energy again gives a rise to the golden solution. One can see in Figure 9 that the boundary layer thickness of the particles gets depreciated when E_a is further motivated to transport the required drug or medicine to the desired target. The Arrhenius equation, which gives the mathematical description of the introduction of activation energy into any system, clearly reveals that the reduction in heat and acceleration of E_a returns a low reaction rate constant.

In the process, this slows down the chemical reaction and results in higher concentration of the particles, which confirms the accumulation of gold nanoparticles at the location of the malign tissue or organ to cured. Finally, the surge in concentration of gold particles is evidenced by the decline in Figures 10–12. The temperature difference ratio brings a remarkable decline in concentration of the heated nanoparticles. As the difference between the ambient fluid temperature and wall temperature widens, the concentration boundary layer thickness expands. This thickness resists the increase in particle concentration displayed in Figure 10.

Similarly, retardation can be witnessed for reaction rate and fitted rate constant. It can be conceived that the rise in these parameters and constants sharpens the chemical reaction, which motivates the concentration gradient at the wall of the inner tube. Hence, a vivid reduction in the concentration of the particles occurs, as is seen in Figures 11 and 12).

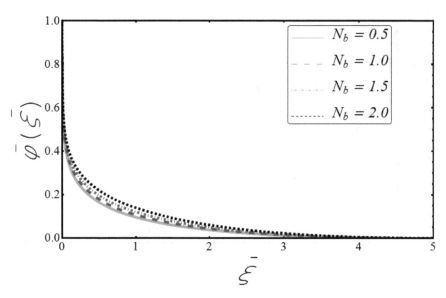

Figure 7. Concentration under the influence of N_b.

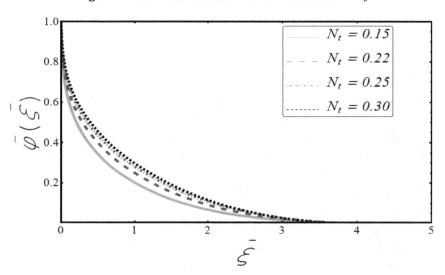

Figure 8. Concentration under the influence of N_t.

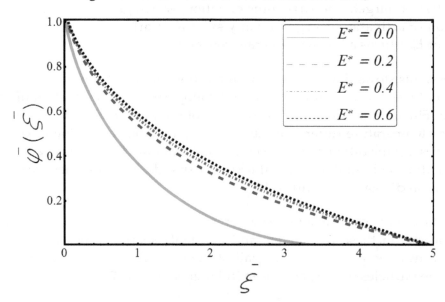

Figure 9. Concentration under the influence of E^*.

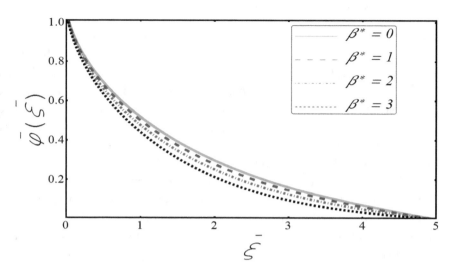

Figure 10. Concentration under the influence of β^*.

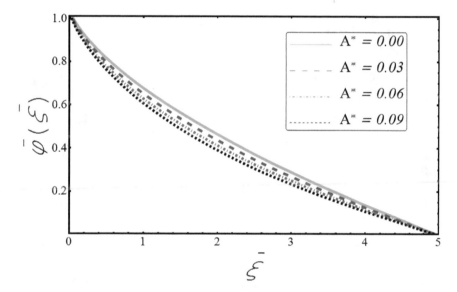

Figure 11. Concentration under the influence of A^*.

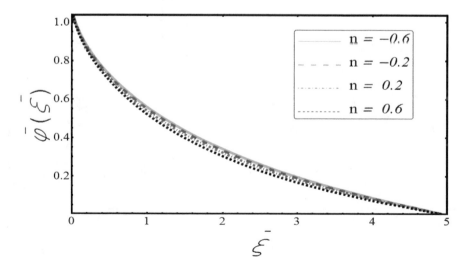

Figure 12. Concentration under the influence of n.

4.4. Trapping Phenomenon/ Streamline Configuration

Finally, the most significant phenomena relevant to any peristaltic motion in a living organism is known as "Trapping". Essentially, this is the appearance of a round closed bolus, which is identified as the hallow cavity, transporting the required medication to the desired tissues or organs, as shown in Figures 13–19. In Figures 13 and 14, one can easily notice that the fluids face less resistance when traveling through the coaxial space, as the contours reduce in size and configuration. In contrast, the couple stress fluid results in shrinking the streamlines and generates the circulating boluses, as depicted in Figure 15. Isotherms of the Brownian motion parameter keep binding closer together, which allows the bolus to expand, as established in Figures 16 and 17, whereas the thermophoresis parameter provides extra potential for isotherms to compress the bolus inwards. Hence, the bolus keeps getting smaller. In the last two graphs, contours are sketched in order to see how concentration is influenced by the reaction rate constant and thermophoresis parameter. One can see in Figure 18 that the bolus bulges out as the reaction rate constant gets stronger, whereas a reverse trend is observed for the thermophoresis parameter in Figure 19.

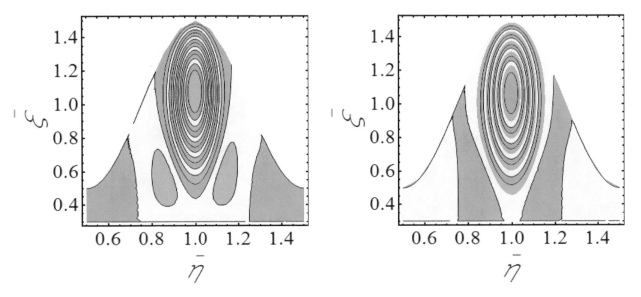

Figure 13. Stream lines for Brownian diffusion constant. (**a**): For $B_r = 0.2$; (**b**): For $B_r = 0.2$.

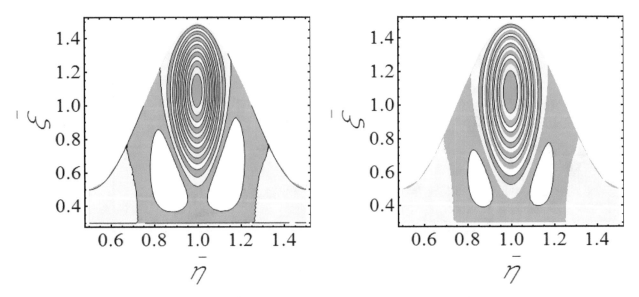

Figure 14. Stream lines for Grashof number. (**a**): For $G_r = 0.1$; (**b**): For $G_r = 0.3$.

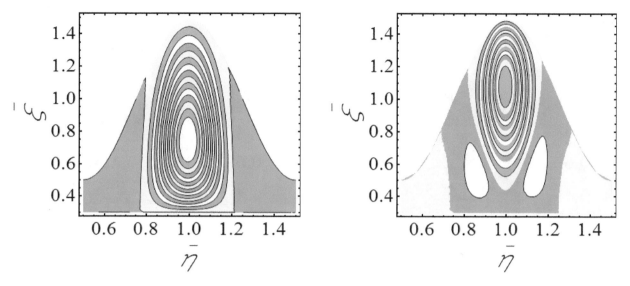

Figure 15. Stream lines for couple stress parameter. (**a**): For $\gamma = 1.0$; (**b**): For $\gamma = 2.0$.

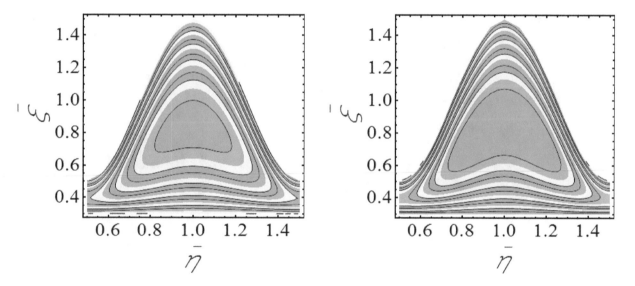

Figure 16. Isotherms for Brownian motion parameter. (**a**): For $N_b = 1.0$; (**b**): For $N_b = 1.5$.

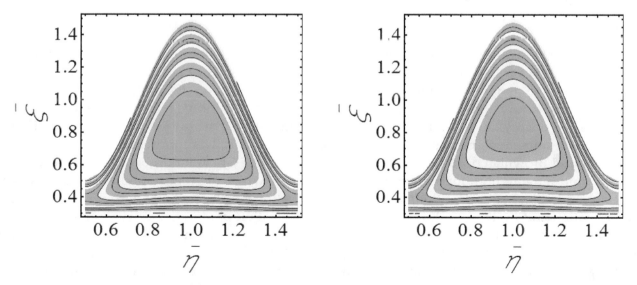

Figure 17. Isotherms for Thermophoresis parameter. (**a**): For $N_t = 0.2$; (**b**): For $N_t = 0.5$.

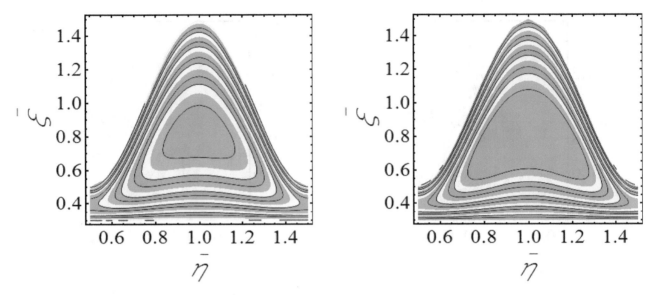

Figure 18. Contour plot for reaction rate constant. (**a**): For $A^* = 0.5$; (**b**): For $A^* = 1.0$.

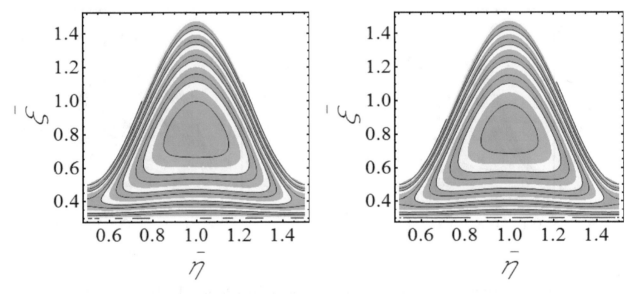

Figure 19. Contour plot for Thermophoresis parameter. (**a**): For $N_t = 0.2$; (**b**): For $N_t = 0.5$.

5. Conclusions

A numerical investigation is carried out for the peristaltic flow of nanofluids between the gap of two coaxial tubes with different configurations and structures. The nanofluid is composed of gold particles, while the couple stress fluid serves as the solvent. To enhance the mutual interaction of gold particles, or the interaction of molecules with the base fluid, additional effects of chemical reaction and activation energy have also been taken into consideration. The performed study reveals very informative results. Such results include that axial velocity is fully supported by the couple stress parameter and Brownian diffusion constant, in contrast to the Grashof number. The temperature of the nanofluid remains high for both involved parameters, which are thermophoresis and Brownian motion parameter. Looking at the graphs of concentrations of the metallic particles, it is inferred that activation energy, thermophoresis, and Brownian motion parameters cause an increase in the concentration of particles, whereas temperature ratio, reaction rate, and fitted rate constants do not support the increase. In the final portion of the graphical study, the number and size of the circulating boluses are depicted. One can easily notice that boluses get enlarged in response to the Brownian motion parameter, couples stress parameter, and reaction rate constant. However, a reverse trend is observed for the Grashof number, thermophoresis parameter, and Brownian diffusion constant. The key finding can be summarized as:

- Strong buoyant force results in retarded axial velocity for the thermophoresis parameter.
- Peristaltic movement of the outer tube enhances the Brownian motion and raises the temperature of the nanofluid.
- Activation energy entering the process maximizes the concentration boundary layer thickness.
- The reaction rate constant increases concentration at the catheter, which decreases the concentration of nanoparticles.
- The thermophoresis parameter shrinks the size of the bolus by strengthening isotherms and closed paths of concentration lines.
- The couple stress parameter and reaction rate constant give freedom to the bolus to swell by binding the stream lines closer to each another.

Author Contributions: Supervision, R.E.; conceptualization, A.Z.; investigation, F.H.; methodology, A.A.

Acknowledgments: R. Ellahi gratefully thanks Prof. Sadiq M. Sait, the Director Office of Research Chair Professors, King Fahd University of Petroleum and Minerals, Dhahran, Saudi Arabia, to honor him with the Chair Professor at KFUPM. F. Hussain is also HEC (Higher Education Commission Pakistan) to provide him the title of indigenous scholar for the pursuance of his Ph.D. studies.

Nomenclature

V	Nanofluid velocity
G	Gravitational acceleration
u	Radial velocity Component (fixed frame)
w	Axial velocity component (Fixed frame)
u^*	Radial velocity component (wave frame)
w^*	Axial velocity component (Wave frame)
$\bar{u}$	Dimensionless radial velocity component
$\bar{w}$	Dimensionless lateral velocity component
d	Amplitude of peristaltic wave
t	Time
k_r	Rate of reaction
c	Propagating velocity of wave
N_t	Thermophoresis parameter
k	Thermal conductivity
N_b	Brownian motion parameter
G_r	Grashof number
D_t	Thermophoretic diffusion coefficient
D_b	Brownian motion coefficient
d	Amplitude of peristaltic wave
t	Time
R_2	Dimensionless radius of outer tube
R_1	Dimensionless radius of inner tube
P^*	Dimensional pressure
B_r	Brownian diffusion constant
A^*	Reaction rate constant
E^*	Activation energy (Dimensionless)
E_a	Activation energy (Dimensional)
n	Fitted rate constant

Greek Symbols

ξ	Radial direction of the flow (Fixed frame)
η	Axial direction of the flow (Fixed frame)
ξ^*	Radial direction of the flow (Wave frame)
η^*	Axial direction of the flow (Wave frame)
$\overline{\xi}$	Radial direction of the flow (Dimensionless)
$\overline{\eta}$	Axial direction of the flow (Dimensionless)
ξ_1	Radius of inner tube (Dimensional)
ξ_2	Radius of outer tube (Dimensional)
$\vec{\varphi}$	Nanoparticle concentration (Fixed frame)
$\vec{\nu}$	Nanofluid temperature (Fixed frame)
φ^*	Nanoparticle concentration (Wave frame)
ν^*	Nanofluid temperature (Wave frame)
$\overline{\varphi}$	Nanoparticle concentration (Dimensionless)
$\overline{\nu}$	Nanofluid temperature (Dimensionless)
γ_1	Couple stress fluid's constant
γ	Couple stress parameter
τ	A ratio defined as $\frac{(\tilde{\rho}c)_f}{(\tilde{\rho}c)_p}$
β^*	Temperature ratio
ρ_p	Density of nanoparticle at reference temperature
ρ_f	Density of nanofluid at reference temperature
$(\rho c)_f$	Heat capacity of base fluid
$(\rho c)_p$	Heat capacity of particle
μ	Dynamic Viscosity
ν	Kinematic viscosity
λ	Wavelength
α	Ratio defined as $\frac{k}{(\rho c)_f}$
$\overline{\epsilon}$	A constant ratio
β_T	Volumetric coefficient of expansion
φ_w	Reference concentration
ν_w	Reference temperature
φ_m	Mass concentration
ν_m	Fluid temperature

Subscripts

f	Base fluid
p	Particle

References

1. Karimipour, A.; Orazio, A.D.; Shadloo, M.S. The effects of different nano particles of Al2O3 and Ag on the MHD nano fluid flow and heat transfer in a microchannel including slip velocity and temperature jump. *Physica E* **2017**, *86*, 146–153. [CrossRef]
2. Tripathi, D.; Beg, O.A. A study on peristaltic flow of nanofluids: Application in drug delivery systems. *Int. J. Heat Mass Transf.* **2014**, *70*, 61–70. [CrossRef]
3. Nabil, T.M.; El-dabe, N.T.M.; Moatimid, G.M.; Hassan, M.A.; Godh, W.A. Wall properties of peristaltic MHD Nanofluid flow through porous channel. *Fluid Mech. Res. Int.* **2018**, *2*, 19. [CrossRef]
4. Swarnalathamma, B.V.; Krishna, M.V. Peristaltic hemodynamic flow of couple stress fluid through a porous medium under the influence of magnetic field with slip effect. *AIP Conf. Proc.* **2016**, *1728*, 020603. [CrossRef]
5. Shit, G.C.; Roy, M. Hydromagnetic effect on inclined peristaltic flow of a couple stress fluid, Hydromagnetic effect on inclined peristaltic flow of a couple stress fluid. *Alex. Eng. J.* **2014**, *53*, 949–958. [CrossRef]
6. Jamalabadi, M.Y.A.; Daqiqshirazi, M.; Nasiri, H.; Safaei, M.R.; Nguyen, T.K. Modeling and analysis of biomagnetic blood Carreau fluid flow through a stenosis artery with magnetic heat transfer: A transient study. *PLoS ONE* **2018**, *13*, e0192138. [CrossRef]

7. Hosseini, S.M.; Safaei, M.R.; Goodarzi, M.; Alrashed, A.A.A.A.; Nguyen, T.K. New temperature, interfacial shell dependent dimensionless model for thermal conductivity of nanofluids. *Int. J. Heat Mass Transf.* **2017**, *114*, 207–210. [CrossRef]

8. Mekheimer, K.; Hasona, W.; Abo-Elkhair, R.E.; Zaher, A. Peristaltic blood flow with gold nanoparticles as a third grade nanofluid in catheter. *Appl. Cancer Ther. Phys. Lett. A* **2018**, *382*, 85–93.

9. Nasiri, H.; Jamalabadi, M.Y.A.; Sadeghi, R.; Safaei, M.R.; Nguyen, T.K.; Shadloo, M.S. A smoothed particle hydrodynamics approach for numerical simulation of nano-fluid flows. *J. Anal. Calorim.* **2018**, 1–9. [CrossRef]

10. Shadloo, M.S.; Kimiaeifar, A. Application of homotopy perturbation method to find an analytical solution for magnetohydrodynamic flows of viscoelastic fluids in converging/diverging channels. *Proc. Mech. Eng. Part C J. Mech. Eng.* **2011**, *225*, 347–353. [CrossRef]

11. Safaei, M.R.; Ahmadi, G.; Goodarzi, M.S.; Shadloo, M.S.; Goshayeshi, H.R.; Dahari, M. Heat transfer and pressure drop in fully developed turbulent flows of graphene nanoplatelets–silver/water nanofluids. *Fluids* **2016**, *1*, 20. [CrossRef]

12. Mohyud-Din, S.T.; Khan, U.; Ahmed, N.; Hassan, S.M. Magnetohydrodynamic flow and heat transfer of nanofluids in stretchable convergent/divergent channels. *Appl. Sci.* **2015**, *5*, 1639–1664. [CrossRef]

13. Chamkha, A.J.; Selimefendigil, F. Forced convection of pulsating nanofluid flow over a backward facing step with various particle shapes. *Energies* **2018**, *11*, 3068. [CrossRef]

14. Xu, Z.; Kleinstreuer, C. Direct nano-drug delivery for tumor targeting subject to shear-augmented diffusion in blood flow. *Med. Biol. Eng. Comput.* **2018**, *56*, 1949–1958. [CrossRef] [PubMed]

15. Rashidi, M.M.; Nasiri, M.; Shadloo, M.S.; Yang, Z. Entropy generation in a circular tube heat exchanger using nanofluids: Effects of different modeling approaches. *Heat Transf. Eng.* **2017**, *38*, 853–866. [CrossRef]

16. Jamalabadi, M.Y.A.; Safaei, M.R.; Alrashed, A.A.A.A.; Nguyen, T.K.; Filho, E.P.B. Entropy generation in thermal radiative loading of structures with distinct heaters. *Entropy* **2017**, *19*, 506. [CrossRef]

17. Sadiq, M.A. MHD stagnation point flow of nanofluid on a plate with anisotropic slip. *Symmetry* **2019**, *11*, 132. [CrossRef]

18. Jawad, M.; Shah, Z.; Islam, S.; Majdoubi, J.; Tlili, I.; Khan, W.; Khan, I. Impact of nonlinear thermal radiation and the viscous dissipation effect on the unsteady three-dimensional rotating flow of single-wall carbon nanotubes with aqueous suspensions. *Symmetry* **2019**, *11*, 207. [CrossRef]

19. Akbarzadeh, O.; Zabidi, N.A.M.; Wahab, Y.A.; Hamizi, N.A.; Chowdhury, Z.Z.; Merican, Z.M.A.; Rahman, M.A.; Akhter, S.; Rasouli, E.; Johan, M.R. Effect of cobalt catalyst confinement in carbon nanotubes support on fischer-tropsch synthesis performance. *Symmetry* **2018**, *10*, 572. [CrossRef]

20. Mustafa, M.; Khan, J.A.; Hayat, T.; Alsaedi, A. Buoyancy effects on the MHD nanofluid flow past a vertical surface with chemical reaction and activation energy. *Int. J. Heat Mass Transf.* **2017**, *108*, 1340–1346. [CrossRef]

21. Xu, H.; Fan, T.; Pop, I. Analysis of mixed convection flow of a nanofluid in a vertical channel with the Buongiorno mathematical model. *Int. Commun. Heat Mass Transf.* **2013**, *44*, 15–22. [CrossRef]

22. Sheikholeslami, M.; Bhatti, M.M. Active method for nanofluid heat transfer enhancement by means of EHD. *Int. J. Heat Mass Transf.* **2017**, *109*, 115–122. [CrossRef]

23. Bhatti, M.M.; Rashidi, M.M. Effects of thermo-diffusion and thermal radiation on Williamson nanofluid over a porous shrinking/stretching sheet. *J. Mol. Liq.* **2016**, *221*, 567–573. [CrossRef]

24. Marin, M. An approach of a heat-flux dependent theory for micropolar porous media. *Meccanica* **2016**, *51*, 127–1133. [CrossRef]

25. Rashidi, S.; Esfahani, J.A.; Ellahi, R. Convective heat transfer and particle motion in an obstructed duct with two side by side obstacles by means of DPM model. *Appl. Sci.* **2017**, *7*, 431. [CrossRef]

26. Zeeshan, A.; Ijaz, N.; Abbas, T.; Ellahi, R. The sustainable characteristic of Bio-bi-phase flow of peristaltic transport of MHD Jeffery fluid in human body. *Sustainability* **2018**, *10*, 2671. [CrossRef]

27. Hussain, F.; Ellahi, R.; Zeeshan, A. Mathematical models of electro magnetohydrodynamic multiphase flows synthesis with nanosized hafnium particles. *Appl. Sci.* **2018**, *8*, 275. [CrossRef]

28. Haq, R.U.; Soomro, F.A.; Hammouch, Z.; Rehman, S. Heat exchange within the partially heated C-shape cavity filled with the water based SWCNTs. *Int. J. Heat Mass Transf.* **2018**, *127*, 506–514. [CrossRef]

29. Kumar, R.V.M.S.S.K.; Kumar, G.V.; Raju, C.S.K.; Shehzad, S.A.; Varma, S.V.K. Analysis of Arrhenius activation energy in magnetohydrodynamic Carreau fluid flow through improved theory of heat diffusion and binary chemical reaction. *J. Phys. Commun.* **2018**, *2*, 35–49. [CrossRef]

30. Anuradha, S.; Yegammai, M. MHD radiative boundary layer flow of nanofluid past a vertical plate with effects of binary chemical reaction and activation energy. *J. Pure Appl. Math.* **2017**, *13*, 6377–6392.

31. Pal, D.; Talukdar, B. Perturbation analysis of unsteady magnetohydrodynamic convective heat and mass transfer in a boundary layer slip flow past a vertical permeable plate with thermal radiation and chemical reaction. *Commun. Nonlinear Sci. Numer. Simul.* **2010**, *15*, 1813–1830. [CrossRef]

32. Hossain, A., Md.; Subba, R.; Gorla, R. Natural convection flow of non-Newtonian power-law fluid from a slotted vertical isothermal surface. *Int. J. Numer. Methods Heat Fluid Flow* **2009**, *19*, 835–846. [CrossRef]

MHD Slip Flow of Casson Fluid along a Nonlinear Permeable Stretching Cylinder Saturated in a Porous Medium with Chemical Reaction, Viscous Dissipation and Heat Generation/Absorption

Imran Ullah [1], Tawfeeq Abdullah Alkanhal [2], Sharidan Shafie [3], Kottakkaran Sooppy Nisar [4], Ilyas Khan [5,*] and Oluwole Daniel Makinde [6]

[1] College of Civil Engineering, National University of Sciences and Technology Islamabad, Islamabad 44000, Pakistan; ullahimran14@gmail.com
[2] Department of Mechatronics and System Engineering, College of Engineering, Majmaah University, Majmaah 11952, Saudi Arabia; t.alkanhal@mu.edu.sa
[3] Department of Mathematical Sciences, Faculty of Science, Universiti Teknologi Malaysia, UTM Johor Bahru 81310, Johor, Malaysia; sharidan@utm.my
[4] Department of Mathematics, College of Arts and Science at Wadi Al-Dawaser, Prince Sattam bin Abdulaziz University, Al Kharj 11991, Saudi Arabia; n.sooppy@psau.edu.sa
[5] Faculty of Mathematics and Statistics, Ton Duc Thang University, Ho Chi Minh City 72915, Vietnam
[6] Faculty of Military Science, Stellenbosch University, Private Bag X2, Saldanha 7395, South Africa; makinded@gmail.com
* Correspondence: ilyaskhan@tdt.edu.vn

Abstract: The aim of the present analysis is to provide local similarity solutions of Casson fluid over a non-isothermal cylinder subject to suction/blowing. The cylinder is placed inside a porous medium and stretched in a nonlinear way. Further, the impact of chemical reaction, viscous dissipation, and heat generation/absorption on flow fields is also investigated. Similarity transformations are employed to convert the nonlinear governing equations to nonlinear ordinary differential equations, and then solved via the Keller box method. Findings demonstrate that the magnitude of the friction factor and mass transfer rate are suppressed with increment in Casson parameter, whereas heat transfer rate is found to be intensified. Increase in the curvature parameter enhanced the flow field distributions. The magnitude of wall shear stress is noticed to be higher with an increase in porosity and suction/blowing parameters.

Keywords: Casson fluid; chemical reaction; cylinder; heat generation; magnetohydrodynamic (MHD); slip

---

1. Introduction

Boundary layer flow on linear or nonlinear stretching surfaces has a wide range of engineering and industrial applications, and has been used in many manufacturing processes, such as extrusion of plastic sheets, glass fiber production, crystal growing, hot rolling, wire drawing, metal and polymer extrusion, and metal spinning. The viscous flow past a stretching surface was first developed by Crane [1]. Later on, this pioneering work was extended by Gupta and Gupta [2] and Chen and Char [3], and the suction/blowing effects on heat transfer flow over a stretching surface were investigated. Gorla and Sidawi [4] analyzed three-dimensional free convection flow over permeable stretching surfaces. Motivated by this, the two-dimensional heat transfer flow of viscous fluid due to a nonlinear stretching sheet was investigated by Vajravelu [5]. The similarity solutions for viscous flow over a nonlinear stretching sheet was obtained by Vajravelu and Cannon [6]. On the other hand, Bachok and Ishak [7] studied the prescribed surface heat flux characteristics on boundary layer flow generated by a stretching cylinder. Hayat et al. [8] analyzed the heat and mass transfer features on two-dimensional flow due to a stretching cylinder placed through a porous media in the presence of convective boundary conditions. The heat transfer analysis in ferromagnetic viscoelastic fluid flow over a stretching sheet was discussed by Majeed et al. [9].

The study of magnetohydrodynamic (MHD) boundary layer flow towards stretching surface has gained considerable attention due to its important practical and engineering applications, such as MHD power generators, cooling or drying of papers, geothermal energy extraction, solar power technology, cooling of nuclear reactors, and boundary layer flow control in aerodynamics. Vyas and Ranjan [10] investigated two-dimensional flow over a nonlinear stretching sheet in the presence of thermal radiation and viscous dissipation. They predicted that stronger radiation boosts the fluid temperature field. The effect of magnetic field on incompressible viscous flow generated due to stretching cylinder was analyzed by Mukhopadhyay [11], and it was observed that a larger curvature parameter allowed more fluid to flow. Fathizadeh et al. [12] studied the MHD effect on viscous fluid due to a sheet stretched in a nonlinear way. Akbar et al. [13] developed laminar boundary layer flow induced by a stretching surface in the presence of a magnetic field. They noticed that the intensity of the magnetic field offered resistance to the fluid flow, because of which, skin friction was enhanced. In another study, Ellahi [14] demonstrated the effects of magnetic field on non-Newtonian nanofluid through a pipe.

The momentum slip at a stretching surface plays an important role in the manufacturing processes of several products, including emulsion, foams, suspensions, and polymer solutions. In recent years, researchers have avoided no-slip conditions and take velocity slip at the wall. The reason is that it has been proven through experiments that momentum slip at the boundary can enhance the heat transfer. Fang et al. [15] obtained the exact solution for two-dimensional slip flow due to stretching surface. The slip effects on stagnation point flow past a stretching sheet were numerically analyzed by Bhattacharyya et al. [16]. The slip effect on viscous flow generated due to a nonlinear stretching surface in the presence of first order chemical reaction and magnetic field was developed by Yazdi et al. [17]. They concluded that velocity slip at the wall reduced the friction factor. Hayat et al. [18] investigated the impact of hydrodynamic slip on incompressible viscous flow over a porous stretching surface under the influence of a magnetic field and thermal radiation. They predicted that suction and slip parameters have the same effect on fluid velocity. Seini and Makinde [19] analyzed the hydromagnetic boundary layer flow of a viscous fluid under the influence of velocity slip at the wall. They noticed that wall shear stress enhanced with the growth of the magnetic parameter. Motivated by this, Rahman et al. [20] discussed the slip mechanisms in boundary layer flow of Jeffery nanofluid through an artery, and the solutions were achieved by the homotopy perturbation method.

In the recent years, the analysis of non-Newtonian fluid past stretching surfaces has gained the attention of investigators due to its wide range practical applications in several industries, for instance, food processes, ground water pollution, crude oil extraction, production of plastic materials, cooling of nuclear reactors, manufacturing of electronic chips, etc. Due to the complex nature of these fluids, different models have been proposed. Among other non-Newtonian model, the Casson fluid model is one of them. The Casson fluid model was originally developed by Casson [21] for the preparation of printing inks and silicon suspensions. Casson fluid has important applications in polymer industries and biomechanics [22]. The Casson fluid model is also suggested as the best rheological model for blood and chocolate [23,24]. For this reason, many authors have considered Casson fluid for different geometries. Shawky [25] analyzed the heat and mass transfer mechanisms in MHD flow of Casson fluid over a linear stretching sheet saturated in a porous medium. Mukhopadhyay [26] and Medikare et al. [27] investigated heat transfer effects on Casson fluid over a nonlinear stretching sheet in the absence and presence of viscous dissipation, respectively. Mythili and Sivaraj [28] considered the geometry of cone and flat plate and studied the impact of chemical reaction on Casson fluid flow with thermal radiation. The impact of magnetic field and heat generation/absorption on heat transfer flow of Casson fluid through a porous medium was presented by Ullah et al. [29]. Imtiaz et al. [30] developed the mixed convection flow of Casson fluid due to a linear stretching cylinder filled with nanofluid with convective boundary conditions.

The above discussion and its engineering applications is the source of motivation to investigate the electrically conductive flow of Casson fluid due to a porous cylinder being stretched in a nonlinear way. It is also clear from the published articles that the mixed convection slip flow of Casson fluid for the geometry of a nonlinear stretching cylinder saturated in a porous medium in the presence of thermal radiation, viscous dissipation, joule heating, and heat generation/absorption has not yet been analyzed. It is worth mentioning that the current problem can be reduced to the flow over a flat plate ($n = 0$ and $\gamma = 0$), linear stretching sheet ($n = 1$ and $\gamma = 0$), nonlinear stretching sheet ($\gamma = 0$), and linear stretching cylinder ($n = 1$). Local similarity transformations are applied to transform the governing equations. The obtained system of equations are then computed numerically using the Keller box method [31] via MATLAB. The variations of flow fields for various pertinent parameters are discussed and displayed graphically. Comparison of the friction factor is made with previous literature results and close agreement is noted. The accuracy achieved has developed our confidence that the present MATLAB code is correct and numerical results are accurate.

2. Mathematical Formulation

Consider a steady, two-dimensional, incompressible mixed convection slip flow of Casson fluid generated due to a nonlinear stretching cylinder in a porous medium in the presence of chemical reaction, slip, and convective boundary conditions. The cylinder is stretched with the velocity of $u_w(x) = cx^n$, where c, n ($n = 1$ represents linear stretching and $n \neq 1$ corresponds to nonlinear stretching) are constants. The x-axis is taken along the axis of the cylinder and the r-axis is measured in the radial direction (see Figure 1). It is worth mentioning here that the momentum boundary layer develops when there is fluid flow over a surface; a thermal boundary layer must develop if the bulk temperature differs from the surface temperature and a concentration boundary layer develops above the surfaces of species in the flow regime.

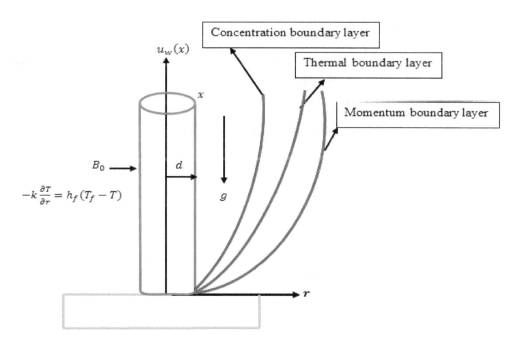

Figure 1. Coordinate system and physical model.

A transverse magnetic field $B(x) = B_0 x^{(n-1)/2}$ is applied in the radial direction with constant B_0. Further, it is also assumed that surface of cylinder is heated by temperature $T_f(x) = T_\infty + Ax^{2n-1}$, in which A is a reference temperature. Concentration is $C_s(x) = C_\infty + B^* x^{2n-1}$, where B^* is the reference concentration. The temperature and concentration at free stream are T_∞ and C_∞, respectively.

The rheological equation of state for an isotropic and incompressible flow of a Casson fluid is

$$\tau_{ij} = \begin{cases} 2\left(\mu_B + p_y/\sqrt{2\pi_1}\right)e_{ij}, & \pi_1 > \pi_c, \\ 2\left(\mu_B + p_y/\sqrt{2\pi_c}\right)e_{ij}, & \pi_1 < \pi_c, \end{cases}$$

Here, $\pi_1 = e_{ij}e_{ij}$ and e_{ij} is the $(i, j) - th$ component of the deformation rate, π_1 is the product of the component of deformation rate with itself, π_c is a critical value of this product based on the non-Newtonian model, μ_B is the plastic dynamic viscosity of the non-Newtonian fluid, and p_y is the yield stress of the fluid.

Under the above assumption, the governing equations for Casson fluid along with the continuity equation are given as

$$\frac{\partial(ru)}{\partial x} + \frac{\partial(rv)}{\partial r} = 0 \tag{1}$$

$$u\frac{\partial u}{\partial x} + v\frac{\partial u}{\partial r} = \nu\left(1 + \frac{1}{\beta}\right)\frac{1}{r}\frac{\partial}{\partial r}\left(r\frac{\partial u}{\partial r}\right) - \left(\frac{\sigma B^2(x)}{\rho} + \left(1 + \frac{1}{\beta}\right)\frac{\nu\phi}{k_1}\right)u + g\beta_T(T - T_\infty) + g\beta_C(C - C_\infty) \tag{2}$$

$$u\frac{\partial T}{\partial x} + v\frac{\partial T}{\partial r} = \alpha\left(1 + \frac{4}{3}R_d\right)\frac{1}{r}\frac{\partial}{\partial r}\left(r\frac{\partial T}{\partial r}\right) - \frac{\nu}{c_p}\left(1 + \frac{1}{\beta}\right)\left(\frac{\partial u}{\partial r}\right)^2 + \frac{\sigma B^2(x)}{\rho c_p}u^2 + \frac{Q}{\rho c_p}(T - T_\infty) \tag{3}$$

$$u\frac{\partial C}{\partial x} + v\frac{\partial C}{\partial r} = D\frac{1}{r}\frac{\partial}{\partial r}\left(r\frac{\partial C}{\partial r}\right) - k_c(C - C_\infty) \tag{4}$$

In the above expressions u and v denote the velocity components in x and r direction, respectively, ν is kinematic viscosity, σ is the electrically conductivity, β is the Casson parameter, ρ is the fluid density, ϕ is the porosity, $k_1(x) = k_0/x^{(n-1)}$ is the variable permeability of porous medium, g is the gravitational force due to acceleration, β_T is the volumetric coefficient of thermal expansion, β_C the coefficient of concentration expansion, $\alpha = \dfrac{k}{\rho c_p}$ is the thermal diffusivity of the Casson fluid, k is the thermal conductivity of fluid, c_p is the heat capacity of the fluid, $R_d = \dfrac{4\sigma^* T_\infty^3}{kk_1^*}$ is the radiation parameter, $Q(x) = Q_0 x^{n-1}$ is heat generation/absorption coefficient, D is the coefficient of mass diffusivity, $k_c(x) = ak_2 x^{n-1}$ is the variable rate of chemical reaction, k_2 is a constant reaction rate and a is the reference length along the flow.

The corresponding boundary conditions are written as follows

$$u = u_w(x) + N_1\nu\left(1 + \frac{1}{\beta}\right)\frac{\partial u}{\partial r}, \ k\frac{\partial T}{\partial r} = -h_f\left(T_f - T\right), \ D\frac{\partial C}{\partial r} = -h_s(C_s - C) \text{ at } r = d \tag{5}$$

$$u \to 0, \ T \to T_\infty, \ C \to C_\infty \text{ as } r \to \infty. \tag{6}$$

Here $N_1(x) = N_0 x^{-\left(\frac{n-1}{2}\right)}$ represents velocity slip with constant N_0, $h_f(x) = h_0 x^{\frac{n-1}{2}}$ and $h_s(x) = h_1 x^{\frac{n-1}{2}}$ represents the convective heat and mass transfer with h_0, h_1 being constants,

Now introduce the stream function ψ, a similar variable η and the following similarity transformations;

$$\psi = \sqrt{\frac{2\nu c}{(n+1)}}x^{\frac{n+1}{2}}df(\eta), \ \eta = \frac{r^2 - d^2}{2d}\sqrt{\frac{(n+1)c}{2\nu}}x^{\frac{n-1}{2}}, \ \theta(\eta) = \frac{T - T_\infty}{T_f - T_\infty}, \varphi(\eta) = \frac{C - C_\infty}{C_s - C_\infty} \tag{7}$$

Equation (1) is identically satisfied by the introduction of the equation

$$u = \frac{1}{r}\frac{\partial \psi}{\partial r}, \quad v = -\frac{1}{r}\frac{\partial \psi}{\partial x} \tag{8}$$

The system of Equations (2)–(4) will take the form

$$\left(1 + \tfrac{1}{\beta}\right)\left[\left(1 + 2\sqrt{\tfrac{2}{n+1}}\gamma\eta\right)f''' + 2\sqrt{\tfrac{2}{n+1}}\gamma f''\right] + ff'' - \tfrac{2n}{n+1}f'^2 - \tfrac{2}{(n+1)}\left(M + \left(1 + \tfrac{1}{\beta}\right)K\right)f' + \tfrac{2}{(n+1)}(Gr\theta + Gm\varphi) = 0 \tag{9}$$

$$\left(1 + \tfrac{4}{3}R_d\right)\left[\left(1 + 2\sqrt{\tfrac{2}{n+1}}\gamma\eta\right)\theta'' + 2\sqrt{\tfrac{2}{n+1}}\gamma\theta'\right] + \Pr f\theta' - 2\left(\tfrac{2n-1}{n+1}\right)\Pr f'\theta + MEcf'^2\left(1 + \tfrac{1}{\beta}\right)\left(1 + 2\sqrt{\tfrac{2}{n+1}}\gamma\eta\right)\Pr Ecf''^2 + \left(\tfrac{2}{n+1}\right)\varepsilon\theta = 0 \tag{10}$$

$$\tfrac{1}{Sc}\left[\left(1 + 2\sqrt{\tfrac{2}{n+1}}\gamma\eta\right)\varphi'' + 2\sqrt{\tfrac{2}{n+1}}\gamma\varphi'\right] + f\varphi' - 2\left(\tfrac{2n-1}{n+1}\right)f'\varphi - \tfrac{2}{(n+1)}R\varphi = 0 \tag{11}$$

The associated boundary conditions in Equations (5) and (6) are transformed as

$$\left.\begin{array}{l}
f(0) = \sqrt{\tfrac{2}{n+1}}S, \quad f'(0) = 1 + \delta\sqrt{\tfrac{n+1}{2}}\left(1 + \tfrac{1}{\beta}\right)f''(0), \quad \theta'(0) = -\left(\sqrt{\tfrac{2}{n+1}}\right)Bi_1[1 - \theta(0)] \\[2mm]
\varphi'(0) = -\left(\sqrt{\tfrac{2}{n+1}}\right)Bi_2[1 - \varphi(0)]
\end{array}\right\}, \tag{12}$$

$$f'(\infty) = 0, \quad \theta(\infty) = 0, \quad \varphi(\infty) = 0. \tag{13}$$

In the above expressions, γ, M, K, Gr, Gm, S ($S > 0$ corresponds to suction and $S < 0$ indicates blowing), δ, $\Pr$, Ec, ε ($\varepsilon > 0$ is for heat generation and $\varepsilon < 0$ denotes heat absorption), Sc, Bi_1, Bi_2, and R ($R > 0$ corresponds to destructive chemical reaction and $R = 0$ represents no chemical reaction) are the curvature parameter, magnetic parameter, porosity parameter, thermal Grashof number, mass Grashof number, suction/blowing parameter, slip parameter, Prandtl number, Eckert number, heat generation/absorption parameter, Schmidt number, Biot numbers and chemical reaction parameter, and are defined as

$$\gamma = \sqrt{\frac{vx^{1-n}}{cd^2}}, \quad M = \frac{\sigma B_0^2}{\rho c}, \quad K = \frac{v\phi}{k_0 c}, \quad Gr = \frac{g\beta_T A}{c^2}, \quad Gm = \frac{g\beta_C B^*}{c^2},$$

$$S = -V_0\sqrt{\frac{1}{cv}}, \quad \delta = N_0\sqrt{cv}, \quad \Pr = \frac{v}{\alpha}, \quad Ec = \frac{u_w^2}{c_p\left(T_f - T_\infty\right)}, \quad \varepsilon = \frac{Q_0}{\rho c_p c},$$

$$Sc = \frac{v}{D}, \quad Bi_1 = \frac{h_0}{k}\left[\frac{v}{c}\right]^{1/2}, \quad Bi_2 = \frac{h_1}{D}\left[\frac{v}{c}\right]^{1/2}, \quad R = \frac{ak_2}{c}$$

The wall skin friction, wall heat flux, and wall mass flux, respectively, are defined by

$$\tau_w = \mu_B\left(1 + \frac{1}{\beta}\right)\left[\frac{\partial u}{\partial r}\right]_{r=d}, \quad q_w = -\left(\left(\alpha + \frac{16\sigma^* T_\infty^3}{3\rho c_p k_1^*}\right)\frac{\partial T}{\partial r}\right)_{r=d} \quad \text{and} \quad q_s = -D\left(\frac{\partial C}{\partial y}\right)_{r=d}$$

The dimensionless skin friction coefficient $Cf_x = \dfrac{\tau_w}{\rho u_w^2}$, the local Nusselt number $Nu_x = \dfrac{xq_w}{\alpha(T_f - T_\infty)}$ and local Sherwood number $Sh_x = \dfrac{xq_s}{D_B(C_w - C_\infty)}$ on the surface along x—direction, local Nusselt number Nu_x and Sherwood number Sh_x are given by

$$(Re_x)^{1/2}Cf_x = \sqrt{\tfrac{n+1}{2}}\left(1 + \tfrac{1}{\beta}\right)f''(0), \quad (Re_x)^{-1/2}Nu_x = -\sqrt{\tfrac{n+1}{2}}\left(1 + \tfrac{4}{3}R_d\right)\theta'(0),$$

$$(Re_x)^{-1/2}Sh_x = -\left(\sqrt{\tfrac{n+1}{2}}\right)\varphi'(0)$$

where $Re_x = \dfrac{cx^{n+1}}{v}$ is the local Reynold number.

3. Results and Discussion

The system of Equations (9)–(11) are solved numerically by using the Keller-box method [31] and numerical computations are carried out for different values of physical parameters including curvature parameter γ, Casson fluid parameter β, nonlinear stretching cylinder parameter n, magnetic parameter M, porosity parameter K, Grashof number Gr, mass Grashof number Gm, Prandtl number Pr, radiation parameter R_d, Eckert number Ec, heat generation/absorption parameter ε, Schmidt number Sc, chemical reaction parameter R, slip parameter δ, and Biot numbers Bi_1, Bi_2. In order to validate the algorithm developed in MATLAB software for the present method, the numerical results for skin friction coefficient are compared with the results of Akbar et al. [13], Fathizadeh et al. [12], Fang et al. [15], and Imtiaz et al. [30], and presented in Table 1. Comparison revealed a close agreement with them.

Table 1. Comparison of skin friction coefficient $f''(0)$ for different values of M with $\beta \to \infty$, $Bi_1 \to \infty$, $Bi_2 \to \infty$, $n = 1$ and $\gamma = M = K = Gr = Gm = S = \delta = R_d = Ec = \varepsilon = R = 0$.

M	$\left(1 + \frac{1}{\beta}\right)f''(0)$				
	Akbar et al. [13]	Fathizadeh et al. [12]	Fang et al. [15]	Imtiaz et al. [30]	Present Results
0	−1	−1	−1	−1	−1
1	−1.4142	−1.4142	−1.4142	−1.4142	−1.4142
5	−2.4495	−2.4494	−2.4494	−2.4494	−2.4495
10	−3.3166	-	-	-	−3.3166

Figures 2–10 are depicted to see the physical behavior of γ, β, n, M, K, Gr, Gm, δ, and S on velocity profile. Figure 2 exhibits the variation of γ on fluid velocity for $n = 1$ (linear stretching) and $n \neq 1$ (nonlinear stretching). It is noticed that fluid velocity is higher for increasing values of γ. Since the increase in γ leads to reduction in the radius of curvature, it also reduces cylinder area. Thus, the cylinder experiences less resistance from the fluid particles and fluid velocity is enhanced. It can also be seen that the momentum boundary layer is thicker with increased γ when $n \neq 1$. The influence of β on velocity profile for different values of S is depicted in Figure 3. In all cases, the fluid velocity is a decreasing function of β. The reason is that the fluid becomes more viscous with the growth of β. Therefore, more resistance is offered which reduces the momentum boundary layer thickness. Figure 4 elucidates the effect of n on velocity profile for $M = 0$ and $M \neq 0$. It is evident that increasing values of n enhance the fluid velocity. Also, this enhancement is more pronounced when $M \neq 0$. The momentum boundary layer is thicker when $n \neq 1$.

The variation of M for $K = 0$ and $K \neq 0$ on the velocity profile is presented in Figure 5. As expected, the strength of the magnetic field lowers the fluid flow. It is an agreement with the fact that increase in M produces Lorentz force that provides resistance to the flow, and apparently thins the momentum boundary layer across the boundary. It can also be seen that the fluid velocity is more influenced with M when $K = 0$. A similar kind of variation is observed on velocity profile for different values of K, as displayed in Figure 6. Since the porosity of porous medium provides resistance to the flow, fluid motion slows down and produces larger friction between the fluid particles and the cylinder surface. The impact of Gr for $M = 0$ and $M \neq 0$ on velocity profile is depicted in Figure 7. The convection inside the fluid rises as the temperature difference $\left(T_f - T_\infty\right)$ enhances due to the growth of Gr. In addition, increase in Gr leads to stronger buoyancy force, in which case, the momentum boundary layer becomes thicker. The same kind of physical explanation can be given for the effect of Gm on velocity profile (see Figure 8). The variation of S on velocity profile for both $n = 1$ and $n \neq 1$ is portrayed in Figure 9. Clearly, the fluid velocity declines when $S > 0$, whereas a reverse trend is noted when $S < 0$. Physically, stronger blowing forces the hot fluid away from the surface, in which case the viscosity reduces and the fluid gets accelerated. On the other hand, wall suction ($S > 0$) exerts a drag force at the surface and hence thinning of the momentum boundary layer. Figure 10 demonstrates the effect of δ on velocity profile for $K = 0$ and $K \neq 0$. It can be easily seen that fluid velocity falls with increase in δ. Since the resistance between the cylinder surface and the fluid particles rises with increase in δ, the momentum boundary layer become thinner.

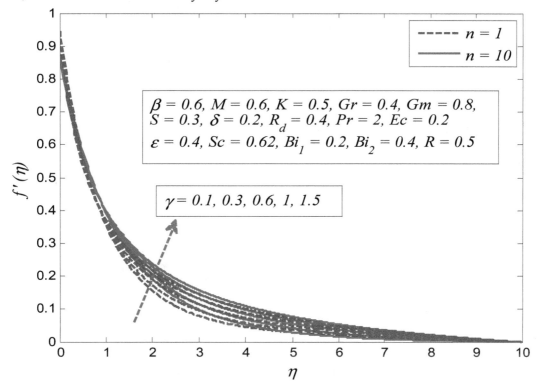

Figure 2. Effect of curvature parameter γ on velocity for linear $(n = 1)$ and nonlinear $n = 10$ stretching parameter.

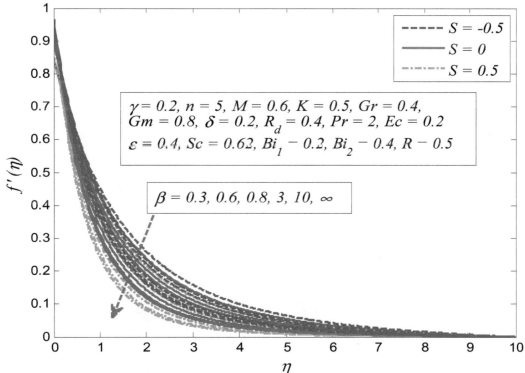

Figure 3. Effect of Casson fluid parameter β on velocity profile for different values of suction/blowing parameter S.

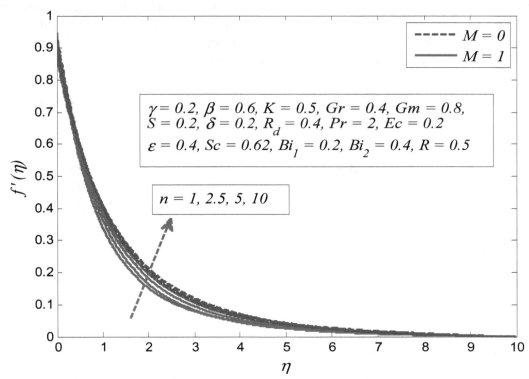

Figure 4. Effect of nonlinear stretching parameter n on velocity profile in the presence and absence of magnetic parameter M.

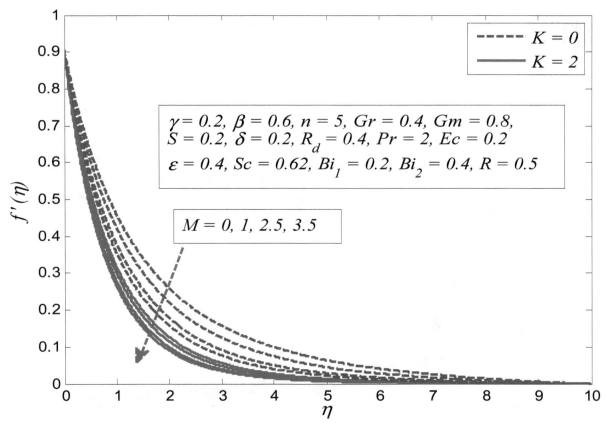

Figure 5. Effect of magnetic parameter M on velocity profile in the presence and absence of porosity parameter K.

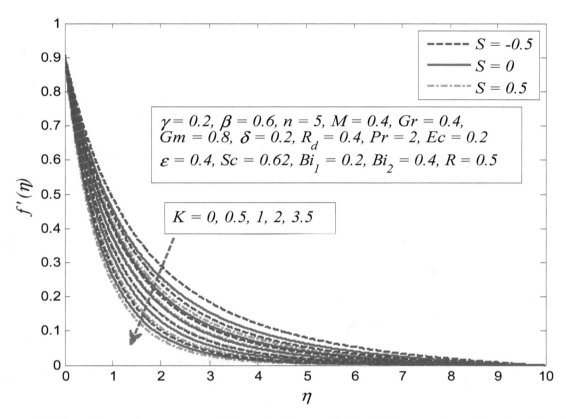

Figure 6. Effect of porosity parameter K on velocity profile for different values of suction/blowing parameter S.

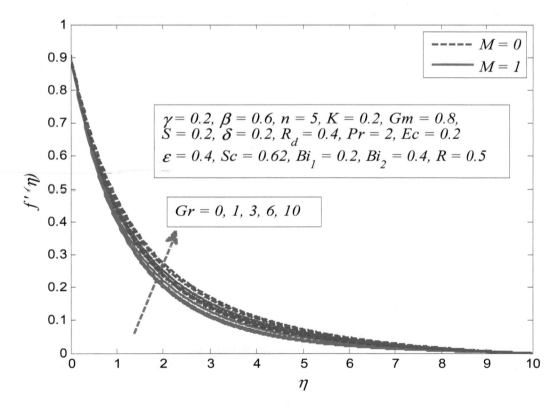

Figure 7. Effect of Grashof number Gr on velocity profile in the presence and absence of magnetic parameter M.

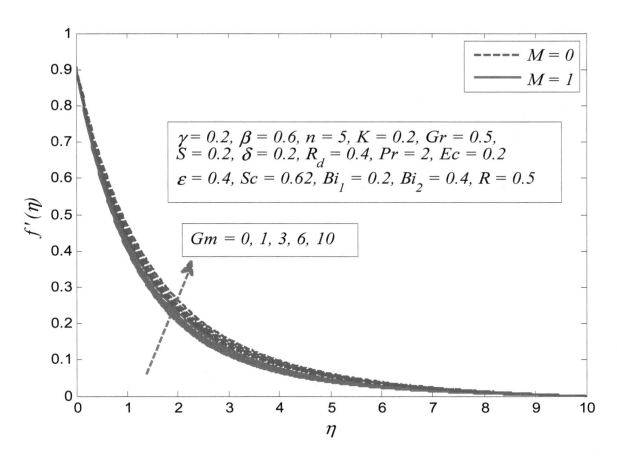

Figure 8. Effect of mass Grashof number Gm on velocity profile in the presence and absence of magnetic parameter M.

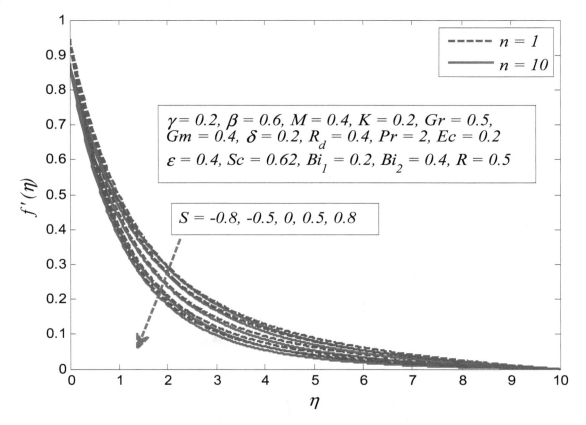

Figure 9. Effect of suction/blowing parameter S on velocity profile for nonlinear stretching parameter n.

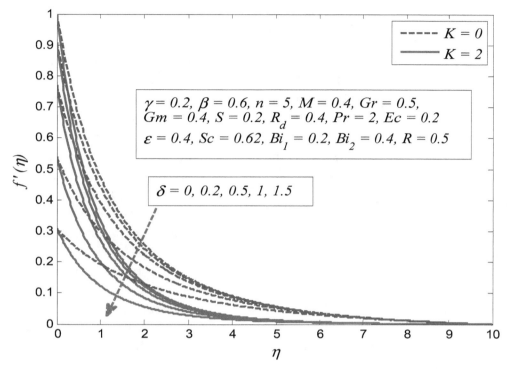

Figure 10. Effect of slip parameter δ on velocity profile in the presence and absence of porosity parameter K.

Figures 11–21 are plotted to get insight on the variation of γ, β, n, M, K, S, Pr, R_d, Ec, ε, and Bi_1 on the temperature profile. Figure 11 illustrates the variation of γ on dimensionless temperature profile for $\beta \to \infty$ (Newtonian fluid) and $\beta = 0.6$ (Casson fluid). It is noticed that temperature rises with increment in γ. A thermal boundary layer thickness is also noted. Figure 12 displays the influence of β on temperature profile for various values of S. It is noticeable that fluid temperature declines with the increase in β for all the three cases of S. The reason is that increase in β implies a reduction in yield stress, and consequently the thickness of the thermal boundary layer reduces. The effect of n on temperature profile for $M = 0$ and $M \neq 0$ is examined in Figure 13. It is clear from this figure that temperature is a decreasing function of n. It is also noticed that the fluid temperature thermal boundary layer is thicker for a linear stretching cylinder ($n = 1$) as compared to nonlinear stretching of the cylinder ($n \neq 1$). Figure 14 shows the variation of M on temperature profile for different values of S. It is noticeable that stronger magnetic field rises the fluid temperature in the vicinity of stretching cylinder. Because increasing M enhances the Lorentz force, this force makes the thermal boundary layer thicker. The same kind of behavior is noticed for the effect of K on dimensionless temperature profile for $\delta = 0$ and $\delta \neq 0$, as presented in Figure 15.

Figure 16 reveals the influence of S on temperature profile for $n = 1$ (linear stretching) and $n \neq 1$ (nonlinear stretching). Clearly, fluid temperature falls when $S > 0$, whereas it rises when $S < 0$. Since the wall suction offers resistance to fluid flow, the thermal boundary layer becomes thinner, and the opposite occurs when $S < 0$. The variation of Pr on dimensionless temperature profile for $Ec = 0$ and $Ec = 0.2$ is depicted in Figure 17. The Prandtl number is defined as the ratio of momentum diffusivity to thermal diffusivity. As expected, fluid temperature drops with the growth of Pr. It is a well-known fact that higher thermal conductivities are associated with lower Prandtl fluids, therefore heat diffuses quickly from the surface as compared to higher Prandtl fluids. Thus, Pr can be utilized to control the rate of cooling in conducting flows. Figure 18 exhibits the effect of R_d on the temperature profile for different values of S. It is noticeable that the strength of R_d boosts the temperature. The larger surface heat flux corresponds to larger values of R_d, causing the fluid to be warmer.

Figure 19 illustrates the influence of Ec on the temperature profile for $K = 0$ and $K \neq 0$. It is noted that the temperature is higher for higher values of Ec. Physically this is true, because viscous dissipation generates heat energy due to friction between fluid particles and thereby thickens the thermal boundary layer structure. It is also observed from this figure that in the presence of porous medium, the strength of Ec effectively enhances the fluid temperature. The influence of ε on temperature profile for $M = 0$ and $M \neq 0$ is displayed in Figure 20. It is clear from this graph that the temperature is enhanced when $\varepsilon > 0$ (heat generation), whereas the opposite trend is observed when $\varepsilon < 0$ (heat absorption). Internal heat generation causes the heat energy to be enhanced. Consequently, the heat transfer rate rises and thickens the thermal boundary layer. Besides, the heat absorption causes a reverse effect, i.e. the heat transfer rate and the thermal boundary layer thickness are reduced. Figure 21 reveals the variation of Bi_1 on the dimensionless temperature profile for $K = 0$ and $K \neq 0$. The Biot number is the ratio of the internal thermal resistance of a solid to the boundary layer thermal resistance. It is noticed that fluid temperature is higher for larger values of Bi_1. The reason is that increment in Bi_1 keeps the convection heat transfer higher and the cylinder thermal resistance lower. It is worth mentioning here that when $Bi_1 < 0.1$, the internal resistance to heat transfer is negligible, representing that the value of k is much larger than h_0, and the internal thermal resistance is noticeably lower than the surface resistance. On the other hand, when $Bi_1 \rightarrow \infty$ the higher Biot number intends that the external resistance to heat transfer reduces, indicating that the surface and the surroundings temperature difference is minor and a noteworthy contribution of temperature to the center comes from the surface of the stretching cylinder.

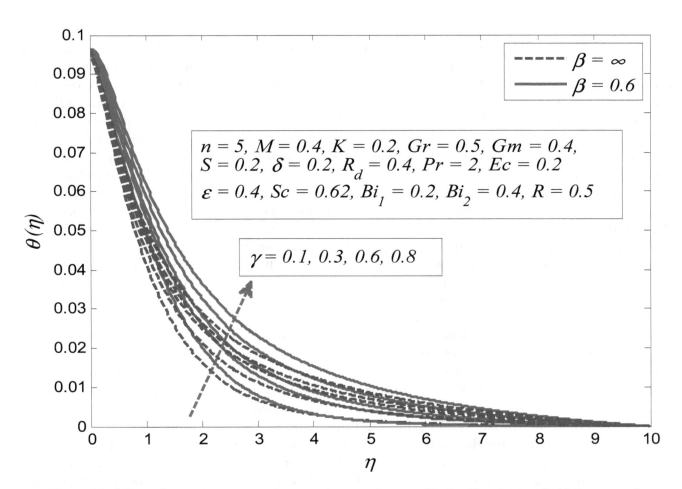

Figure 11. Effect of curvature parameter γ on temperature profile for Newtonian fluid $\beta = \infty$ and Casson fluid $\beta = 0.6$.

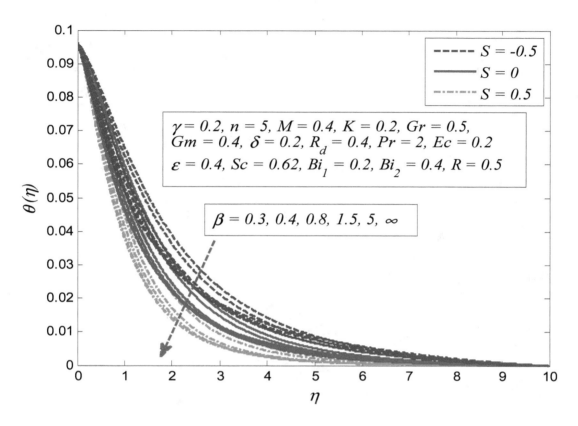

Figure 12.   Effect of Casson fluid parameter β on temperature profile for different values of suction/blowing parameter S.

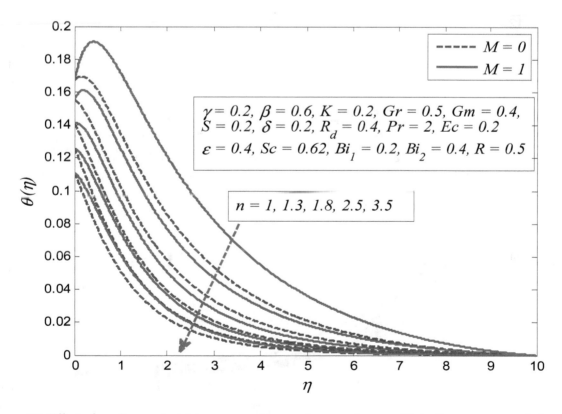

Figure 13. Effect of nonlinear stretching parameter n on temperature profile in the presence and absence of magnetic parameter M.

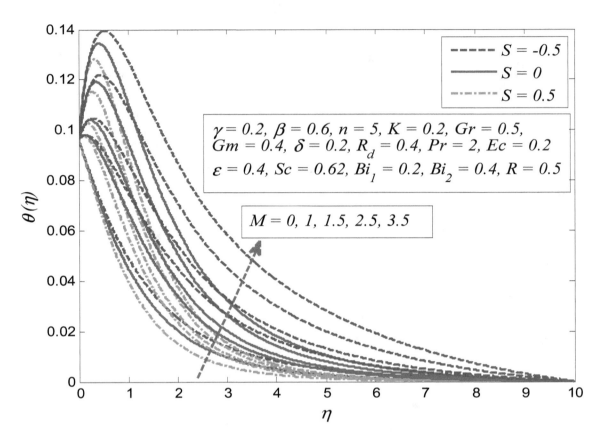

Figure 14. Effect of magnetic parameter M on temperature profile for different values of suction/blowing parameter S.

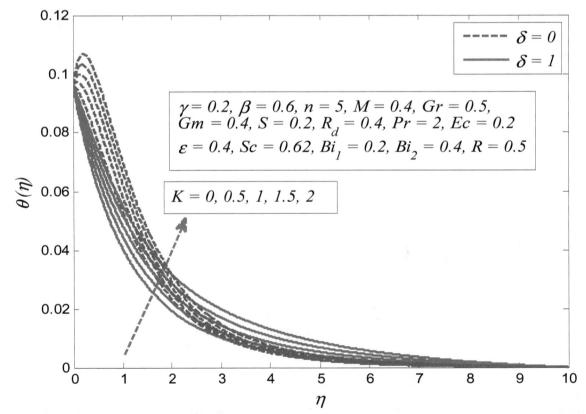

Figure 15. Effect of porosity parameter K on temperature profile in the presence and absence of slip parameter δ.

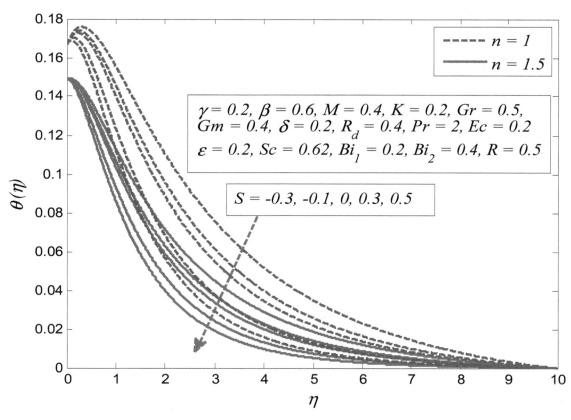

Figure 16. Effect of suction/blowing parameter S on temperature profile for different values of nonlinear stretching parameter n.

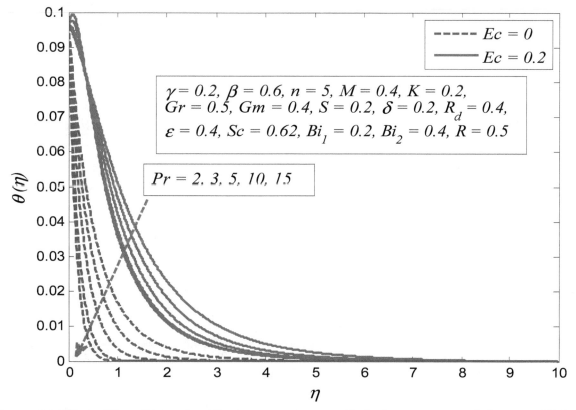

Figure 17. Effect of Prandtl number Pr on temperature profile in the presence and absence of Eckert number Ec.

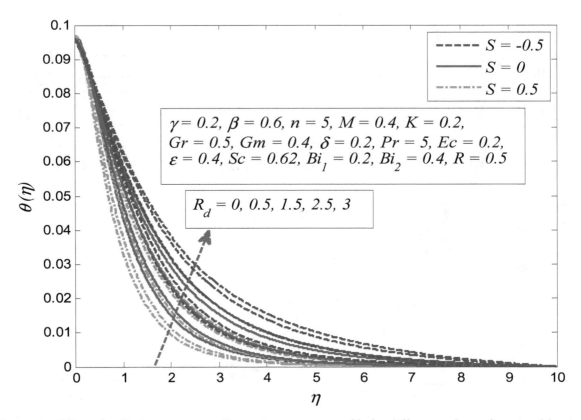

Figure 18. Effect of radiation parameter R_d on temperature profile for different values of suction/blowing parameter S.

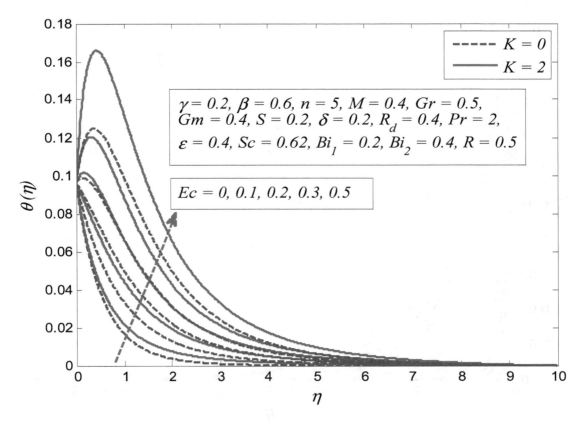

Figure 19. Effect of Eckert number Ec on temperature profile in the presence and absence of porosity parameter K.

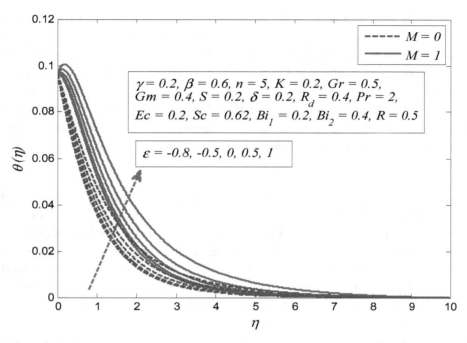

Figure 20. Effect of heat generation/absorption parameter ε on temperature profile in the presence and absence of magnetic parameter M.

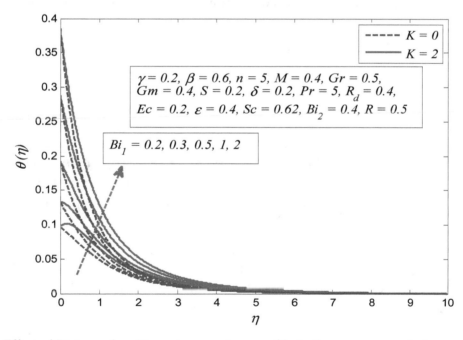

Figure 21. Effect of Biot number Bi_1 on temperature profile in the presence and absence of porosity parameter K.

Figures 22–31 display the variation of γ, β, n, M, K, δ, S, Sc, R, and Bi_1 on concentration profile, respectively. Figure 22 elucidates the effect of γ on concentration profile for $\beta \to \infty$ (Newtonian fluid) and $\beta = 0.6$ (Casson fluid). It is found that increasing values of γ enhances the fluid concentration and associated boundary layer thickness. Figure 23 demonstrates the influence of β on concentration profile for $M = 0$ and $M \neq 0$. It is noted that fluid concentration is higher as β grows. The viscosity of the fluid increases with increasing β, in which case the concentration rises and the concentration boundary layer becomes thicker. The opposite behavior is noticed for the effect of n on concentration profile for various values of S (see Figure 24). It is also observed that thickness of concentration boundary layer shortens for large n. Figure 25 determines the variation of M on the dimensionless concentration profile for $K = 0$ and $K \neq 0$. It is seen that fluid concentration is higher for higher values of M. As mentioned earlier for velocity and temperature profiles, fluid motion reduces due to magnetic field and results in an enhancement in thermal and concentration boundary layer thicknesses. A similar trend is observed for the effect of K and δ on the concentration profile, as plotted in Figures 26 and 27, respectively. The growth of both parameters offers resistance to the fluid particles and the concentration boundary layer becomes thicker. Figure 28 shows that fluid concentration reduces when $S > 0$, while it is enhanced when $S < 0$. Indeed, when mass suction occurs, some of the fluid is sucked through the wall which thins the boundary layer; on the contrary, blowing thickens the concentration boundary layer structure.

Figure 29 examines the variation of Sc (Sc = 0.30, 0.62, 0.78, 0.94, 2.57 corresponds to hydrogen, helium, water vapor, hydrogen sulphide, and propyl Benzene) on the dimensionless concentration profile when $\beta \rightarrow \infty$ (Newtonian fluid) and $\beta = 0.6$ (Casson fluid). For both fluids, an increase in Sc reduces the fluid concentration. Since higher values of Sc lead to higher mass transfer rate, the thickness of the concentration boundary layer declines. The effect of R on the concentration distribution for different values of S is depicted in Figure 30. It is clear that fluid concentration drops with the growth of R. Physically this makes sense, because the decomposition rate of reactant species enhances in the destructive chemical reaction ($R > 0$). Consequently, the mass transfer rate grows and thickens the concentration boundary layer. Figure 31 exhibits the variation of Bi_2 on concentration distribution for $M = 0$ and $M \neq 0$. It is noticeable that fluid concentration rises with increasing Bi_2. As increase in Biot number enhances the temperature field, the concentration field excites, making the solutal boundary layer thicker.

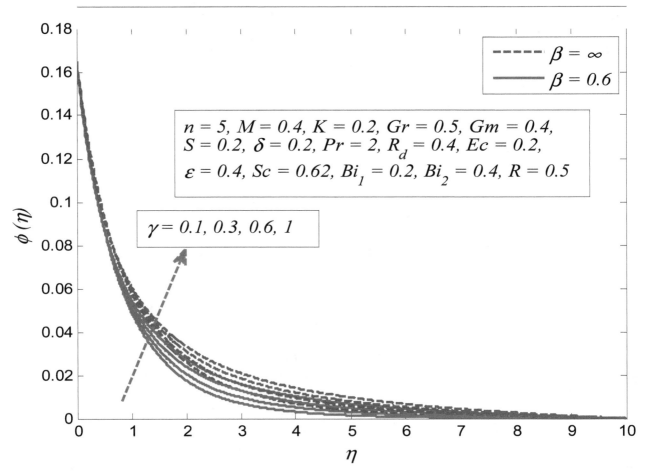

Figure 22. Effect of curvature parameter γ on concentration profile for Newtonian fluid $\beta = \infty$ and Casson fluid $\beta = 0.6$.

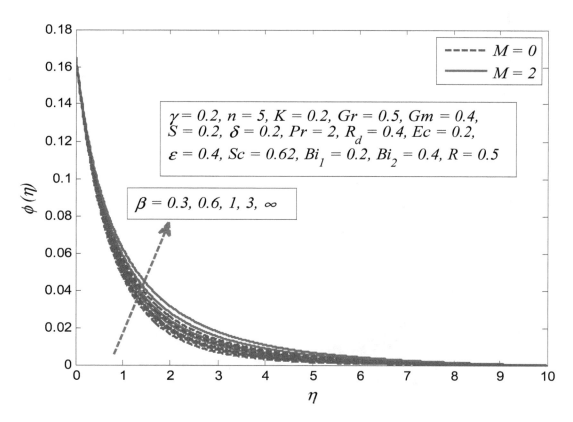

Figure 23. Effect of Casson parameter β on concentration profile in the presence and absence of magnetic parameter M.

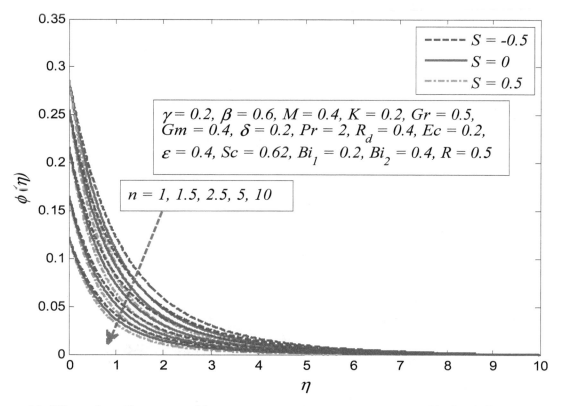

Figure 24. Effect of nonlinear stretching parameter n on concentration profile for different values of suction/blowing parameter S.

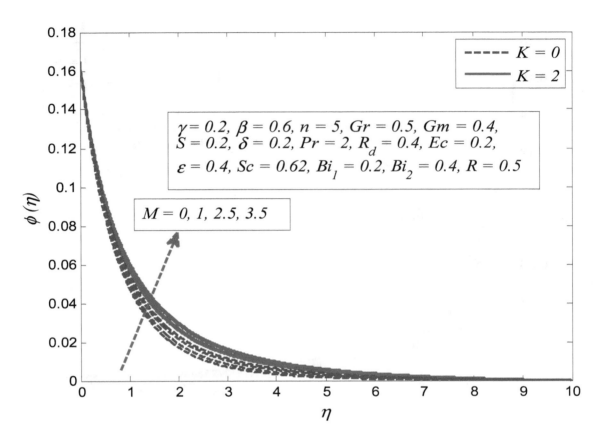

Figure 25. Effect of magnetic parameter M on concentration profile in the presence and absence of porosity parameter K.

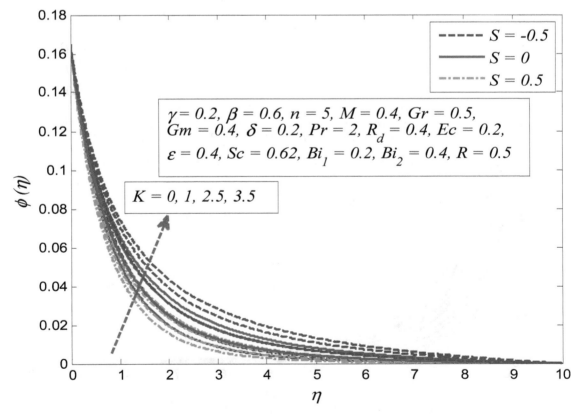

Figure 26. Effect of porosity parameter K on concentration profile for different values of suction/blowing parameter S.

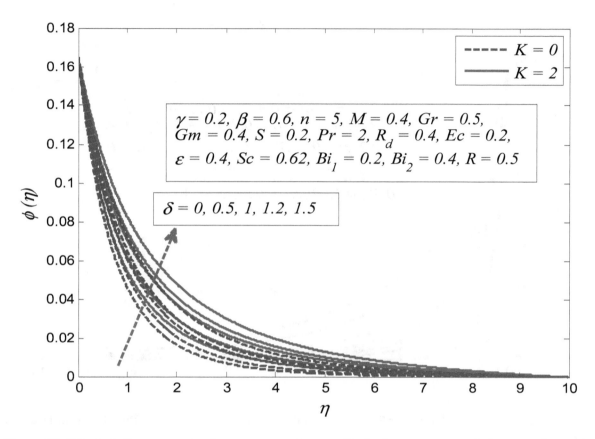

Figure 27. Effect of slip parameter δ on concentration profile in the presence and absence of porosity parameter K.

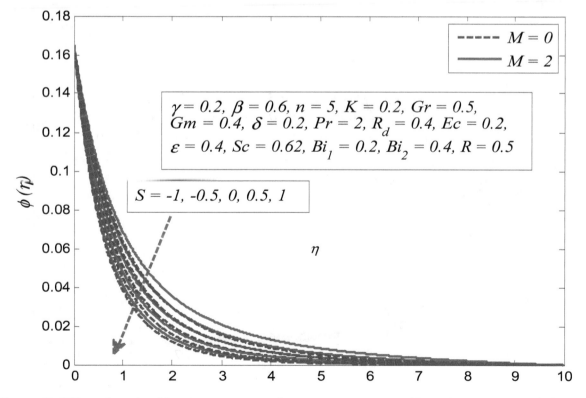

Figure 28. Effect of suction/blowing parameter S on concentration profile in the presence and absence of magnetic parameter M.

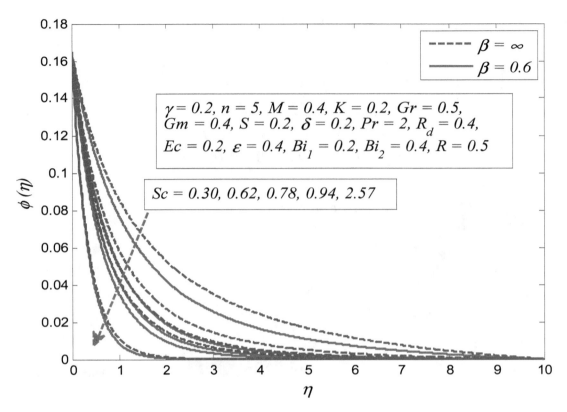

Figure 29. Effect of Schmidt number Sc on concentration profile for Newtonian fluid $\beta = \infty$ and Casson fluid $\beta = 0.6$.

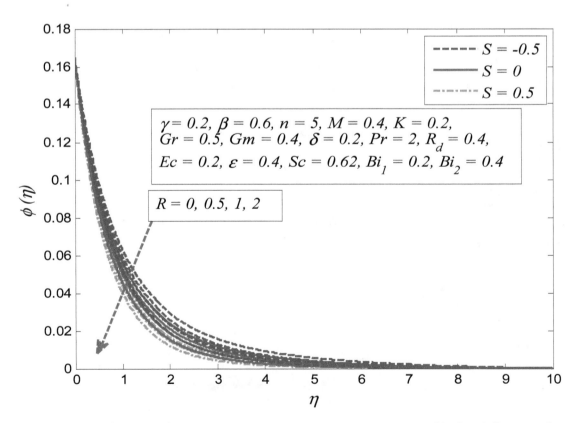

Figure 30. Effect of chemical reaction parameter R on concentration profile for different values of suction/blowing parameter S.

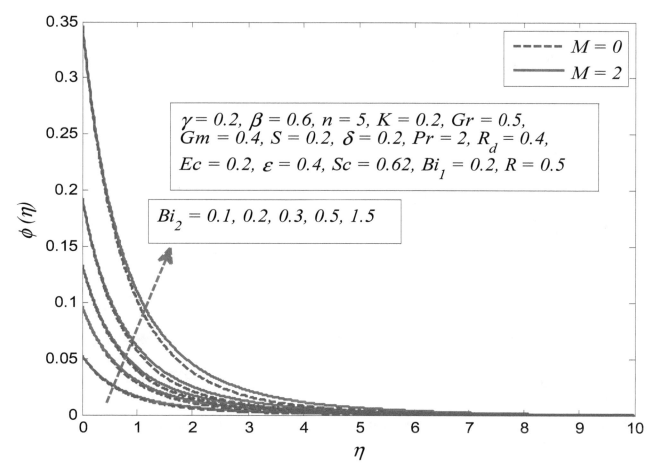

Figure 31. Effect of Biot number Bi_1 on concentration profile in the presence and absence of magnetic parameter M.

Figures 32–35 depict the effect of the skin friction coefficient, Nusselt number, and Sherwood number for different values of γ, β, M, K, n, S, Ec, and R, respectively. Figure 32 reveals the variation of wall shear stress for various values of γ, β, and M. It is noted that the absolute values of wall shear stress increase as γ and M increase, whereas the opposite is observed for the effect of β. It is also noticeable that the values of friction factor are negative, which shows that the stretching cylinder experiences a drag force from the fluid particles. Moreover, the effect of γ on wall shear stress is more pronounced for Casson fluid. The effect of K, n, and S on dimensionless skin friction coefficient is examined in Figure 33. This figure shows that friction factor absolute values decline as K, n, and S increase. Figure 34 portrays the variation of Nusselt number for various values of γ, β, and Ec. It is shown that heat transfer rate drops as γ and Ec increase, whereas they increase for larger values of β. However, the heat transfer rate is more influenced for Casson fluid. It is also noted that heat transfer rate is negative for higher values of Ec. These negative values show that heat is transferred from the working fluid to the stretching surface. Finally, the effect of Sherwood number for various γ, β, and R is illustrated in Figure 35. It is found that the mass transfer rate is an increasing function of γ and R and a decreasing function of β.

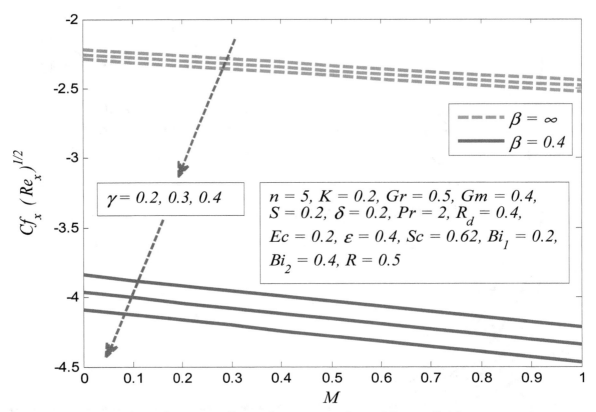

Figure 32. Variation of skin friction coefficient for various values of Casson fluid parameter β, curvature parameter γ, and magnetic parameter M.

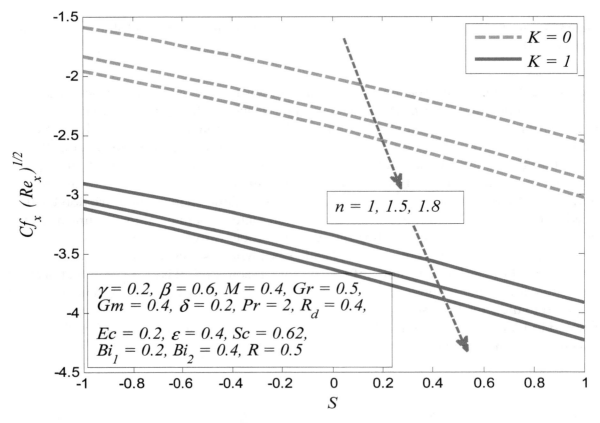

Figure 33. Variation of skin friction coefficient for various values of nonlinear stretching parameter n, porosity parameter K, and suction/blowing parameter S.

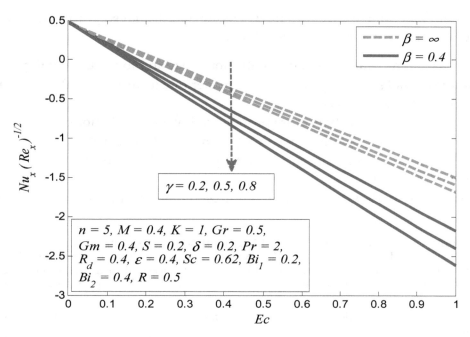

Figure 34. Variation of Nusselt number for various values of Casson parameter β, curvature parameter γ, and Eckert number Ec.

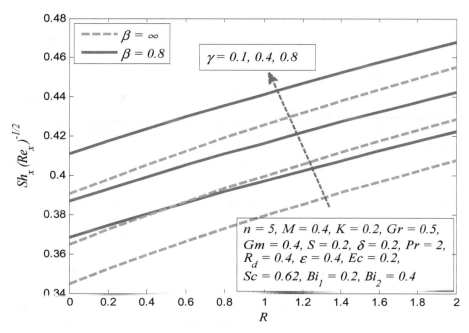

Figure 35. Variation of Sherwood number for various values of Casson fluid parameter β, curvature parameter γ, and chemical reaction parameter R.

4. Conclusions

In the present study, the influence of chemical reaction on MHD slip flow of Casson fluid due to nonlinear cylinder stretching was investigated numerically. Similarity solutions for velocity, temperature, and concentration distributions are achieved via the Keller box method. The numerical results of wall shear stress and heat transfer rate are also compared as a limiting case. The effect of physical parameters, namely, unsteadiness parameter γ, Casson parameter β, nonlinear stretching parameter n, magnetic parameter M, porosity parameter K, thermal Grashof number Gr, mass Grashof number Gm, Prandtl number Pr, radiation parameter R_d, Eckert number Ec, heat generation/absorption parameter ε, Schmidt number Sc, chemical reaction parameter R, suction/blowing parameter S, slip parameter δ, and Biot numbers Bi_1, Bi_2 are discussed and displayed graphically. Some interesting observations from the present analysis are as follow:

1. The fluid velocity, temperature, and concentration are found to increase with γ.
2. The magnitude of wall shear stress and mass transfer rate increase with the growth of β, whereas the heat transfer rate is enhanced.
3. The effect of M on fluid velocity is more pronounced when $K = 0$ (nonporous medium).
4. The temperature field is more influenced with increasing Ec when $K \neq 0$.
5. The velocity, temperature, and concentration distributions decrease when $S > 0$, while the reverse trend is seen when $S < 0$.
6. The concentration boundary layer is observed to be thinner during destructive chemical reaction.

Author Contributions: Conceptualization, I.U. and I.K.; methodology, I.U. and I.K.; software, I.U.; validation, I.U., I.K. and S.S.; formal analysis, I.U., O.D.M., T.A.A., S.S. and K.S.N.; investigation, I.U., T.A.A., K.S.N. and O.D.M; resources, I.U., I.K., and S.S; data curation, I.U.; writing—original draft preparation, I.U. and O.D.M.; writing—review and editing, I.U.; visualization, T.A.A., K.S.N., I.U. and I.K.; supervision, I.U. and S.S.

Acknowledgments: The authors would like to thank the reviewers and editor for their constructive and insightful comments in relation to this work.

References

1. Crane, L.J. Flow past a stretching plate. *Z. Angew. Math. Phys.* **1970**, *21*, 645–647. [CrossRef]
2. Gupta, P.S.; Gupta, A.S. Heat and mass transfer on a stretching sheet with suction or blowing. *Can. J. Chem. Eng.* **1977**, *55*, 744–746. [CrossRef]
3. Chen, C.K.; Char, M.I. Heat transfer of a continuous, stretching surface with suction or blowing. *J. Math. Anal. Appl.* **1988**, *135*, 568–580. [CrossRef]
4. Gorla, I.; Sidawi, R.S.R. Free Convection on a Vertical Stretching Surface with Suction and Blowing. *Appl. Sci. Res.* **1994**, *52*, 247–257. [CrossRef]
5. Vajravelu, K. Viscous flow over a nonlinearly stretching sheet. *Appl. Math. Comput.* **2001**, *124*, 281–288. [CrossRef]
6. Vajravelu, K.; Cannon, J.R. Fluid flow over a nonlinearly stretching sheet. *Appl. Math. Comput.* **2006**, *181*, 609–618. [CrossRef]
7. Bachok, N.; Ishak, A. Flow and heat transfer over a stretching cylinder with prescribed surface heat flux. *Malays. J. Math. Sci.* **2010**, *4*, 159–169.
8. Hayat, T.; Saeed, Y.; Asad, S.; Alsaedi, A. Convective heat and mass transfer in flow by an inclined stretching cylinder. *J. Mol. Liq.* **2016**, *220*, 573–580. [CrossRef]
9. Majeed, A.; Zeeshan, A.; Alamri, S.Z.; Ellahi, R. Heat transfer analysis in ferromagnetic viscoelastic fluid flow over a stretching sheet with suction. *Neur. Comp. Appl.* **2018**, *30*, 1947–1955. [CrossRef]
10. Vyas, A.; Ranjan, P. Dissipative mhd boundary layer flow in a porous medium over a sheet stretching nonlinearly in the presence of radiation. *Appl. Math. Sci.* **2010**, *4*, 3133–3142.
11. Mukhopadhyay, S. MHD boundary layer flow along a stretching cylinder. *Ain Shams Eng. J.* **2013**, *4*, 317–324. [CrossRef]
12. Fathizadeh, M.; Madani, M.; Khan, Y.; Faraz, N.; Yildirim, A.; Tutkun, S. An effective modification of the homotopy perturbation method for MHD viscous flow over a stretching sheet. *J. King Saud Univ. Sci.* **2013**, *25*, 107–113. [CrossRef]
13. Akbar, N.S.; Ebaid, A.; Khan, Z.H. Numerical analysis of magnetic field effects on Eyring-Powell fluid flow towards a stretching sheet. *J. Magn. Magn. Mater.* **2015**, *382*, 355–358. [CrossRef]
14. Ellahi, R. The effects of MHD and temperature dependent viscosity on the flow of non-Newtonian nanofluid in a pipe: Analytical solutions. *Appl. Math. Model.* **2013**, *37*, 1451–1467. [CrossRef]
15. Fang, T.; Zhang, J.; Yao, S. Slip MHD viscous flow over a stretching sheet—An exact solution. *Commun. Nonlinear Sci. Numer. Simul.* **2009**, *14*, 3731–3737. [CrossRef]

16. Bhattacharyya, K.; Mukhopadhyay, S.; Layek, G.C. Slip effects on an unsteady boundary layer stagnation-point flow and heat transfer towards a stretching sheet. *Chin. Phys. Lett.* **2011**, *28*, 094702. [CrossRef]

17. Yazdi, M.H.; Abdullah, S.; Hashim, I.; Sopian, K. Slip MHD liquid flow and heat transfer over non-linear permeable stretching surface with chemical reaction. *Int. J. Heat Mass Transf.* **2011**, *54*, 3214–3225. [CrossRef]

18. Hayat, T.; Qasim, M.; Mesloub, S. MHD flow and heat transfer over permeable stretching sheet with slip conditions. *Int. J. Numer. Methods Fluids* **2011**, *66*, 963–975. [CrossRef]

19. Seini, I.Y.; Makinde, O.D. Boundary layer flow near stagnation-points on a vertical surface with slip in the presence of transverse magnetic field. *Int. J. Numer. Methods Heat Fluid Flow* **2014**, *24*, 643–653. [CrossRef]

20. Rahman, S.U.; Ellahi, R.; Nadeem, S.; Zia, Q.Z. Simultaneous effects of nanoparticles and slip on Jeffrey fluid through tapered artery with mild stenosis. *J. Mol. Liq.* **2016**, *218*, 484–493. [CrossRef]

21. Casson, N. *A Flow Equation for Pigment-Oil Suspensions of the Printing Ink Type*; Mill, C.C., Ed.; Rheol. Disperse Syst. Pergamon Press: Oxford, UK, 1959; pp. 84–104.

22. Nandy, S.K. Analytical Solution of MHD Stagnation-Point Flow and Heat Transfer of Casson Fluid over a Stretching Sheet with. *Thermodynamics* **2013**. [CrossRef]

23. Singh, S. Clinical significance of aspirin on blood flow through stenotic blood vessels. *J. Biomim. Biomater. Tissue Eng.* **2011**, *10*, 17–24.

24. Mukhopadhyay, S.; Bhattacharyya, K.; Hayat, T. Exact solutions for the flow of Casson fluid over a stretching surface with transpiration and heat transfer effects. *Chin. Phys. B* **2013**, *22*, 114701. [CrossRef]

25. Shawky, H.M. Magnetohydrodynamic Casson fluid flow with heat and mass transfer through a porous medium over a stretching sheet. *J. Porous Media* **2012**, *15*, 393–401. [CrossRef]

26. Mukhopadhyay, S. Casson fluid flow and heat transfer over a nonlinearly stretching surface. *Chin. Phys. B* **2013**, *22*, 074701. [CrossRef]

27. Medikare, M.; Joga, S.; Chidem, K.K. MHD Stagnation Point Flow of a Casson Fluid over a Nonlinearly Stretching Sheet with Viscous Dissipation. *Am. J. Comput. Math.* **2016**, 37–48. [CrossRef]

28. Mythili, D.; Sivaraj, R. Influence of higher order chemical reaction and non-uniform heat source/sink on Casson fluid flow over a vertical cone and flat plate. *J. Mol. Liq.* **2016**, *216*, 466–475. [CrossRef]

29. Ullah, I.; Khan, I.; Shafie, S. Hydromagnetic Falkner-Skan flow of Casson fluid past a moving wedge with heat transfer. *Alex. Eng. J.* **2016**, *55*, 2139–2148. [CrossRef]

30. Imtiaz, M.; Hayat, T.; Alsaedi, A. Mixed convection flow of Casson nanofluid over a stretching cylinder with convective boundary conditions. *Adv. Powder Technol.* **2016**, *27*, 2245–2256. [CrossRef]

31. Cebeci, T.; Bradshaw, P. *Physical and Computational Aspects of Convective Heat Transfer*, 1st ed.; Springer: New York, NY, USA, 1988.

12

Numerical Solution of Non-Newtonian Fluid Flow Due to Rotatory Rigid Disk

Khalil Ur Rehman [1,*], **M. Y. Malik** [2], **Waqar A Khan** [3], **Ilyas Khan** [4] **and S. O. Alharbi** [4]

[1]   Department of Mathematics, Air University, PAF Complex E-9, Islamabad 44000, Pakistan
[2]   Department of Mathematics, College of Sciences, King Khalid University, Abha 61413, Saudi Arabia; drmymalik@qau.edu.pk
[3]   Department of Mechanical Engineering, College of Engineering, Prince Mohammad Bin Fahd University, Al Khobar 31952, Kingdom of Saudi Arabia; wkhan@pmu.edu.sa
[4]   Department of Mathematics, College of Science Al-Zulfi, Majmaah University, Al-Majmaah 11952, Saudi Arabia; i.said@mu.edu.sa (I.K.); so.alharbi@mu.edu.sa (S.O.A.)
*   Correspondence: krehman@math.qau.edu.pk

Abstract: In this article, the non-Newtonian fluid model named Casson fluid is considered. The semi-infinite domain of disk is fitted out with magnetized Casson liquid. The role of both thermophoresis and Brownian motion is inspected by considering nanosized particles in a Casson liquid spaced above the rotating disk. The magnetized flow field is framed with Navier's slip assumption. The Von Karman scheme is adopted to transform flow narrating equations in terms of reduced system. For better depiction a self-coded computational algorithm is executed rather than to move-on with build-in array. Numerical observations via magnetic, Lewis numbers, Casson, slip, Brownian motion, and thermophoresis parameters subject to radial, tangential velocities, temperature, and nanoparticles concentration are reported. The validation of numerical method being used is given through comparison with existing work. Comparative values of local Nusselt number and local Sherwood number are provided for involved flow controlling parameters.

Keywords: Casson fluid model; rotating rigid disk; nanoparticles; Magnetohydrodynamics (MHD)

1. Introduction

The examination of non-Newtonian fluids has received remarkable attention from researchers and scientists because of their extensive use in industrial and technological areas. For instance, paints, synthetic lubricants, sugar solutions, certain oils, clay coating, drilling muds, and blood as a biological fluid are common examples of non-Newtonian fluids, just to mention a few. The fundamental mathematical equations given by Navier–Stokes cannot briefly delineate the flow field characteristics of non-Newtonian fluids because of the complex mathematical expression involved in the formulation of flow problem. In addition, the relation between strain rate and shear stress is non-linear so the single constitutive expressions are fruitless to report complete description of flows subject to non-Newtonian fluids. Numerous non-Newtonian fluid models are exposed to explore rheological characteristics, namely Bingham Herschel–Bulkley fluid model, Seely, Carreau Carreau–Yasuda, Sisko, Eyring, Cross, Ellis, Williamson, tangent hyperbolic, Generalized Burgers, Burgers, Oldroyd-8 constants, Oldroyd-A, Oldroyd-B fluid model, Maxwell, Jeffrey, Casson fluid model, etc. Researchers discussed flow characteristics of non-Newtonian fluid models via stretching surfaces by incorporating pertinent physical effects. Among these, Casson fluid model has many advantages as compared to rest of fluid models. This model can be used to approximate the properties of blood and daily life suspensions. One can assessed recent developments in this direction in References [1–15].

The centrifugal filtration, gas turbine rotors, rotating air cleaning machines, food processing, medical equipment, system of electric-power generation, crystal growth processes, and many others are the practical applications of rotational fluids flow. Therefore, analysis of flows due to rotation of solid surfaces is widely recognized by scientists, and researchers like Karman [16] firstly report viscous fluid flow induced by rotating solid disk. A special transformation named as Karman transformation given by him for the first time in this attempt. These transformations are utilized for conversion of fundamental equations termed as Naviers–Stokes equations in terms of ordinary differential system. Later on, a number of studies were given by researchers to depict the flow characteristics of both Newtonian and non-Newtonian fluids model over a rotating disk. Preceding these analyses in 2013, the extension of Karman problem was given by Turkyilmazoglu and Senel [17]. In this attempt they discussed numerical results for heat transfer properties of rotating partial slip fluid flow. In 2014, the magnetized slip flow via porous disk was reported by Rashidi et al. [18]. In addition, they discussed entropy measurements for this case. The flow properties in the presence of nano-size particles were discussed by Turkyilmazoglu [19]. He used numerical algorithm for solution purpose. In fact, he dealt comparative execution to report the impact of various nanoparticles suspended in fluid flow regime. Afterwards, tremendous attempts are given in this direction by way of both analytical and numerical approach. One can find the concern developments on rotating flows in References [20–31].

The present article contains analysis of Casson liquid towards rotating rigid disk. The Casson flow field is magnetized and has nanoparticles. Further, slip effects are also taken into account. The physical model is translated in terms of mathematical model. For solution purposes, the van Karman way of study is adopted. A computational algorithm is applied and the obtained results of involved parameters of concerned quantities are discussed via graphs and tables. Further, the current attempt is compared with existing literature and we found a good agreement which leads to the surety of the present work.

2. Problem Formulation

The Casson liquid is quipped above the disk for $\bar{z} > 0$. The constant frequency $(\bar{\Omega})$ is constant. The semi bounded magnetized flow regime contains suspended nanoparticles. The surface is taken with velocity slip condition. The quantities $(\bar{u}, \bar{v}, \bar{w})$ are in $(\bar{r}, \bar{\phi}, \bar{z})$ directions. The ultimate differential system of said problem is:

$$\frac{\partial \bar{w}}{\partial \bar{z}} + \frac{\partial \bar{u}}{\partial \bar{r}} + \frac{\bar{u}}{\bar{r}} = 0, \tag{1}$$

$$\bar{w}\frac{\partial \bar{u}}{\partial \bar{z}} + \bar{u}\frac{\partial \bar{u}}{\partial \bar{r}} - \frac{\bar{v}^2}{\bar{r}} = \nu\left(1 + \frac{1}{\lambda}\right)\left(\frac{\partial^2 \bar{u}}{\partial \bar{z}^2} + \frac{1}{\bar{r}}\frac{\partial \bar{u}}{\partial \bar{r}} + \frac{\partial^2 \bar{u}}{\partial \bar{r}^2} - \frac{\bar{u}}{\bar{r}^2}\right) - \frac{\sigma B_0^2}{\rho_f}\bar{u}, \tag{2}$$

$$\bar{w}\frac{\partial \bar{v}}{\partial \bar{z}} + \bar{u}\frac{\partial \bar{v}}{\partial \bar{r}} + \frac{\bar{u}\bar{v}}{\bar{r}} = \nu\left(1 + \frac{1}{\lambda}\right)\left(\frac{\partial^2 \bar{v}}{\partial \bar{z}^2} + \frac{\partial^2 \bar{v}}{\partial \bar{r}^2} - \frac{\bar{v}}{\bar{r}^2} + \frac{1}{\bar{r}}\frac{\partial \bar{v}}{\partial \bar{r}}\right) - \frac{\sigma B_0^2}{\rho_f}\bar{v}, \tag{3}$$

$$\bar{w}\frac{\partial \bar{w}}{\partial \bar{z}} + \bar{u}\frac{\partial \bar{w}}{\partial \bar{r}} = \nu\left(1 + \frac{1}{\lambda}\right)\left(\frac{\partial^2 \bar{w}}{\partial \bar{z}^2} + \frac{1}{\bar{r}}\frac{\partial \bar{w}}{\partial \bar{r}} + \frac{\partial^2 \bar{w}}{\partial \bar{r}^2}\right), \tag{4}$$

$$\bar{w}\frac{\partial \bar{T}}{\partial \bar{z}} + \bar{u}\frac{\partial \bar{T}}{\partial \bar{r}} = \alpha\left(\frac{\partial^2 \bar{T}}{\partial \bar{z}^2} + \frac{1}{\bar{r}}\frac{\partial \bar{T}}{\partial \bar{r}} + \frac{\partial^2 \bar{T}}{\partial \bar{r}^2}\right) + \frac{(\rho c)_p}{(\rho c)_f}\left[D_B\left(\frac{\partial \bar{T}}{\partial \bar{z}}\frac{\partial \bar{C}}{\partial \bar{z}} + \frac{\partial \bar{T}}{\partial \bar{r}}\frac{\partial \bar{C}}{\partial \bar{r}}\right)\right]$$

$$+ \frac{(\rho c)_p}{(\rho c)_f}\left[\frac{D_T}{\bar{T}_\infty}\left(\left(\frac{\partial \bar{T}}{\partial \bar{z}}\right)^2 + \left(\frac{\partial \bar{T}}{\partial \bar{r}}\right)^2\right)\right], \tag{5}$$

$$\bar{w}\frac{\partial \bar{C}}{\partial \bar{z}} + \bar{u}\frac{\partial \bar{C}}{\partial \bar{r}} = D_B\left(\frac{\partial^2 \bar{C}}{\partial \bar{z}^2} + \frac{1}{\bar{r}}\frac{\partial \bar{C}}{\partial \bar{r}} + \frac{\partial^2 \bar{C}}{\partial \bar{r}^2}\right) + \frac{D_T}{\bar{T}_\infty}\left[\frac{\partial^2 \bar{T}}{\partial \bar{z}^2} + \frac{1}{\bar{r}}\frac{\partial \bar{T}}{\partial \bar{r}} + \frac{\partial^2 \bar{T}}{\partial \bar{r}^2}\right], \tag{6}$$

$$\bar{u} = L\frac{\partial \bar{u}}{\partial \bar{z}}, \bar{v} = \bar{r}\bar{\Omega} + L\frac{\partial \bar{v}}{\partial \bar{z}}, \bar{w} = 0, \bar{T} = \bar{T}_w, \bar{C} = \bar{C}_w \text{ at } \bar{z} = 0, \tag{7}$$

$$\bar{u} \to 0, \bar{v} \to 0, \bar{T} \to \bar{T}_\infty, \bar{C} \to \bar{C}_\infty \text{ as } \bar{z} \to \infty, \tag{8}$$

for order reduction one can use the variables [16],

$$\bar{u} = \bar{r}\overline{\Omega}\frac{dF(\xi)}{d\xi}, \bar{v} = G(\xi)\bar{r}\overline{\Omega}, \overline{w} = -F(\xi)\sqrt{2\overline{\Omega}\nu},$$

$$C(\xi) = \frac{\overline{C}-\overline{C}_\infty}{\overline{C}_w-\overline{C}_\infty}, T(\xi) = \frac{\overline{T}-\overline{T}_\infty}{\overline{T}_w-\overline{T}_\infty}, \xi = \sqrt{\frac{2\overline{\Omega}}{\nu}}\bar{z}.$$

(9)

We get:

$$2\frac{d^3F(\xi)}{d\xi^3}\left(1+\frac{1}{\lambda}\right) + 2F(\xi)\frac{d^2F(\xi)}{d\xi^2} - \left(\frac{dF(\xi)}{d\xi}\right)^2 + (G(\xi))^2 - \gamma\frac{dF(\xi)}{d\xi} = 0,$$

(10)

$$2\frac{d^2G(\xi)}{d\xi^2}\left(1+\frac{1}{\lambda}\right) + 2F(\xi)\frac{dG(\xi)}{d\xi} - 2G(\xi)\frac{dF(\xi)}{d\xi} - \gamma G(\xi) = 0,$$

(11)

$$\frac{d^2T(\xi)}{d\xi^2} + \Pr\left(F(\xi)\frac{dT(\xi)}{d\xi} + N_B\frac{dT(\xi)}{d\xi}\frac{dC(\xi)}{d\xi} + N_T\left(\frac{dT(\xi)}{d\xi}\right)^2\right) = 0,$$

(12)

$$\frac{d^2C(\xi)}{d\xi^2} + Le\Pr F(\xi)\frac{dC(\xi)}{d\xi} + \frac{N_T}{N_B}\frac{d^2T(\xi)}{d\xi^2} = 0,$$

(13)

$$F(\xi) = 0, \frac{dF(\xi)}{d\xi} = \beta\frac{d^2F(\xi)}{d\xi^2}, G(\xi) = 1 + \beta\frac{dG(\xi)}{d\xi}, T(\xi) = 1, C(\xi) = 1, \text{at } \xi = 0,$$

$$\frac{dF(\xi)}{d\xi} \to 0, G(\xi) \to 0, T(\xi) \to 0, C(\xi) \to 0, \text{as}\, \xi \to \infty.$$

(14)

and:

$$\gamma = \sqrt{\frac{\sigma B_0{}^2}{\rho_f\overline{\Omega}}}, \beta = L\sqrt{\frac{2\overline{\Omega}}{\nu}}, N_B = \frac{(\rho c)_p}{(\rho c)_f}\frac{(\overline{T}_w-\overline{T}_\infty)D_T}{\overline{T}_\infty\nu},$$

(15)

$$Le = \frac{\alpha}{D_B}, \Pr = \frac{\nu}{\alpha}, N_T = \frac{(\rho c)_p}{(\rho c)_f}\frac{(\overline{C}_w-\overline{C}_\infty)D_B}{\nu},$$

the surface quantities are defined as:

$$\sqrt{\mathrm{Re}_{\bar{r}}}C_F = \left(1+\frac{1}{\lambda}\right)\frac{d^2F(0)}{d\xi^2}, \sqrt{\mathrm{Re}_{\bar{r}}}C_G = \left(1+\frac{1}{\lambda}\right)\frac{dG(0)}{d\xi},$$

$$\frac{Nu}{\sqrt{\mathrm{Re}_{\bar{r}}}} = -\frac{dT(0)}{d\xi}, \frac{Sh}{\sqrt{\mathrm{Re}_{\bar{r}}}} = -\frac{dC(0)}{d\xi},$$

(16)

3. Computational Outline

To transform the system of Equations (10)–(13) into an initial value problem one can use the dummy substitutions:

$Y_2 = F'(\xi), Y_3 = F'_2 = F''(\xi), Y_5 = G'(\xi), Y_7 = T'(\xi), Y_9 = C'(\xi),$ so we have

$$\begin{bmatrix} Y'_1 \\ Y'_2 \\ Y'_3 \\ Y'_4 \\ Y'_5 \\ Y'_6 \\ Y'_7 \\ Y'_8 \\ Y'_9 \end{bmatrix} = \begin{bmatrix} Y_2 \\ Y_3 \\ \frac{\gamma Y_2 + (Y_2)^2 - 2Y_1Y_3 - (Y_4)^2}{2\left(1+\frac{1}{\lambda}\right)} \\ Y_5 \\ \frac{2Y_2Y_4 + \gamma Y_4 - 2Y_1Y_5}{2\left(1+\frac{1}{\lambda}\right)} \\ Y_7 \\ -\Pr\left[Y_1Y_7 + N_BY_7Y_9 + N_TY_7{}^2\right] \\ Y_9 \\ -Le\Pr Y_9 + \frac{N_T}{N_B}Y'_7 \end{bmatrix}$$

(17)

$$Y_1(\xi) = 0, Y_2(\xi) = \beta F''(\xi) = \beta\alpha_1, Y_3(\xi) = F''(\xi), Y_4(\xi) = 1 + \beta G'(\xi) = 1 + \beta\alpha_2,$$

$$Y_5(\xi) = G'(\xi), Y_6(\xi) = 1, Y_7(\xi) = \alpha_3, Y_8(\xi) = 1, Y_9(\xi) = \alpha_4, \text{when } \xi \to 0,$$

(18)

with

$$Y_2(\xi) = 0, \ Y_4(\xi) = 0, Y_6(\xi) = 0, \ Y_8(\xi) = 0, \ \text{when } \xi \to \infty, \quad (19)$$

here,α_1, α_2, α_3 and α_4 are initial guess values.

4. Analysis

The Casson fluid (CF) flow is considered on a rigid disk. The flow field is magnetized with suspended nanoparticles. The said problem is controlled mathematically and a numerical solution is offered through the shooting method. In detail, Figures 1–6 are used to highlight the variations of both CF velocities ($F'(\xi)$ and $G(\xi)$) via physical parameters, namely λ, γ, and β. Figures 1 and 2 are plotted to examine the impact of λ on CF velocity. It is clear from Figures 1 and 2 that the CF velocity decreases against λ.

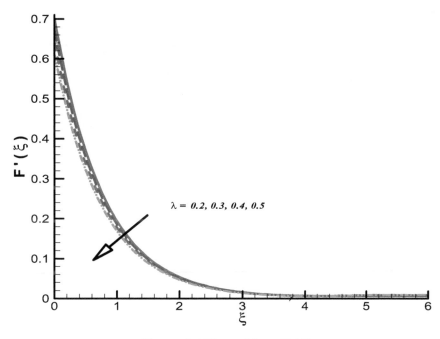

Figure 1. Effect of λ on $F(\xi)$.

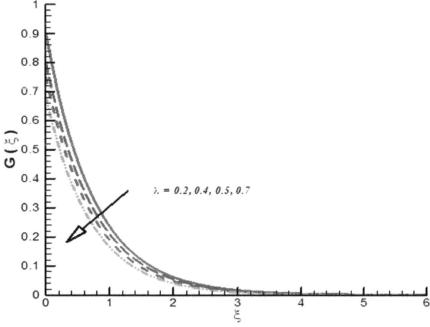

Figure 2. Effect of λ on $G(\xi)$.

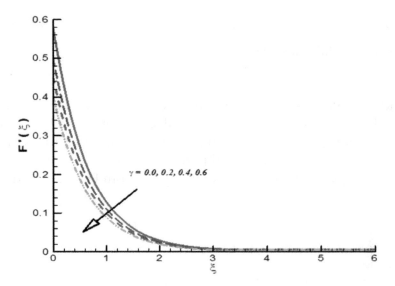

Figure 3. Effect of γ on $F'(\xi)$.

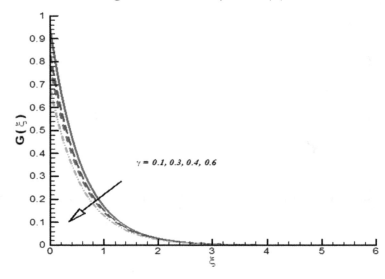

Figure 4. Effect of γ on $G(\xi)$.

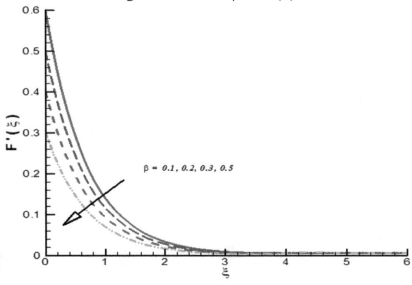

Figure 5. Effect of β on $F'(\xi)$.

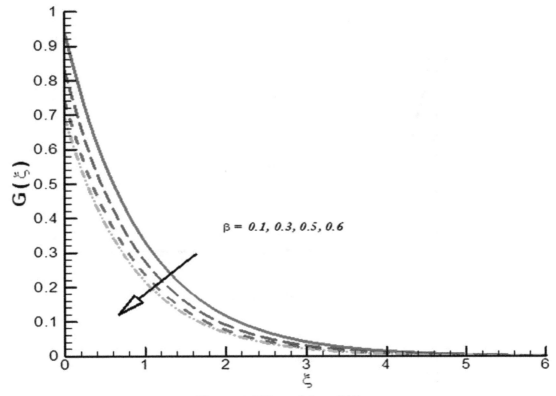

Figure 6. Effect of β on $G(\xi)$.

The impact of γ on CF velocity is examined and provided via Figure 3. The CF velocity decreases for higher values of γ. This is due to activation of Lorentz force via increasing γ. Similarly, the effect of γ on tangential velocity $G(\xi)$ is examined and given by means of Figure 4. It is important to note that the tangential velocity decreases for γ like radial one. The effect of β on radial velocity is offer in Figure 5. It is noticed that the radial velocity reflects a diminishing nature for positive values of β and the corresponding momentum boundary layer is also effected and admits decline values. Figure 6 gives the effect of β on tangential velocity of Casson fluid parameter. It is observed that the tangential velocity decreases for slip parameter. The Casson fluid temperature is examined and provided via Figures 7–9. Particularly, Figure 7 is plotted against N_T while Figure 8 is used to identify the influence of Pr on $T(\xi)$. Figure 9 reports influence of N_B on $T(\xi)$. From these figures we observed that Casson fluid temperature increases towards N_T, N_B but opposite trend is testified for Pr. Figures 10–12 reports the impact of Le, N_B and N_t on $C(\xi)$. In detail, Figure 10 paints the effect of Le on $C(\xi)$. The Casson concertation decreases for positive variations in Le. The $C(\xi)$ effected significantly towards N_B. Figure 11 is evident that the N_B results decline values in $C(\xi)$ for both zero and non-zero values of β. Such decreasing trend is due to higher values of Brownian force. The change in $C(\xi)$ is observed towards N_t and offer in Figure 12. The higher values of N_B corresponds increasing trends in $C(\xi)$ and related momentum boundary layer. In this attempt the MHD Casson nanofluid flow brought by rotating solid disk in the presence of slip conditions is examined. For comparison purpose, when Casson fluid parameter approaches to infinity our problem absolutely match with Hayat et al. [32]. In this work they studied nanoparticle aspects on viscous fluid flow due to rotating disk along with slip effects numerically. We have compared the variation of both Nusselt and Sherwood numbers with their findings as shown in Tables 1 and 2. One can see from these tables our finding match with existing values in a limiting sense. The trifling difference is due to choice of numerical method used in both attempts. Their values are obtained by build in command in Mathematica while we have used self-coded algorithm (shooting method with R-K scheme) subject to Casson nanofluid flow induced by solid rotating disk. Beside this one can extend idea to computational fluid dynamics in context of industrial and standpoints, see References [32–42].

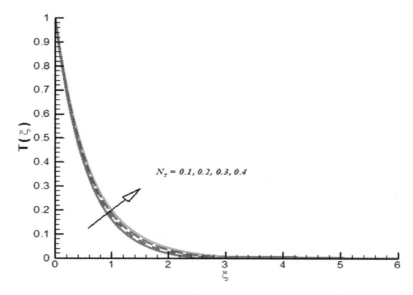

Figure 7. Effect of N_T on $T(\xi)$.

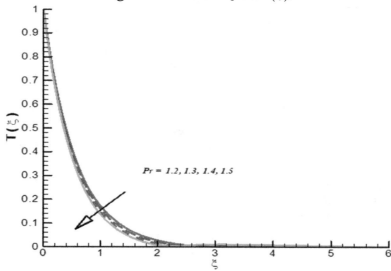

Figure 8. Effect of Pr on $T(\xi)$.

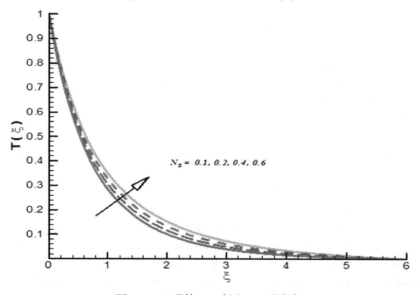

Figure 9. Effect of N_B on $T(\xi)$.

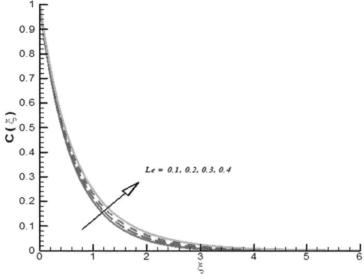

Figure 10. Effect of Le on $C(\xi)$.

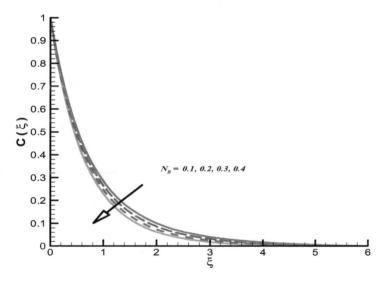

Figure 11. Effect of N_B on $C(\xi)$.

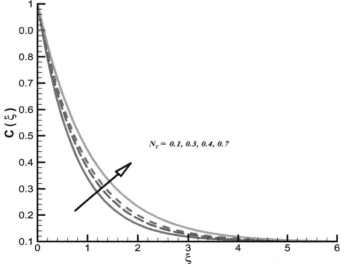

Figure 12. Effect of N_T on $C(\xi)$.

Table 1. Local Nusselt number comparison with Hayat et al. [32].

						$\frac{Nu}{\sqrt{Re_{\bar{r}}}} = -\frac{dT(0)}{d\xi}$	
β	γ	N_T	Le	Pr	N_B	Hayat et al. [32]	Present values
0.2	-	-	-	-	-	0.32655	0.326600
0.5	-	-	-	-	-	0.30360	0.30363
0.8	-	-	-	-	-	0.28715	0.28724
-	0.0	-	-	-	-	0.30494	0.30502
-	0.7	-	-	-	-	0.24421	0.24434
-	1.4	-	-	-	-	0.17566	0.17575
-	-	0.5	-	-	-	0.25913	0.25916
-	-	0.7	-	-	-	0.23865	0.23879
-	-	1.0	-	-	-	0.21010	0.21025
-	-	-	0.5	-	-	0.29633	0.29642
-	-	-	1.0	-	-	0.28954	0.28963
-	-	-	1.5	-	-	0.28395	0.28398
-	-	-	-	0.5	-	0.24989	0.24999
-	-	-	-	1.0	-	0.29211	0.29224
-	-	-	-	1.5	-	0.32286	0.32294
-	-	-	-	-	0.5	0.26341	0.26358
-	-	-	-	-	0.7	0.23677	0.23687
-	-	-	-	-	1.0	0.20056	0.20068

Table 2. Local Sherwood number comparison with Hayat et al. [32].

						$\frac{Sh}{\sqrt{Re_{\bar{r}}}} = -\frac{dC(0)}{d\xi}$	
β	γ	N_T	Le	Pr	N_B	Hayat et al. [32]	Present values
0.2	-	-	-	-	-	0.27583	0.27593
0.5	-	-	-	-	-	0.26933	0.26945
0.8	-	-	-	-	-	0.26493	0.26498
-	0.0	-	-	-	-	0.27000	0.27012
-	0.7	-	-	-	-	0.25387	0.25394
-	1.4	-	-	-	-	0.23722	0.23735
-	-	0.5	-	-	-	0.22206	0.22215
-	-	0.7	-	-	-	0.22539	0.22564
-	-	1.0	-	-	-	0.22285	0.22288
-	-	-	0.5	-	-	0.21373	0.21380
-	-	-	1.0	-	-	0.30132	0.30145
-	-	-	1.5	-	-	0.38690	0.38696
-	-	-	-	0.5	-	0.22934	0.22944
-	-	-	-	1.0	-	0.26624	0.26636
-	-	-	-	1.5	-	0.31262	0.31276
-	-	-	-	-	0.5	0.30338	0.30342
-	-	-	-	-	0.7	0..31875	0..31887
-	-	-	-	-	1.0	0.32959	0.32978

5. Closing Remarks

A Casson fluid (CF) flow yield by rotating rigid disk is considered. **Both the** Brownian and thermophoresis aspects are entertained by incorporating nanoparticles. The flow characteristics are reported numerically with the support of computational algorithm. The summary is as follows:

- CF velocities which includes $[G(\xi), F'(\xi)]$ reflects decline trend towards β.

- CF velocities are decreasing function of λ and γ.

- CFT $[T(\xi)]$ admits inciting nature towards both N_T and N_B but opposite trend is observed for Pr.

- CFC $[C(\xi)]$ shows decline values for both Le, and N_B.

- CFC $[C(\xi)]$ reflect inciting trend for N_T.

- Comparative values of HTR and MTR are provided for involved flow controlling parameters.

Author Contributions: Conceptualization, K.U.R. and M.Y.M.; methodology, W.A.K.; software, I.K.; validation K.U.R. and M.Y.M. and W.A.K.; formal analysis, I.K.; investigation, W.A.K.; resources, K.U.R. and M.Y.M; writing original draft preparation, W.A.K. and I.K.; writing review and editing, I.K.; visualization, K.U.R. and M.Y.M; supervision, W.A.K.; project administration, W.A.K; funding acquisition, S.O.A.

Acknowledgments: The authors extend their appreciation to the Deanship of Scientific Research at King Khalid University, Abha 61413, Saudi Arabia for funding this work through research groups program under grant number R.G.P-2/29/40.

Nomenclature

$V = (\overline{u}, \overline{v}, \overline{w})$	**Velocity field**
$(\overline{r}, \overline{\phi}, \overline{z})$	Polar coordinates
ν	Kinematic viscosity
λ	Casson fluid parameter
ρ_f	Fluid density
σ	Electrical conductivity
B_0	Uniform applied magnetic field
α	Thermal diffusivity
D_B	Brownian diffusion coefficient
D_T	Thermophoretic diffusion coefficient
$\overline{T}_\infty$	Ambient temperature
L	Velocity slip parameter
$\overline{T}_w$	Surface temperature
$\overline{C}_w$	Surface concentration
$\overline{C}$	Concentration
$F'(\xi), G(\xi)$	Dimensionless velocities
$T(\xi)$	Dimensionless temperature
$C(\xi)$	Dimensionless concentration
γ	Magnetic field parameter
Pr	Prandtl number
N_B	Brownian motion parameter
N_T	Thermophoresis parameter
Le	Lewis number
β	Velocity slip parameter
Re_r	Reynolds number

References

1. Mustafa, M.; Hayat, T.; Pop, I.; Aziz, A. Unsteady boundary layer flow of a Casson fluid due to an impulsively started moving flat plate. *Heat Transf. Asian Res.* **2011**, *6*, 563–576. [CrossRef]
2. Nadeem, S.; Rizwan, U.H.; Lee, C. MHD flow of a Casson fluid over an exponentially shrinking sheet. *Sci. Iran.* **2012**, *19*, 1550–1553. [CrossRef]
3. Mustafa, M.; Tasawar, H.; Pop, I.; Awatif, H. Stagnation-point flow and heat transfer of a Casson fluid towards a stretching sheet. *Z. Naturforsch. A* **2012**, *67*, 70–76. [CrossRef]
4. Mukhopadhyay, S. Casson fluid flow and heat transfer over a nonlinearly stretching surface. *Chin. Phys. B* **2013**, *22*, 074701. [CrossRef]
5. Mukhopadhyay, S.; Iswar, C.M.; Tasawar, H. MHD boundary layer flow of Casson fluid passing through an exponentially stretching permeable surface with thermal radiation. *Chin. Phys. B* **2014**, *23*, 104701. [CrossRef]
6. Mustafa, M.; Junaid, A.K. Model for flow of Casson nanofluid past a non-linearly stretching sheet considering magnetic field effects. *AIP Adv.* **2015**, *5*, 077148. [CrossRef]
7. Ramesh, K.; Devakar, M. Some analytical solutions for flows of Casson fluid with slip boundary conditions. *Ain Shams Eng. J.* **2015**, *6*, 967–975. [CrossRef]
8. Sandeep, N.; Olubode, K.K.; Isaac, L.A. Modified kinematic viscosity model for 3D-Casson fluid flow within boundary layer formed on a surface at absolute zero. *J. Mol. Liq.* **2016**, *221*, 1197–1206. [CrossRef]
9. Qing, J.; Muhammad, M.B.; Munawwar, A.A.; Mohammad, M.R.; Mohamed, E.-S.A. Entropy generation on MHD Casson nanofluid flow over a porous stretching/shrinking surface. *Entropy* **2016**, *18*, 123. [CrossRef]
10. Ali, M.E.; Sandeep, N. Cattaneo-christov model for radiative heat transfer of magnetohydrodynamic Casson-ferrofluid: A numerical study. *Results Phys.* **2017**, *7*, 21–30. [CrossRef]
11. Reddy, J.V.R.; Sugunamma, V.; Sandeep, N. Enhanced heat transfer in the flow of dissipative non-Newtonian Casson fluid flow over a convectively heated upper surface of a paraboloid of revolution. *J. Mol. Liq.* **2017**, *229*, 380–388. [CrossRef]
12. Rehman, K.U.; Aneeqa, A.M.; Malik, M.Y.; Sandeep, N.; Saba, N.U. Numerical study of double stratification in Casson fluid flow in the presence of mixed convection and chemical reaction. *Results Phys.* **2017**, *7*, 2997–3006. [CrossRef]
13. Kumaran, G.; Sandeep, N. Thermophoresis and Brownian moment effects on parabolic flow of MHD Casson and Williamson fluids with cross diffusion. *J. Mol. Liq.* **2017**, *233*, 262–269. [CrossRef]
14. Ali, F.; Nadeem, A.S.; Ilyas, K.; Muhammad, S. Magnetic field effect on blood flow of Casson fluid in axisymmetric cylindrical tube: A fractional model. *J. Magn. Magn. Mater.* **2017**, *423*, 327–336. [CrossRef]
15. Raju, C.S.K.; Mohammad, M.H.; Sivasankar, T. Radiative flow of Casson fluid over a moving wedge filled with gyrotactic microorganisms. *Adv. Powder Technol.* **2017**, *28*, 575–583. [CrossRef]
16. Kármán, T.V. Über laminare und turbulente Reibung. *ZAMM* **1921**, *1*, 233–252. [CrossRef]
17. Turkyilmazoglu, M.; Senel, P. Heat and mass transfer of the flow due to a rotating rough and porous disk. *Int. J. Therm. Sci.* **2013**, *63*, 146–158. [CrossRef]
18. Rashidi, M.M.; Kavyani, N.; Abelman, S. Investigation of entropy generation in MHD and slip flow over a rotating porous disk with variable properties. *Int. J. Heat Mass Transf.* **2014**, *70*, 892–917. [CrossRef]
19. Turkyilmazoglu, M. Nanofluid flow and heat transfer due to a rotating disk. *Comput. Fluids* **2014**, *94*, 139–146. [CrossRef]
20. Griffiths, P.T.; Stephen, J.G.; Stephen, S.O. The neutral curve for stationary disturbances in rotating disk flow for power-law fluids. *J. Non Newton. Fluid Mech.* **2014**, *213*, 73–81. [CrossRef]
21. Mustafa, M.; Junaid, A.K.; Hayat, T.; Alsaedi, A. On Bödewadt flow and heat transfer of nanofluids over a stretching stationary disk. *J. Mol. Liq.* **2015**, *211*, 119–125. [CrossRef]
22. Sheikholeslami, M.; Hatami, M.; Ganji, D.D. Numerical investigation of nanofluid spraying on an inclined rotating disk for cooling process. *J. Mol. Liq.* **2015**, *211*, 577–583. [CrossRef]
23. Xun, S.; Zhao, J.; Zheng, L.; Chen, X.; Zhang, X. Flow and heat transfer of Ostwald-de Waele fluid over a variable thickness rotating disk with index decreasing. *Int. J. Heat Mass Transf.* **2016**, *103*, 1214–1224. [CrossRef]
24. Latiff, N.A.; Uddin, M.J.; Ismail, A.M. Stefan blowing effect on bioconvective flow of nanofluid over a solid rotating stretchable disk. *Propuls. Power Res.* **2016**, *5*, 267–278. [CrossRef]
25. Ming, C.; Zheng, L.; Zhang, X.; Liu, F.; Anh, V. Flow and heat transfer of power-law fluid over a rotating disk with generalized diffusion. *Int. Commun. Heat Mass Transf.* **2016**, *79*, 81–88. [CrossRef]

26. Imtiaz, M.; Tasawar, H.; Ahmed, A.; Saleem, A. Slip flow by a variable thickness rotating disk subject to magnetohydrodynamics. *Results Phys.* **2017**, *7*, 503–509. [CrossRef]
27. Doh, D.H.; Muthtamilselvan, M. Thermophoretic particle deposition on magnetohydrodynamic flow of micropolar fluid due to a rotating disk. *Int. J. Mech. Sci.* **2017**, *130*, 350–359. [CrossRef]
28. Hayat, T.; Madiha, R.; Maria, I.; Ahmed, A. Nanofluid flow due to rotating disk with variable thickness and homogeneous-heterogeneous reactions. *Int. J. Heat Mass Transf.* **2017**, *113*, 96–105. [CrossRef]
29. Devi, M.; Chitra, L.; Rajendran, A.B.Y.; Fernandez, C. Non-linear differential equations and rotating disc electrodes: Padé approximationtechnique. *Electrochim. Acta* **2017**, *243*, 1–6. [CrossRef]
30. Guha, A.; Sayantan, S. Non-linear interaction of buoyancy with von Kármán's swirling flow in mixed convection above a heated rotating disc. *Int. J. Heat Mass Transf.* **2017**, *108*, 402–416. [CrossRef]
31. Ellahi, R.; Ahmed, Z.; Farooq, H.; Tehseen, A. Study of shiny film coating on multi-fluid flows of a rotating disk suspended with nano-sized silver and gold particles: A comparative analysis. *Coatings* **2018**, *8*, 422. [CrossRef]
32. Hayat, T.; Taseer, M.; Sabir, A.S.; Ahmed, A. On magnetohydrodynamic flow of nanofluid due to a rotating disk with slip effect: A numerical study. *Comput. Methods Appl. Mech. Eng.* **2017**, *315*, 467–477. [CrossRef]
33. Vo, T.Q.; Park, B.S.; Park, C.H.; Kim, B.H. Nano-scale liquid film sheared between strong wetting surfaces: Effects of interface region on the flow. *J. Mech. Sci. Technol.* **2015**, *29*, 1681–1688. [CrossRef]
34. Kherbeet, A.; Sh, H.A.; Mohammed, B.H.; Salman, H.E.; Ahmed, O.; Alawi, A.; Rashidi, M.M. Experimental study of nanofluid flow and heat transfer over microscale backward-and forward-facing steps. *Exp. Therm. Fluid Sci.* **2015**, *65*, 13–21. [CrossRef]
35. Abbas, T.; Muhammad, A.; Muhammad, B.; Mohammad, R.; Mohamed, A. Entropy generation on nanofluid flow through a horizontal Riga plate. *Entropy* **2016**, *18*, 223. [CrossRef]
36. Bhatti, M.M.; Abbas, T.; Rashidi, M.M. Numerical study of entropy generation with nonlinear thermal radiation on magnetohydrodynamics non-Newtonian nanofluid through a porous shrinking sheet. *J. Magn.* **2016**, *21*, 468–475. [CrossRef]
37. Ghorbanian, J.; Alper, T.C.; Beskok, A. A phenomenological continuum model for force-driven nano-channel liquid flows. *J. Chem. Phys.* **2016**, *145*, 184109. [CrossRef]
38. Bao, L.; Priezjev, N.V.; Hu, H.; Luo, K. Effects of viscous heating and wall-fluid interaction energy on rate-dependent slip behavior of simple fluids. *Phys. Rev. E* **2017**, *96*, 033110. [CrossRef]
39. Ghorbanian, J.; Beskok, A. Temperature profiles and heat fluxes observed in molecular dynamics simulations of force-driven liquid flows. *Phys. Chem. Chem. Phys.* **2017**, *19*, 10317–10325. [CrossRef]
40. Mohebbi, R.; Rashidi, M.M.; Mohsen, I.; Nor, A.C.S.; Hong, W.X. Forced convection of nanofluids in an extended surfaces channel using lattice Boltzmann method. *Int. J. Heat Mass Transf.* **2018**, *117*, 1291–1303. [CrossRef]
41. Rehman, K.U.; Malik, M.Y.; Iffat, Z.; Alqarni, M.S. Group theoretical analysis for MHD flow fields: A numerical result. *J. Braz. Soc. Mech. Sci. Eng.* **2019**, *41*, 156. [CrossRef]
42. Rehman, K.U.; Malik, M.Y.; Mahmood, R.; Kousar, N.; Zehra, I. A potential alternative CFD simulation for steady Carreau–Bird law-based shear thickening model: Part-I. *J. Braz. Soc. Mech. Sci. Eng.* **2019**, *41*, 176. [CrossRef]

13

MHD Flow and Heat Transfer over Vertical Stretching Sheet with Heat Sink or Source Effect

Ibrahim M. Alarifi [1], Ahmed G. Abokhalil [2,3], M. Osman [1,4], Liaquat Ali Lund [5], Mossaad Ben Ayed [6,7], Hafedh Belmabrouk [8,9] and Iskander Tlili [1,*

[1]   Department of Mechanical and Industrial Engineering, College of Engineering, Majmaah University, Al-Majmaah 11952, Saudi Arabia; i.alarifi@mu.edu.sa (I.M.A.); m.othman@mu.edu.sa (M.O.)
[2]   Department of Electrical Engineering, College of Engineering, Majmaah University, Al-Majmaah 11952, Saudi Arabia; a.abokhalil@mu.edu.sa
[3]   Electrical Engineering Department, Assiut University, Assiut 71515, Egypt
[4]   Mechanical Design Department, Faculty of Engineering Mataria, Helwan University, Cairo El-Mataria 11724, Egypt
[5]   Sindh Agriculture University, Tandojam Sindh 70060, Pakistan; balochliaqatali@gmail.com
[6]   Computer Science Department, College of Science and Humanities at Alghat, Majmaah University, Al-Majmaah 11952, Saudi Arabia; mm.ayed@mu.edu.sa
[7]   Computer and Embedded System Laboratory, Sfax University, Sfax 3011, Tunisia
[8]   Electronics and Microelectronics Laboratory, Faculty of Science of Monastir, University of Monastir, Monastir 5019, Tunisia; ha.belmabrouk@mu.edu.sa
[9]   Department of Physics, College of Science at Zulfi, Majmaah University, Al Zulfi 11932, Saudi Arabia
*   Correspondence: l.tlili@mu.edu.sa

Abstract: A steady laminar flow over a vertical stretching sheet with the existence of viscous dissipation, heat source/sink, and magnetic fields has been numerically inspected through a shooting scheme based Runge—Kutta–Fehlberg-integration algorithm. The governing equation and boundary layer balance are expressed and then converted into a nonlinear normal system of differential equations using suitable transformations. The impact of the physical parameters on the dimensionless velocity, temperature, the local Nusselt, and skin friction coefficient are described. Results show good agreement with recent researches. Findings reveal that the Nusselt number at the sheet surface augments, since the Hartmann number, stretching velocity ratio A, and Hartmann number Ha increase. Nevertheless, it reduces with respect to the heat generation/absorption coefficient δ.

Keywords: steady laminar flow; nanofluid; heat source/sink; magnetic field; stretching sheet

1. Introduction

The steady laminar flow and heat transfer of a viscous fluid over a vertical stretching sheet with the existence of heat source/sink and magnetic fields has gained significant interest because of its various usages in engineering procedures like geothermal energy extraction, glass fiber, and plasma studies, etc. Several researchers investigated numerically MHD mixed convective stagnation point flow lengthways a perpendicular widening piece in the presence of a heat source/sink in order to evaluate the impacts of relevant physical parameters especially Hartmann number, Stretching velocity ratio and Biot number on velocity and temperature profiles besides to skin friction and heat transfer properties. Using the Runge-Kutta-Fehlberg methods joined with shooting technique [1–4]. P.R. Sharma et al. [3] analyzed numerically the impacts of an external magnetic field. Tarek M. A. El-Mistikawy [5] focused on the flow resulting from a linearly stretching sheet with a transverse magnetic field. A. Mohammadeina et al. [1] investigated the impacts of thermal radiation and magnetic field on flow of CuO-water nanofluid past a stretching sheet characterized by forced-convection with a stagnation point with suction/injection. S.S. Ghadikolaei et al. [4] investigated the impact of thermal radiation and Joule heating. Wubshet Ibrahim [2] studied numerically the melting heat transfer and magneto hydrodynamic stagnation point flow of a nanofluid pasta extending piece.

Homotopy technique and shooting method have been widely used in many studies to obtain exact and wide-ranging analytic solution [6–9]. Arif Hussain et al. [6] focused on the thermal and physical properties of features of MHD hyperbolic refraction of fluid flow over a non-linear widening sheet taking into account convective boundary conditions and viscous dissipation. Tasawar Hayat et al. [9] studied convective stream of ferrofluid owing to nonlinear widening curved slip while M. N. Tufail et al. [7] analyzed numerically the heat transmission ended an unsteady widening sheet for an MHD Casson fluid through viscous dissipation impacts using the homotopy method. M. Y. Malik et al. [10] examined numerically an MHD flow of the Carreau fluid over a stretching sheet with inconstant thickness with the Keller box method. T.M. Agbaje et al. [11] proposed the spectral perturbation technique, which is a sequence development method which spreads the usage of the typical trepidation methods. When joined with the Chebyshev pseudo-spectral process, the SPM can provide higher order approximate mathematical resolutions for intricate increases faced in perturbation patterns. Siti Khuzaimah Soida et al. [12] analyzed numerically a stable MHD flow through a centrifugally widening or lessening floppy using the boundary value problem solver in Matlab software.

The temperature profile behavior was investigated by several researchers. U.S. Mahabaleshwar et al. [13] studied an MHD couple stress liquid caused by a perforated sheet experiencing lined widening with radiation. It was found that the temperature rises as the heat source/sink NI parameter and Chandrasekhar number Q augment [13]. M. Ferdows et al. [14] studied a steady two-dimensional free convective flow of a viscous incompressible fluid lengthways a perpendicular stretching sheet. The temperature profiles increase as the combined effect of porous diffusivity and magnetic field R, perturbation parameter increase. The temperature profile augments with the Eckert number Ec [15] and Biot number [6,13] and the magnetic field parameter, radiation and viscous dissipation [7,8,14–17]. In fact, when variable thickness exists, the radiative heat transfer and viscous dissipation enhance the nanofluid temperature [18]. When the unsteadiness of the stretching sheet extends, temperature rises [19]. The fluid temperature increases with increasing of Brownian motion, thermophoresis [20,21], temperature ratio [1] and heat source/sink parameter [3]. In the other hand, temperature drops when the Casson fluid parameter augments [7] and when the Prandtl number increases [13,20] and as the power law increases [6]. G.S. Seth et al. [19] reported that temperature is reduced when the stretching sheet is nonlinear. The fluid temperature declines with higher values of velocity ratio parameter, suction/injection parameter [1,16], Hartmann number Ha, Prandtl number Pr, thermal stratification parameter [18] or mixed convection parameter k [3]. Whereas, Kai-Long Hsiao [15] found that the temperature impact is improved in parallel with the rise of Prandtl number.

The impacts of significant factors on the solitary velocity for stretching sheet circumstance were discussed in many studies. Decrease in the velocity profile has been reported verses the increase of the Casson fluid parameter [7], non-linearity restriction, flow comportment index, the magnetic field, power law index and Weissenberg number [1,6–8]. However, the impacts of power law index and magnetic field parameter are more significant than nonlinearity parameter and Weissenberg number [6]. The dimensionless velocity is reduced when values of volume fraction and dimensionless velocity slip parameters augment [22]. It declines also when mass concentration parameter and permeability parameter augment [4]. Velocity profile is reduced by the combined effect of porous diffusivity [14], with the increase of Hartmann number and suction/injection parameter [23]. Greater values of suction and stretching parameter m and wall thickness parameter reduce velocity profile while lead to the rise of the volume fraction and porosity parameter [17]. The fluid velocity is improved in parallel with increase of velocity ratio, curvature parameter, electric field parameter and material parameter [4].

M. Ferdows et al. [14] reported that velocity profile is enhanced with the increase of the proportion of the flow velocity constraint to stretching sheet indicator and viscosity ratio. G.S. Seth et al. [24] reported that nanofluid velocity is reduced when the stretching sheet is nonlinear whereas when the unsteadiness of the stretching sheet augments velocity is decreased. M. N. Tufail et al. [7] found that as the unsteadiness parameter growths, the velocity decreases close the sheet and increases distant to the sheet. The higher of the slight order derivative causes the quicker velocity of viscoelastic fluids close the platter [25–32]. The temperature and velocity profile behavior was investigated by several researchers with the effect of heat generation/absorption on MHD flow though rare of them trait the effect of an external magnetic field and heat generation/absorption on mixed convective flow lengthways a perpendicular extending piece [33–37].

An inclusive analysis of the works about nanofluids is presented by Wang et al. [19,20]. The inactivity point flow of a nanofluid near a stretching sheet has been explored by Khan and Pop [21], Mustafa et al. [22], Nazar et al. [23], and Ibrahim et al. [24]. Nadeem and Haq [29] inspected MHD three-dimensional flow of nanofluids past a shrinking sheet with thermal radiation. They employed Boungiorno model and considered the influences of thermophoresis and Brownian motion on the local Nusselt ad Sherwood numbers.

This study includes a numerically solution of MHD boundary layer flow over a vertical stretching sheet with company of heat sink/ source and magnetic fields effect. The governing equations for the problematic have been clearly described with some appropriate changes and then explained numerically via shooting arrangement based RKFI procedure. In this work the effects of relevant physical indicators essentially Hartmann number, Stretching velocity ratio and Biot number on velocity and temperature distributions in addition to the skin friction and heat transfer properties have been investigated. Comparison of the results with research results [3,26–30] shows a good agreement as shown in Table 1.

Table 1. Numerical results of Nusselt number—$\theta(0)$ for diverse Prandtl number when $A = 1$, $A = 0$, $Ha = 0$, and $\delta = 0$.

Pr	Ramchandran et al. [26]	Hassanien and Gorla [30]	Lok et al. [27]	Ishak et al. [28]	Ali et al. [29]	Sharma et al. [3]	Present
1	-	-	-	0.8708	0.8708	0.8707	0.87078
10	-	1.9446	-	1.9446	1.9448	1.94463	1.94465
20	2.4576	-	2.4577	2.4576	2.4579	2.4576	2.4577
40	3.1011	-	3.1023	3.1011	3.1017	3.1011	3.1015
60	3.5514	-	3.556	3.5514	3.5524	3.55142	3.55148
80	3.9055	-	3.9195	3.9095	3.9108	3.90949	3.90919
100	4.2116	4.2337	4.2289	4.2116	4.2133	4.21163	4.21135

2. Problem Description

The present study deals with a viscous incompressible fluid that was analyzed through steady laminar flow two-dimensional condition lengthways on an upright stretching sheet that was positioned in the x track and the y axis is perpendicular to the plan of this sheet.

u is the velocity component in x direction.

v is the velocity component in y direction.

Taking into consideration that mutually c then a are positive constants; $u = u_e(x) = ax$ represents the unrestricted stream velocity, while $u = u_w(x) = cx$ represents the velocity when there is stretching on the sheet.

When a heat source/sink is present, H_0 is an outside magnetic field that is practically perpendicular to the sheet.

The principal equations of continuity, momentum, and energy are written as

$$\frac{\partial u}{\partial x} + \frac{\partial v}{\partial y} = 0, \tag{1}$$

$$u\frac{\partial u}{\partial x} + v\frac{\partial u}{\partial y} = -\frac{1}{\rho}\frac{dp}{dx} + v\frac{\partial^2 u}{\partial y^2} - \frac{\sigma\mu_e^2 H_0^2}{\rho}u + g\beta(T - T_\infty), \tag{2}$$

$$u\frac{\partial T}{\partial x} + v\frac{\partial T}{\partial y} = \frac{k}{\rho C_p}\frac{\partial^2 T}{\partial y^2}P + \frac{Q}{\rho C_p}(T - T_\infty), \tag{3}$$

where

σ is the electrical conductivity,

μ_e is the magnetic permeability,

T_∞ is the temperature of free stream,

g is acceleration due to gravity,

β is the volumetric coefficient of thermal expansion,

k is the thermal conductivity,

$v(= \mu/\rho)$ is the kinematic viscosity, and

$T_w = T_\infty + bx$ is the temperature of the sheet.

When the superficial is heated $b > 0$ so that cfr4 whereas for cooled superficial $b < 0$ and $T_w \langle T_\infty$. The boundary conditions are

$$v = 0, \; u = u_w(x) = cx, \; -k\frac{\partial T}{\partial y} = h_f(T_f - T) \; at \; y = 0, \tag{4}$$
$$u = u_e(x) = ax, \; T = T_\infty \, as \, y \to \infty,$$

The forces will be in equilibrium because of the presence of the hydrostatic and magnetic pressure gradient, as mentioned below

$$-\frac{1}{\rho}\frac{dp}{dx} = u_e\frac{du_e}{dx} + \frac{\sigma\mu_e^2 H_0^2}{\rho}u_e, \tag{5}$$

Therefore, the momentum equation turns into

$$u\frac{\partial u}{\partial x} + v\frac{\partial u}{\partial y} = u_e\frac{du_e}{dx} - \frac{\sigma\mu_e^2 H_0^2}{\rho}(u - u_e) + v\frac{\partial^2 u}{\partial y^2} + g\beta(T - T_\infty). \tag{6}$$

3. Scheme Analysis

The subsequent change and dimensionless quantities are used into equations, while taking into account the mentioned boundary condition (4) within the solution of Equations (1) to (3), Equation (7), obtained as:

$$\eta = \sqrt{\frac{a}{v}}y, = \psi = x\sqrt{av}f(\eta), \; \theta(\eta) = \frac{T - T_\infty}{T_w - T_\infty} and u = \frac{\partial\psi}{\partial y}, \; v = -\frac{\partial\psi}{\partial x} \tag{7}$$

Therefore, the equation of continuity is approved.
The momentum and energy equation becomes

$$f''' + ff'' - (f')^2 + 1 + H_a^2(1 - f') + \lambda\theta = 0 \tag{8}$$

$$\theta'' + P_r(f\theta' - f'\theta + \delta\theta) = 0 \tag{9}$$

where major represents the derivative according to η,

$Ha\left(= \mu_e H_0\sqrt{\frac{\sigma}{\rho a}}\right)$ is the Hartmann number,

$\lambda\left(= \frac{Gr_x}{Re_x^2}\right)$ is mixed convection parameter,

$Gr_x\left(= g\beta(T_w - T_\infty)\frac{x^3}{v^2}\right)$ is the local Grashof number,
$Re_x\left(= u_e(x)\frac{x}{v}\right)$ and $P_r\left(\frac{v}{x}\right)$, and
$\delta\left(= \frac{Q}{\rho a C_p}\right)$ represent the factor of heat generation/absorption.
The related boundary circumstances are limited to

$$f = 0, \ f' = c/a = A, \ \theta' = B(\theta - 1) \text{ at } \eta = 0$$
$$f' = 1, \ \theta = 0 \text{ as } \eta \to \infty \tag{10}$$

where $A\left(= \frac{c}{a}\right)$ represent the rate of velocity

Also, $B = \frac{h_f}{k}\sqrt{\frac{v}{a}}$.

Therefore, the Nusselt number and Skin friction are expressed by

$$C_f = \frac{2\tau_w}{\rho u_e^2} = Re_x^{-1/2}f''(0), \ Nu_x = \frac{xq_w}{k(T_w - T_\infty)} = -Re_x^{1/2}\theta'(0). \tag{11}$$

Additionally, respectively, the wall shear stress τ_w and the heat flux q_w are illustrated, as below:

$$\tau_w = \mu\left(\frac{\partial u}{\partial y}\right)_{y=0}, \ q_w = -k\left(\frac{\partial T}{\partial y}\right)_{y=0}$$

It is intricate to get the locked procedure explanations because calculations (8) and (9) are significantly nonlinear. Consequently, schemes are obtained by replacing

$$f = f_1, f' = f_2, f'' = f^3, f''' = f_3', \theta' = f_5, \theta'' = f_5'$$

Thus, the scheme of calculations becomes

$$f_1' = f_2, f_2' = f_3, f_3' = f_2^2 - f_1f_3 - 1 + Ha^2(f_2 - 1) - \lambda f_4$$
$$f_4' = f_5, f_5' = P_r(f_2f_4 - f_1f_5 - \delta f_4)$$

Depending on the next conditions

$$f_1(0) = 0, f_2(0) = A, f_3(0) = s_1, f_4(0) = 1, f_5(0) = s_2 \text{ and } f_2(\infty) = 1, f_4(\infty) = 0$$

In order to obtain stage-by-stage integration and scheming, which are determined by relying on MATLAB software; the Runge–Kutta fourth order method with the shooting technique is actually used.

4. Stability Analysis

When there exists more than one solution in any fluid flow problem, the stability of solution is necessary to perform in that problem. In order to perform stability analysis, we adopt the algorithm of Merkin [33], Weidman et al. [36], and Rosca and Pop [37].

Step 1: To convert the governing Equations (2) and (3) of fluid flow in unsteady form, we have

$$\frac{\partial u}{\partial t} + u\frac{\partial u}{\partial x} + v\frac{\partial u}{\partial y} = -\frac{1}{\rho}\frac{dP}{dx} + v\frac{\partial^2 u}{\partial y^2} - g\beta(T - T_\infty) - \frac{\sigma\mu_e H_0^2 u}{\rho} \tag{12}$$

$$\frac{\partial T}{\partial t} + u\frac{\partial T}{\partial x} + v\frac{\partial T}{\partial y} = \frac{k}{\rho C_p}\frac{\partial^2 T}{\partial y^2} + \frac{k}{\rho C_p}(T - T_\infty) \tag{13}$$

Step 2: To introduce a new non-dimensional time variable $\tau = a.t$, and all other similarity variables are also a function of τ, can be written as,

$$\psi = x\sqrt{av}f(\eta,\tau); \eta = y\sqrt{\frac{a}{v}}; \theta(\eta,\tau) = \frac{(T-T_\infty)}{(T_w-T_\infty)};$$ (14)

Step 3: By applying Equation (14) on Equations (12) and (13), we have

$$\frac{\partial^3 f(\eta,\tau)}{\partial \eta^3} + f\frac{\partial^2 f(\eta,\tau)}{\partial \eta^2} - \left(\frac{\partial f(\eta,\tau)}{\partial \eta}\right)^2 + 1 + Ha^2\left(1-\frac{\partial f(\eta,\tau)}{\partial \eta}\right) - \lambda\theta(\eta,\tau)$$
$$-\frac{\partial^2 f(\eta,\tau)}{\partial\tau\partial\eta} = 0$$ (15)

$$\frac{1}{Pr}\frac{\partial^2\theta(\eta,\tau)}{\partial\eta^2} + f(\eta,\tau)\frac{\partial\theta(\eta,\tau)}{\partial\eta} - \frac{\partial f(\eta,\tau)}{\partial\eta}\theta(\eta,\tau) + \delta\theta(\eta,\tau) - \frac{\partial\theta(\eta,\tau)}{\partial\tau} = 0$$ (16)

and the related boundary conditions are

$$f(0,\tau) = 0; \frac{\partial f(0,\tau)}{\partial\eta} = A; \frac{\partial\theta(0,\tau)}{\partial\eta} = B(\theta(0,\tau)-1),$$
$$\frac{\partial f(\eta,\tau)}{\partial\eta} \to 1; \theta(\eta,\tau) \to 0 \text{ as } \eta \to \infty$$ (17)

Step 4: To check the stability of steady flow solutions $f(\eta) = f_0(\eta)$ and $\theta(\eta) = \theta_0(\eta)$ will satisfy the basic model by introducing the following functions

$$f(\eta,\tau) = f_0(\eta) + e^{-\tau}F(\eta,\tau); \theta(\eta,\tau) = \theta_0(\eta) + e^{-\tau}G(\eta,\tau)$$ (18)

Here, $F(\eta,\tau)$, and $G(\eta,\tau)$ are small relative to $f_0(\eta)$, and $\theta_0(\eta)$. The unknown eigenvalue is γ, which is to be found out.

Step 5: By putting Equation (18) into Equations (15) and (16) and keeping $\tau = 0$, we have

$$F_0''' + f_0(\eta)F_0'' + F_0f_0'' - 2f_0'F_0' - \lambda G_0 - Ha^2F_0' + \gamma F_0' = 0$$ (19)

$$\frac{1}{Pr}G_0'' + f_0G_0' + F_0\theta_0' - f_0'G_0 - F_0'\theta_0 + \delta G_0 + \gamma G_0 = 0$$ (20)

Along with boundary conditions

$$F_0(0) = 0, \ F_0'(0) = 0, \ G_0'(0) = BG_0$$
$$F_0'(\eta) \to 0, \ G_0(\eta) \to 0, \text{ as } \eta \to \infty$$ (21)

Step 6: To relax one boundary condition into an initial condition, as suggested by Weidman et al. [36] and Harris et al. [34]. In this problem, we relaxed $G_0(\eta) \to 0$, as $\eta \to \infty$ into $G_0'((0) = 1$. We have to solve Equation (19) and (20) with boundary and relaxed initial condition in order to find the values of smallest eigenvalue γ.

It is worth mentioning that the negative values of γ indicate the growth of disturbance and the flow becomes unstable. On the other hand, if the values of γ are positive, which means that the flow is stable and shows an initial decay. The values of smallest eigenvalue are given in Table 2, which indicate that only first (second) solution is stable (unstable).

Table 2. Smallest eigenvalue γ when $\lambda = -0.2$, $Pr = 1$, $A < 0$ (for Shrinking surface) and $A > 0$ (for Stretching surface).

ε	Ha	γ	-
		1st Solution	**2nd Solution**
0.5	0.3	0.97533	−0.09572
	0.5	0.65753	−0.06946
−0.5	0	1.34857	−0.75392
-	0.5	1.02349	−0.58327

The bvp4c solver function has performed stability analysis. According to Rahman et al. [35], "this collocation formula and the collocation polynomial provides a C^1 continuous solution that is fourth order accurate uniformly in [a,b]. Mesh selection and error control are based on the residual of the continuous solution". As we know, only the first solution is the stable and only the stable solution has physical meaning. In these regards, the various effect of different physical parameters on velocity and temperature profiles have been demonstrated for the first solution only. Finally, from Figure 1a,b, we draw some graphs in order to show the existence of multiple solutions for the opposing flow case.

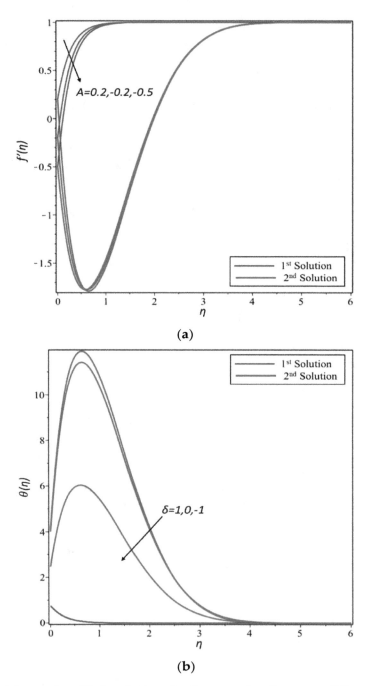

Figure 1. (a) The existence of multiple solutions for the opposing flow case different parameter A; and, (b) the existence of multiple solutions for the opposing flow case different parameter δ.

5. Results and Discussion

A steady laminar flow above a vertical stretching sheet with the existence of viscous dissipation, heat sink/source, and magnetic fields has been numerically explored by using of RKFI process through shooting scheme. In this work, the effects of pertinent physical parameters are essentially Hartmann number, Stretching velocity ratio, and Biot number on temperature and velocity distributions; also, the skin friction and heat transfer properties have been examined.

The effect of Hartmann number on dimensionless velocity for mutually opposing and assisting flow are shown in Figure 2a,b respectively, with supplementary parameters are usually the regularly of velocity ratio parameter $A = 1$, Heat generation/absorption coefficient $\delta = 1$, Prandtl number $Pr = 1$, and Biot number $B_i = 1$. It is perceived that, for assisting flow ($\lambda > 0$), the dimensionless velocity is the maximum at the superficial of the vertical stretching sheet and it increasingly reduces to the minimum value $f' = 1$ as it changes to gone after the superficial, while for opposing flow ($\lambda < 0$), the dimensionless velocity is the lowest at the surface of the vertical stretching sheet and gradually increases to the maximum value $f' = 1$, as it changes away from the superficial, this effect is mathematically obvious in Equation (10). It is additional observed that the velocity profile decreases with the Hartmann number for the assisting flow, whereas, it increases with the Hartmann number for the opposing flow. Consequently, the hydrodynamic boundary layer thickness also depends upon Ha. It is well known that the Hartmann number represents the proportion of electromagnetic force to the viscous force, thus, in the case of assisting flow, increasing the Hartmann number means that electromagnetic force was enhanced when compared to viscous force, which in turn Lorentz force augments, then opposes the flow, and then reduces the velocity profile. Nonetheless, in the situation of opposing flow, there is a reverse effect of the Hartmann number on dimensionless velocity. It should be pointed out that, in the case of assisting flow, ($\lambda > 0$) means the heating of the fluid, therefore the thermal buoyancy forces were enhanced. It can be interpreted on this fact that the highest value of dimensionless velocity is near the stretched surface; however, for opposing flow ($\lambda > 0$), which means that the fluid is consequently cooled; the thermal buoyancy forces decreases and then we realize the lowest value of dimensionless velocity near the stretched surface. It is worthwhile to note that the velocity profile increases with the mixed convection parameter λ for mutually case opposing and assisting flow due to an increasing of the thermal buoyancy forces. It can be seen that, for buoyancy opposed ($\lambda < 0$, opposing), the velocity profile will be significantly affected.

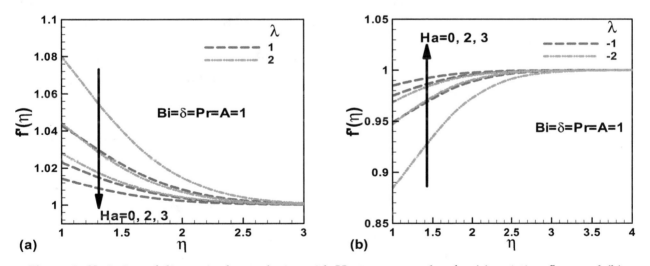

Figure 2. Variation of dimensionless velocity with Hartmann number for (**a**) assisting flow and (**b**) opposing flow.

Figure 3a,b, respectively, show the effects of stretching velocity ratio A when stretching in the flow and in the opposite direction, with further parameters sustaining the constant Heat generation/absorption coefficient $\delta = -1$, Prandtl number $Pr = 1$, Biot number $B_i = 1$, and Hartmann number $Ha = 1$. It can be seen that the velocity profile increases with stretching velocity ratio when stretching in the flow direction. Whereas, the velocity profile decreases with stretching velocity ratio when stretching in the opposite direction, this can be attributed to the significant enhancement in pressure on the sheet. Furthermore, it is remarked that the velocity profile augment with mixed convection parameter λ for both stretching in the flow direction and stretching in the opposite direction, as proven in Figure 1. The physical reason behind this is that, by augmenting the mixed convection parameter, the thermal buoyancy forces rise and help to push the flow in y direction, which in turn increases the velocity profile.

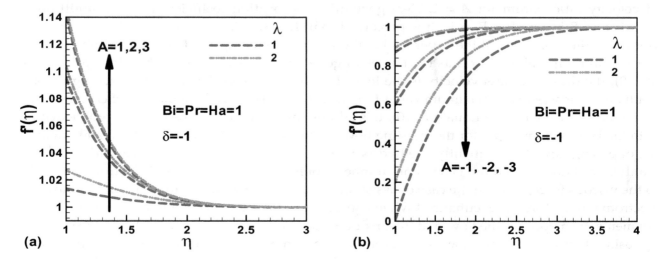

Figure 3. Variation of dimensionless velocity with stretching velocity ratio when (**a**) stretching in the flow direction and (**b**) stretching in the opposite direction.

The effects of Biot number in the presence of heat source on the dimensionless temperature for assisting and opposing flow are shown in Figure 4a,b, respectively at $P_r = H_a = A = 1$. It is clear that the temperature profile is greater at the stretching sheet surface and then exponentially lessens along the streamwise path up to the zero value for both assisting and opposing flow; this effect is proved and designated by the choice of boundary conditions and it is mathematically noticeable in Equation (10). It is worthwhile to note that the dimensionless temperature augment with both Biot number Bi and Heat generation/absorption coefficient δ for both case assisting and opposing flow; therefore, the thermal boundary layer thickness increases. It is well known that the Biot number signifies the proportion of heat convection to heat conduction; therefore increasing the Biot number leads to more heat will be released to the fluid flow, which in turn augments the temperature profile. Similarly, augmenting heat source leads to applying more heat to the fluid flow and it results in enhancing the dimensionless temperature.

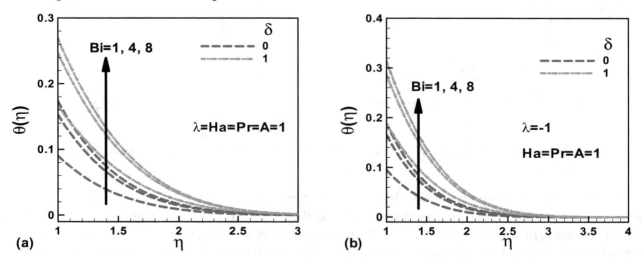

Figure 4. Variation of dimensionless temperature with Biot number in the presence of heat source for (**a**) assisting flow and (**b**) opposing flow.

Figure 5a,b, respectively, illustrate the result of Biot number in the company of heat sink on the dimensionless temperature for assisting and opposing flow at $P_r = H_a = A = 1$. As expected, the temperature profile satisfies the boundary conditions, starting with a higher value at the surface of the sheet and then meaningfully declining to zero value when η increases. Furthermore, it is remarked that dimensionless temperature increases with Biot number, even in the company of heat sink, as shown in Figure 3 in the circumstance of heat basis, so it can be concluded that the temperature profile increase with biot number independent of Heat generation/absorption coefficient δ. It should be pointed out that, for positive values of Heat generation/absorption coefficient, δ acts as a heat source, but for a negative value of Heat generation/absorption coefficient, δ acts as sink source, this signifies that the dimensionless temperature reduced in case of a negative value of δ (heat sink) when compared to a positive value of δ (heat source), as presented in Figure 3. Similarly, the thermal boundary layer thickness decreases.

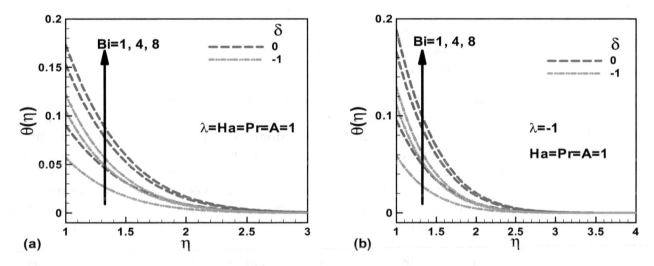

Figure 5. Variation of dimensionless temperature with Biot number in the presence of heat sink for (**a**) assisting flow and (**b**) opposing flow.

Figure 6a,b, respectively, display the effects of stretching velocity ratio and mixed convection parameter on dimensionless skin friction for both assisting and opposing flow. It is found that the skin friction increases with the mixed convection parameter and Hartmann number Ha, although it drops with the stretching velocity ratio for mutually opposing and assisting flow. This can be attributed to the result of velocity profile in the boundary layer and consequently disturbs the boundary layer thickness, as shown by Figures 1 and 2. On the other hand, the decreasing of skin friction with stretching velocity ratio can be associated to the augmentation of the velocity; therefore, the velocity boundary layer increases.

It is well recognized that augmenting velocity means growing the Reynold number which in turn leads to lessening viscous force regarding inertial force, consequently the dropping in viscous force will reduce skin friction. It is valuable to mention that the effect of assisting flows on dimensionless skin friction is slightly more pronounced than that of opposing flow, because the pressure near to the surface is greater than not near to the surface. As observed, the skin friction continually and significantly increases with the Hartmann number, since it represents the ratio of electromagnetic force to the viscous force, therefore the magnetic field will increase and accordingly the Lorentz force will oppose and push the flow to the surface, which in turn augments the skin friction.

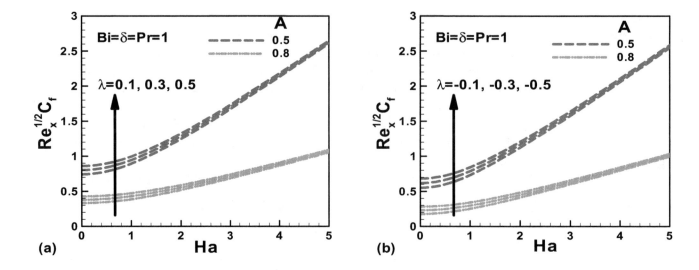

Figure 6. Variation of dimensionless skin friction with several parameters in the presence of heat source for (**a**) assisting flow and (**b**) opposing flow.

The effects of Biot number Bi, Heat generation/absorption coefficient δ, and mixed convection parameter λ on the dimensionless heat transfer rate for both assisting and opposing flow are illustrated, respectively, in Figure 7a,b. It is indicates that the dimensionless heat transfer rate increase with the Biot number and mixed convection parameter; whereas, it decreases with heat generation/absorption coefficient for both assisting and opposing flow. Physically, this can be attributed to the increase of temperature gradient with respect to the Biot number and mixed convection parameter, as presented by Figures 3 and 4. However, in an unexpected and perplexing result, it can be seen that the dimensionless heat transfer rate decrease with heat generation/absorption coefficient δ, it is well known that Nusselt numbers represent the proportion of convection to conduction heat transfer and in both situation assisting and opposing flow the effect of heat generation/absorption coefficient is more pronounced in conduction more than the convection heat transfer. It is clear that the effect of the mixed convection parameter on the heat transfer rate is slightly perceptible. Finally, the Nusselt number at the sheet surface augments, because the Hartmann number, stretching velocity ratio A, Hartmann number Ha, and mixed convection parameter λ increase. Though, it declines with respect to heat generation/absorption coefficient δ.

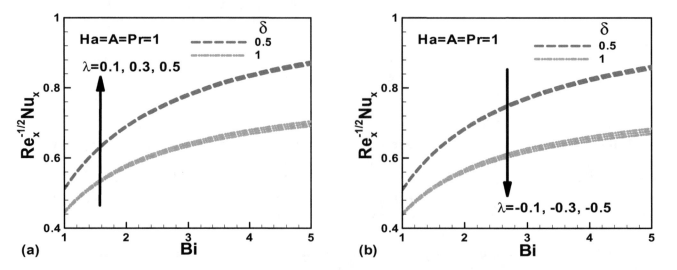

Figure 7. Variation of dimensionless heat transfer rate with several parameters in the presence of heat source for (**a**) assisting flow and (**b**) opposing flow.

6. Conclusions

A steady laminar flow over a vertical stretching sheet with the existence of viscous dissipation, heat source/sink, and magnetic fields has been mathematically explored through shooting arrangement based RKFI procedure. A comparison of the results with research results [3,26–29] demonstrates good agreement. The influence of relevant involved indicators on temperature and velocity, as well as the Nusselt number and skin friction coefficient are perceived. The main results that are enclosed in this work are concluded, as follows:

1. It is perceived that, for assisting flow ($\lambda > 0$), the dimensionless velocity is the maximum at the superficial of the vertical stretching sheet and it gradually lessens to the minimum value $f' = 1$, as it transfers to gone after the superficial, while for opposing flow ($\lambda < 0$), the dimensionless velocity is the lowest at the superficial of the vertical stretching sheet and gradually increases to the maximum value $f' = 1$.
2. The velocity profile augments with mixed convection parameter λ for both stretching in the flow direction and stretching in the opposite direction.
3. The temperature profile is greater at the stretching sheet superficial then exponentially lessens along the streamwise path up to the zero value for both assisting and opposing flow.
4. Skin friction increase with a mixed convection parameter and Hartmann number Ha, though it declines by stretching speed ratio for together opposing and assisting flow.
5. The effect of mixed convection parameter on the heat transfer rate is slightly perceptible. The Nusselt number at the sheet surface augments, because the Hartmann number, Hartmann number Ha, stretching velocity ratio A, and mixed convection parameter λ increase. Though, it declines according to heat generation/absorption coefficient δ.

Author Contributions: All authors contribute equally in this work in all parts and in all steps.

Acknowledgments: Ibrahim M. Alarifi would like to thank Deanship of Scientific Research at Majmaah University for supporting this work under the Project Number No. 1440-50.

References

1. Mohammadein, S.A.; Raslan, K.; Abdel-Wahed, M.S.; Abedel-Aal, E.M. KKL-model of MHD CuO-nanofluid flow over a stagnation point stretching sheet with nonlinear thermal radiation and suction/injection. *Results Phys.* **2018**, *10*, 194–199. [CrossRef]
2. Ibrahim, W. Magnetohydrodynamic (MHD) boundary layer stagnation point flow and heat transfer of a nanofluid past a stretching sheet with melting. *Propuls. Power Res.* **2017**, *6*, 214–222. [CrossRef]
3. Sharma, P.R.; Sinha, S.; Yadav, R.S.; Filippov, A.N. MHD mixed convective stagnation point flow along a vertical stretching sheet with heat source/sink. *Int. J. Heat Mass Transf.* **2018**, *117*, 780–786. [CrossRef]
4. SGhadikolaei, S.; Hosseinzadeh, K.H.; Ganji, D.D. MHD raviative boundary layer analysis of micropolar dusty fluid with graphene oxide (Go)-engine oil nanoparticles in a porous medium over a stretching sheet with joule heating effect. *Powder Technol.* **2018**. [CrossRef]
5. El-Mistikawy, T.M.A. MHD flow due to a linearly stretching sheet with induced magnetic field. *Acta Mech.* **2016**, *227*, 3049–3053. [CrossRef]
6. Hussain, A.; Malik, M.Y.; Salahuddin, T.; Rubab, A.; Khan, M. Effects of viscous dissipation on MHD tangent hyperbolic fluid over a nonlinear stretching sheet with convective boundary conditions. *Results Phys.* **2017**, *7*, 3502–3509. [CrossRef]
7. Tufail, M.N.; Butt, A.S.; Ali, A. Computational modeling of an MHD flow of a non-newtonian fluid over an unsteady stretching sheet with viscous dissipation effects. *J. Appl. Mech. Tech. Phys.* **2016**, *57*, 900–907. [CrossRef]
8. Mabood, F.; Khan, W.A.; Ismail, A.I.M. MHD flow over exponential radiating stretching sheet using homotopy analysis method. *J. King Saud Univ. Eng. Sci.* **2017**, *29*, 68–74. [CrossRef]

9. Hayat, T.; Qayyum, S.; Alsaedi, A.; Ahmad, B. Magnetohydrodynamic (MHD) nonlinear convective flow of Walters-B nanofluid over a nonlinear stretching sheet with variable thickness. *Int. J. Heat Mass Transf.* **2017**, *110*, 506–514. [CrossRef]

10. Malik, M.Y.; Khan, M.; Salahuddin, T. Study of an MHD flow of the carreau fluid flow over a stretching sheet with a variable thickness by using an implicit finite difference scheme. *J. Appl. Mech. Tech. Phys.* **2017**, *58*, 1033–1039. [CrossRef]

11. Agbaje, T.M.; Mondal, S.; Makukula, Z.G.; Motsa, S.S.; Sibanda, P. A new numerical approach to MHD stagnation point flow and heat transfer towards a stretching sheet. *Ain Shams Eng. J.* **2018**, *9*, 233–243. [CrossRef]

12. Soid, S.K.; Ishak, A.; Pop, I. MHD flow and heat transfer over a radially stretching/shrinking disk. *Chin. J. Phys.* **2018**, *56*, 58–66. [CrossRef]

13. Mahabaleshwar, U.S.; Sarris, I.E.; Hill, A.A.; Lorenzini, G.; Pop, I. An MHD couple stress fluid due to a perforated sheet undergoing linear stretching with heat transfer. *Int. J. Heat Mass Transf.* **2017**, *105*, 157–167. [CrossRef]

14. Ferdows, M.; Khalequ, T.S.; Tzirtzilakis, E.E.; Sun, S. Effects of Radiation and Thermal Conductivity on MHD Boundary Layer Flow with Heat Transfer along a Vertical Stretching Sheet in a Porous Medium. *J. Eng. Thermophys.* **2017**, *26*, 96–106. [CrossRef]

15. Hsiao, K.-L. Micropolar nanofluid flow with MHD and viscous dissipation effects towards a stretching sheet with multimedia feature. *Int. J. Heat Mass Transf.* **2017**, *112*, 983–990. [CrossRef]

16. Daniel, Y.S.; Aziz, Z.A.; Ismail, Z.; Salah, F. Thermal stratification effects on MHD radiative flow of nanofluid over nonlinear stretching sheet with variable thickness. *J. Comput. Des. Eng.* **2018**, *5*, 232–242. [CrossRef]

17. Hayat, T.; Rashid, M.; Alsaedi, A. MHD convective flow of magnetite-Fe_3O_4 nanoparticles by curved stretching sheet. *Results Phys.* **2017**, *7*, 3107–3115. [CrossRef]

18. Daniel, Y.S.; Aziz, Z.A.; Ismail, Z.; Salah, F. Thermal radiation on unsteady electrical MHD flow of nanofluid over stretching sheet with chemical reaction. *J. King Saud Univ. Sci.* **2017**, in press. [CrossRef]

19. Seth, G.S.; Singha, A.K.; Mandal, M.S.; Banerjee, A.; Bhattacharyya, K. MHD stagnation-point flow and heat transfer past a non-isothermal shrinking/stretching sheet in porous medium with heat sink or source effect. *Int. J. Mech. Sci.* **2017**, *134*, 98–111. [CrossRef]

20. Khan, M.; Hussain, A.; Malik, M.Y.; Salahuddin, T.; Khan, F. Boundary layer flow of MHD tangent hyperbolic nanofluid over a stretching sheet: A numerical investigation. *Results Phys.* **2017**, *7*, 2837–2844. [CrossRef]

21. Madhu, M.; Kishan, N. MHD boundary-layer flow of a non-newtonian nanofluid past a stretching sheet with a heat source/sink. *J. Appl. Mech. Tech. Phys.* **2016**, *57*, 908–915. [CrossRef]

22. Babu, M.J.; Sandeep, N. Three-dimensional MHD slip flow of nanofluids over a slendering stretching sheet with thermophoresis and Brownian motion effects. *Adv. Powder Technol.* **2016**, *27*, 2039–2050. [CrossRef]

23. Khan, I.; Malik, M.Y.; Hussain, A.; Salahuddin, T. Effect of homogenous-heterogeneous reactions on MHD Prandtl fluid flow over a stretching sheet. *Results Phys.* **2017**, *7*, 4226–4231. [CrossRef]

24. Seth, G.S.; Mishra, M.K. Analysis of transient flow of MHD nanofluid past a non-linear stretching sheet considering Navier's slip boundary condition. *Adv. Powder Technol.* **2017**, *28*, 375–384. [CrossRef]

25. Chen, X.; Ye, Y.; Zhang, X.; Zheng, L. Lie-group similarity solution and analysis for fractional viscoelastic MHD fluid over a stretching sheet. *Comput. Math. Appl.* **2018**, *75*, 3002–3011. [CrossRef]

26. Ramchandran, N.; Chen, T.S.; Armaly, B.F. Mixed convection in stagnation flows adjacent to vertical surfaces. *J. Heat Transf.* **1988**, *110*, 373–377. [CrossRef]

27. Lok, Y.Y.; Amin, N.; Pop, I. Unsteady mixed convection flow of a micropolar fluid near the stagnation point on a vertical surface. *Int. J. Therm. Sci.* **2006**, *45*, 1149–1152. [CrossRef]

28. Ishak, A.; Nazar, R.; Pop, I. Dual solutions in mixed convection flow near a stagnation point on a vertical porous plate. *Int. J. Therm. Sci.* **2008**, *47*, 417–422. [CrossRef]

29. Ali, F.M.; Nazar, R.; Arifin, N.M.; Pop, I. Mixed convection stagnation point flow on vertical stretching sheet with external magnetic field. *Appl. Math. Mech. Engl. Ed.* **2014**, *35*, 155–166. [CrossRef]

30. Hassanien, I.A.; Gorla, R. Nonsimilar Solutions for Natural Convection in Micropolar Fluids on a Vertical Plate. *Int. J. Fluid Mech. Res.* **2003**, *30*, 4–14. [CrossRef]

31. Li, J.; Zheng, L.; Liu, L. MHD viscoelastic flow and heat transfer over a vertical stretching sheet with Cattaneo-Christov heat flux effects. *J. Mol. Liq.* **2016**, *221*, 19–25. [CrossRef]

32. Freidoonimehr, N.; Rahimi, A.B. Exact-solution of entropy generation for MHD nanofluid flow induced by a stretching/shrinking sheet with transpiration: Dual solution. *Adv. Powder Technol.* **2017**, *28*, 671–685. [CrossRef]

33. Merkin, J.H. On dual solutions occurring in mixed convection in a porous medium. *J. Eng. Math.* **1986**, *20*, 171–179. [CrossRef]

34. Harris, S.D.; Ingham, D.B.; Pop, I. Mixed convection boundary-layer flow near the stagnation point on a vertical surface in a porous medium: Brinkman model with slip. *Transp. Porous Media* **2009**, *77*, 267–285. [CrossRef]

35. Rahman, M.M.; Roşca, A.V.; Pop, I. Boundary layer flow of a nanofluid past a permeable exponentially shrinking/stretching surface with second order slip using Buongiorno's model. *Int. J. Heat Mass Transf.* **2014**, *77*, 1133–1143. [CrossRef]

36. Weidman, P.D.; Kubitschek, D.G.; Davis, A.M.J. The effect of transpiration on self-similar boundary layer flow over moving surfaces. *Int. J. Eng. Sci.* **2006**, *44*, 730–737. [CrossRef]

37. Roşca, A.V.; Pop, I. Flow and heat transfer over a vertical permeable stretching/shrinking sheet with a second order slip. *Int. J. Heat Mass Transf.* **2013**, *60*, 355–364. [CrossRef]

MHD Nanofluids in a Permeable Channel with Porosity

Ilyas Khan [1] and Aisha M. Alqahtani [2,*]

[1]   Faculty of Mathematics and Statistics, Ton Duc Thang University, Ho Chi Minh 72915, Vietnam; ilyaskhan@tdt.edu.vn

[2]   Department of Mathematics, Princess Nourah bint Abdulrahman University, Riyadh 11564, Saudi Arabia

*   Correspondence: alqahtani@pnu.edu.sa

Abstract: This paper introduces a mathematical model of a convection flow of magnetohydrodynamic (MHD) nanofluid in a channel embedded in a porous medium. The flow along the walls, characterized by a non-uniform temperature, is under the effect of the uniform magnetic field acting transversely to the flow direction. The walls of the channel are permeable. The flow is due to convection combined with uniform suction/injection at the boundary. The model is formulated in terms of unsteady, one-dimensional partial differential equations (PDEs) with imposed physical conditions. The cluster effect of nanoparticles is demonstrated in the $C_2H_6O_2$, and H_2O base fluids. The perturbation technique is used to obtain a closed-form solution for the velocity and temperature distributions. Based on numerical experiments, it is concluded that both the velocity and temperature profiles are significantly affected by ϕ. Moreover, the magnetic parameter retards the nanofluid motion whereas porosity accelerates it. Each H_2O-based and $C_2H_6O_2$-based nanofluid in the suction case have a higher magnitude of velocity as compared to the injections case.

Keywords: Permeable walls; suction/injection; nanofluids; porous medium; mixed convection; magnetohydrodynamic (MHD)

---

1. Introduction

Heat transport in unsteady laminar flows has numerous real-world applications, particularly flows in a porous channel with permeable walls, which include medical devices, aerodynamic heating, chemical industry, electrostatic precipitation, petroleum industry, nuclear energy, and polymer technology. Based on this motivation, many researchers have considered the porous channel problem with suction and injection under different physical conditions. In earlier studies, Torda [1] studied the boundary layer flow with the suction/injection effect. Berman [2] derived an exact solution for the channel flow taking into consideration the uniform suction/injection at the boundary wall of the channel. The suction and injection and the combined effect of heat and mass transfer on a moving continuous flat surface were analyzed by Erickson et al. [3]. Alamri et al. [4] studied the Poiseuille flow of nanofluid in a channel under Stefan blowing and the second-order slip effect. Zeeshan et al. [5] reported analytical solutions for the Poiseuille flow of nanofluid in a porous wavy channel. Hassan et al. [6] investigated the flow of H_2O based nanofluid on a wavy surface. Ellahi et al. [7] studied the boundary layer Poiseuille plan flow of kerosene oil based nanofluid fluid with variable thermal conductivity. Ijaz et al. [8] presented a comprehensive study on the interaction of nanoparticles in the flow of nanofluid in a finite symmetric channel. Some recent important and interesting studies can be found in [9–12].

Magnetohydrodynamic (MHD) is referred to as the magnetic properties of the fluids under the influence of an electromagnetic force. MHD flows have numerous applications in MHD bearings and

MHD pumps. Many studies have been carried out on MHD flow in the literature. Abbas et al. [13] investigated the MHD flow of Maxwell fluid in a porous channel. The convective MHD flow of second-grade fluid was reported by Hayat and Abbas [14]. The effect of a transverse magnetic field on different flows in a semi-porous channel was presented by Sheikholeslami et al. [15]. Ravikumar et al. [16] studied three dimensional MHD due to the pressure gradient over the porous plate. Batti et al. [17] analyzed the heat transfer flow of nanofluid in a channel. They studied the effect of thermal radiation and the MHD effect by using Roseland's approximation, Ohm's law, and Maxwell equations. Ma et al. [18] study the MHD flow of nanofluid in a U-shaped enclosure using the Koo–Kleinstreuer–Li (KKL) correlation approximation for the effective thermal conductivity. Opreti [19] studied water-based silver nanofluid over a stretching sheet. They considered the effect of MHD, suction/injection, and heat generation/absorption in their study. Hosseinzadeh et al. [20] investigated the MHD squeezing flow of nonfluid in a channel. They presented analytical solutions by using similarity transformation and the perturbation technique. Narayana et al. [21] developed a mathematical model for the MHD stagnation point flow of Watler's-B fluid nanofluid.

Nano-sized particles of (Ag) nanoparticles inside H_2O-based fluids are commonly known as silver-based nanofluids. The viscosity of the nanofluids containing metallic nanoparticles has a much higher thermal conductivity than the nanofluids containing metallic oxide and non-metallic nanoparticles. Because of this, the interest of researchers in investigating nanofluids containing metallic nanoparticles has increased recently. The first exact solutions for different types of nanofluid were developed by Loganathan et al. [22]. Qasim et al. [23] reported numerical solutions for MHD ferrofluid in a stretching cylinder. Amsa et al. [24] investigated nanofluid flow near a vertical plate containing five different nanoparticles. The radiative heat transfer in the natural convection flow of oxide nanofluid was studied by Das and Jana [25]. Dhanai et al. [26]. Numerically studied the MHD mixed convection flow of nanofluid in a cylindrical coordinate system. The MHD rotational flow of nanofluid taking into consideration the effect of a porous medium, thermal radiation, and the chemical reaction was presented by Reddy et al. [27]. For some other interesting studies, readers are referred to [28–40].

Motivated by the above-discussed literature, the present study focused on the MHD channel flow of nanofluid in a porous medium with the suction and injection effect. The flow of electrically conducting nanofluid is considered under the influence of a transverse magnetic field. The analytical solutions for the proposed model are developed by using the perturbation method. The solutions are numerically computed, and the influence of various flow parameters is studied graphically.

2. Problem Description

Consider a porous channel of a width, d, filled with incompressible H_2O and $C_2H_6O_2$ based nanofluids with Ag nanoparticles. The channel walls are stationary with isothermal temperature conditions. The flow in the x-direction due to the temperature gradient is shown in Figure 1. Under the assumption of [11], the governing equations are as follows:

$$\rho_{nf}\left(\frac{\partial v}{\partial t} - v_\omega \frac{\partial v}{\partial y}\right) = -\frac{\partial p}{\partial x} + \mu_{nf}\frac{\partial^2 v}{\partial y^2} - \left(\sigma_{nf}B_0^2 + \frac{\mu_{nf}}{k_1}\right)u + (\rho\beta)_{nf}g(T - T_0), \tag{1}$$

$$(\rho C_p)_{nf}\left(\frac{\partial T}{\partial t} - v_\omega \frac{\partial T}{\partial y}\right) = k_{nf}\frac{\partial^2 T}{\partial y^2} - \frac{\partial q}{\partial y}, \tag{2}$$

together with the following physical conditions:

$$v(0,t) = 0, \ v(d,t) = 0, \tag{3}$$

$$T(0,t) = T_0, \ T(d,t) = T_w, \tag{4}$$

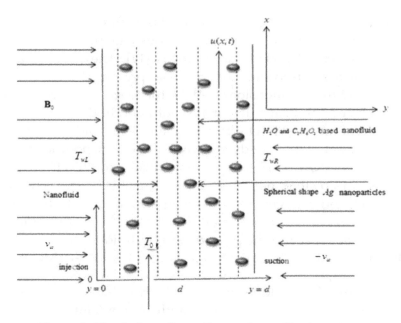

Figure 1. Physical configuration and coordinate system.

By using Xuan et al.'s [28] model, the effective thermal conductive, k_{nf}, and dynamic viscosity, μ_{nf}, of nanofluids are defined as:

$$k_{nf} = k_{static} + k_{Brownian},$$

$$k_{static} = k_f\left[\frac{(k_s + 2k_f) - 2\phi(k_f - k_s)}{(k_s + 2k_f) + \phi(k_f - k_s)}\right], \quad k_{brownian} = \frac{\rho_s\phi c_{pf}}{2k_s}\sqrt{\frac{k_b T}{3\pi r_c \mu_f}}, \tag{5}$$

$$\mu_{nf} = \mu_{static} + \mu_{Brownian},$$

$$\mu_{static} = \frac{\mu_f}{(1-\phi)^{2.5}}, \quad \mu_{Brownian} = \frac{\phi\rho_s(c_p)_s}{2k_s}\sqrt{\frac{k_b T}{3\pi r_c \mu_f}}, \tag{6}$$

where $k_b = 1.3807 \times 10^{-23} JK^{-1}$ and $300K > T > 325K$ are used, ϕ is the nanoparticles' volume fraction, and r_c is the radius of gyration for a number of particles. The static part in the effective thermal conductivity is derived from Maxwell's [29] model and the effective viscosity is derived from Brinkman's [30] model. Xuan et al.'s [28] model:

$$\rho_{nf} = (1-\phi)\rho_f + \phi\rho_s, \quad (\rho\beta)_{nf} = (1-\phi)(\rho\beta)_f + \phi(\rho\beta)_s,$$
$$(\rho c_p)_{nf} = (1-\phi)(\rho c_p)_f + \phi(\rho c_p)_s, \quad \sigma_{nf} = \alpha_{nf}(\rho c_p)_{nf},$$
$$\sigma_{nf} = \sigma_f\left[1 + \frac{3(\sigma-1)\phi}{(\sigma+2)-(\sigma-1)\phi}\right], \sigma = \frac{\sigma_s}{\sigma_f}, \tag{7}$$

where the numerical values of the thermo-physical of base fluid and nanoparticles are given in Table 1 [11,31]. The radiative heat flux is given by:

$$-\frac{\partial q}{\partial y} = 4\alpha^2(T - T_0), \tag{8}$$

Substituting Equation (8) into Equation (2), gives:

$$(\rho c_p)_{nf}\left(\frac{\partial T}{\partial t} - v_\omega\frac{\partial T}{\partial y}\right) = k_{nf}\frac{\partial^2 T}{\partial y^2} + 4\alpha^2(T - T_0), \tag{9}$$

The dimensionless variables:

$$x^* = \frac{x}{d}, \; y^* = \frac{y}{d}, \; u^* = \frac{u}{U_0}, \; t^* = \frac{tU_0}{d}, \; T^* = \frac{T - T_0}{T_w - T_0},$$
$$p^* = \frac{d}{\mu U_0} p, \; \omega^* = \frac{d\omega_1}{U_0}, \; v_0 = \frac{v_w}{U_0}, \tag{10}$$

Table 1. Thermo-physical properties of base fluid and nanoparticles.

Model	C_P (kg^{-1} K^{-1})	ρ (kg m^{-3})	k (Wm^{-1} K^{-1})	$\beta \times 10^{-5}$ (K^{-1})	(σ S/m)
Water (H_2O)	4179	997.1	0.613	21	5.5×10^{-6}
EG ($C_2H_6O_2$)	0.58	1.115	0.1490	6.5	1.07×10^{-6}
Alumina (Al_2O_3)	756	3970	40	0.85	1.07×10^{-6}
Silver (Ag)	235	10,500	429	1.89	6.30×10^7
Copper (Cu)	385	8933	401	1.67	59.6×10^6
Titanium Dioxide (TiO_2)	686.2	4250	8.9528	0.9	2.6×10^6

Are introduced into Equations (1) and (9), we get:

$$a_0 \left(\frac{\partial u}{\partial t} - v_0 \frac{\partial u}{\partial y} \right) = \lambda \varepsilon \exp(i\omega t) + \phi_2 \frac{\partial^2 u}{\partial y^2} - m_0^2 u + a_1 T, \tag{11}$$

$$u(0,t) = 0; \; u(1,t) = 0; \; t > 0, \tag{12}$$

$$b_0 \left(\frac{\partial T}{\partial t} - v_0 \frac{\partial T}{\partial y} \right) = \frac{\partial^2 T}{\partial y^2} + b_1 T, \tag{13}$$

$$T(0,t) = 0; T(1,t) = 1; t > 0, \tag{14}$$

where:

$$a_0 = \phi_1 Re, \; \phi_1 = (1 - \phi) + \phi \frac{\rho_s}{\rho_f}, \; Re = \frac{U_0 d}{v}, \; \phi_2 = \frac{1}{(1-\phi)^{2.5}}, \; m_0^2 = \phi_5 M^2,$$

$$\phi_5 = \left[1 + \frac{3(\sigma - 1)\phi}{(\sigma + 2) - (\sigma - 1)\phi} \right], \; M^2 = \frac{\sigma_f B_0^2 d^2}{\mu_f}, \; a_1 = \phi_3 Gr, \; \phi_3 = (1 - \phi)\rho_f + \phi \frac{(\rho\beta)_s}{\beta_f},$$

$$Gr = \frac{g\beta_f d^2 (T_w - T_0)}{v_f U_0}, \; b_0^2 = \frac{Pe\phi_4}{\lambda_n}, \; Pe = \frac{U_0 d (\rho c_p)_f}{k_f},$$

$$\lambda_n = \frac{k_{nf}}{k_f} = \frac{(k_s + 2k_f) - 2\phi(k_f - k_s)}{(k_s - 2k_f) + \phi(k_f - k_s)}, \; \phi_4 = \left[(1 - \phi) + \phi \frac{(\rho c_p)_s}{(\rho c_p)_f} \right], \; b_1^2 = \frac{N^2}{\lambda_n}, \; N^2 = \frac{4d^2 \alpha_0^2}{k_f}.$$

The following general perturbed solutions are considered for Equations (11)—(14), the following type of solutions are assumed:

$$u(y,t) = [u_0(y) + \varepsilon \exp(i\omega t) u_1(y)], \tag{15}$$

$$T(y,t) = [T_0(y) + \varepsilon \exp(i\omega t) T_1(y)]. \tag{16}$$

Which lead to the following solutions:

$$\frac{d^2 u_0(y)}{dy^2} + \frac{a_0 v_o}{\phi_2} \frac{\partial u(y)}{\partial y} - \frac{m_0^2}{\phi_2} u_0(y) = -a_2 T_0, \tag{17}$$

$$u_0(0) = 0; u_0(1) = 0, \tag{18}$$

$$\frac{d^2 u_1(y)}{dy^2} + \frac{v_0}{\phi_2} \frac{\partial u_1(y)}{\partial y} - m_2^2 u_1(y) = -\frac{\lambda}{\phi_2}, \tag{19}$$

$$u_1(0) = 0; u_1(1) = 0, \tag{20}$$

$$\frac{d^2 T_0(y)}{dy^2} + b_0 v_0 \frac{\partial T_0(y)}{\partial y} + b_1^2 T_0(y) = 0, \tag{21}$$

$$T_0(0) = 0; T_0(1) = 1, \tag{22}$$

$$\frac{d^2 T_1(y)}{dy^2} + v_0 \frac{\partial T_1(y)}{\partial y} + (b_1 - b_0 i\omega) T_1(y) = 0, \tag{23}$$

$$T_1(0) = 0; T_1(1) = 0, \tag{24}$$

where:

$$m_1 = \sqrt{\frac{m_0^2}{\phi_2}}, a_2 = \frac{a_1}{\phi_2}, m_2 = \sqrt{\frac{m_0^2 + i\omega a_0}{\phi_2}}, m_3 = \sqrt{b_1 - i\omega b_0}.$$

The solutions of Equations (21) and (23) under the boundary conditions, (22) and (24), are obtained as:

$$T_0(y) = e^{-\alpha y} e^\alpha \frac{\sin(\beta y)}{\sin(\beta)}, \tag{25}$$

$$T_1(y) = 0, \tag{26}$$

where:

$$\alpha = \frac{b_0 v_0}{2}, \beta = \frac{1}{2}\sqrt{b_0 v_0 - 4b_1}.$$

Using Equations (25) and (26), Equation (16) becomes:

$$T(y,t) = T(y) = e^{-\alpha y} e^\alpha \frac{\sin(\beta y)}{\sin(\beta)}. \tag{27}$$

The solutions of Equations (17) and (19) after substituting Equation (25) under the boundary conditions, (18) and (20), are obtained as:

$$u_0(y) = \begin{aligned} & e^{-\alpha_2 y} (c_5 \sinh(\beta_2 y) + c_6 \cosh(\beta_2 y)) \\ & + a_1 e^{-\alpha y} e^\alpha \frac{[A\sin(\beta y) - B\cos(\beta y)]}{[A^2 + B^2]}, \end{aligned} \tag{28}$$

$$u_1(y) = e^{-\alpha_3 y} (c_7 \sinh(\beta_3 y) + c_8 \cosh(\beta_3 y)) + \frac{\lambda}{(m_2^2 \phi_2)}, \tag{29}$$

With:

$$
\begin{aligned}
&\alpha_2 = \frac{a_0 v_0}{2\phi_2}, \beta_2 = \frac{1}{2}\sqrt{\frac{a_0^2 b_0^2}{\phi_2^2} + \frac{4m_0^2}{\phi_2}}, A = \alpha^2 - \beta^2 - \alpha \frac{a_0 v_0}{\phi_2} - \frac{m_0^2}{\phi_2}, \\
&B = -2\alpha\beta - \beta \frac{a_0 v_0}{(\phi_2)_3}, \alpha_3 = \frac{v_0}{2\phi_2}, \beta_3 = \frac{1}{2}\sqrt{\frac{v_0^2}{\phi_2^2} + 4m_0^2}, \\
&c_5 = \frac{1}{\sinh(\beta_2)}\left[\left(\frac{a_1 e^\alpha \beta}{[A^2 + B^2]}\right)\cosh(\beta_2) + \frac{e^{\alpha_2}[A\sin(\beta) - B\cos(\beta)]}{[A^2 + B^2]}\right] \\
&c_6 = -\frac{a_1 e^\alpha \beta}{[A^2 + B^2]}, c_7 = \frac{\lambda}{(m_2^2 \phi_2)\sinh(\beta_3)}\cosh(\beta_3) - \frac{\lambda e^{\alpha_3}}{(m_2^2 \phi_2)}\frac{1}{\sinh(\beta_3)}, \\
&c_8 = -\frac{\lambda}{(m_2^2 \phi_2)}.
\end{aligned} \tag{30}
$$

Finally, substituting Equations (28) to (30) into Equation (16), we get:

$$
\begin{aligned}
u(y,t) = \ & e^{-\alpha_2 y}\left(\left(\frac{\sinh(\beta_2 y)}{\sinh(\beta_2)}\right)\left(\frac{\left(\frac{a_1 e^{\alpha}\beta}{[A^2+B^2]}\right)\cosh(\beta_2)}{+\frac{e^{\alpha_2}[A\sin(\beta)-B\cos(\beta)]}{[A^2+B^2]}}\right)\right.\\
& \left.-\left(\frac{a_1 e^{\alpha}\beta}{[A^2+B^2]}\right)\cosh(\beta_2 y)\right)\\
& -a_1 e^{-\alpha y}e^{\alpha}\frac{[A\sin(\beta)-B\cos(\beta)]}{[A^2+B^2]}\\
& +\exp(i\omega t)\left[e^{-\alpha_3 y}\left(\left(\frac{\frac{\lambda}{(m_2^2\phi_2)\sinh(\beta_3)}\cosh(\beta_3)}{-\frac{\lambda e^{\alpha_3}}{(m_2^2\phi_2)}\frac{1}{\sinh(\beta_3)}}\right)\sinh(\beta_3 y)\right.\right.\\
& \left.\left.-\left(\frac{\lambda}{(m_2^2\phi_2)}\right)\cosh(\beta_3 y)\right)+\frac{\lambda}{(m_2^2\phi_2)}\right].
\end{aligned}
\tag{31}
$$

3. Nusselt Number

The dimensionless expression for the Nusselt number is given by:

$$
Nu = \frac{\beta_1 e^{\alpha}}{\sin(\beta_1)},
\tag{32}
$$

4. Skin-Friction

From Equation (31), the skin friction is calculated as:

$$
\begin{aligned}
\tau_t(t) = \ & \frac{a_1 e^{\alpha}\beta_1\beta_2\cosh(\beta_2)}{[A^2+B^2]\sinh(\beta_2)} - \frac{\alpha_2 e^{\alpha_2}[A\sin(\beta_1)-B\cos(\beta_1)]}{[A^2+B^2]} + \frac{\alpha_2 a_1 e^{\alpha}\beta_1}{[A^2+B^2]}\\
& -a_1 e^{\alpha}\frac{[\beta_1 A+\alpha B]}{[A^2+B^2]} + \exp(i\omega t)\left(\frac{\frac{\lambda\beta_3}{\left((m_2)_3^2(\phi_2)_3\right)\sinh(\beta_3)}\cosh(\beta_3)-}{\frac{\lambda e^{\alpha_3}\beta_3}{\left((m_2)_3^2(\phi_2)_3\right)}\frac{1}{\sinh(\beta_3)}-\frac{\alpha_3\lambda}{\left((m_2)_3^2(\phi_2)_3\right)}}\right).
\end{aligned}
\tag{33}
$$

5. Results and Discussion

In this section, the graphs of the velocity and temperature for H_2O and $C_2H_6O_2$ based nanofluids containing Ag nanoparticles were plotted for different values of volume fraction, ϕ, and buoyancy parameter, Gr, permeability parameter, K, magnetic parameter, M, and radiation parameter, N, for both cases of suction and injection. The thermophysical properties of the base fluids and Ag nanoparticles are mentioned in Table 1. For this purpose, Figures 2–19 were plotted. Figures 2–5 were prepared to study the effects of the velocity for the cases of suction and injection of Ag in H_2O and $C_2H_6O_2$ based nanofluids, respectively. It was found that the velocity increases with increasing ϕ for both cases of suction and injection. However, no variation is observed in the velocity of Ag in $C_2H_6O_2$ based nanofluids in the case of injection. This behavior of velocity is found to be similar qualitatively to the results of Hajmohammadi et al. [32], however, they used Cu in water-based nanofluids.

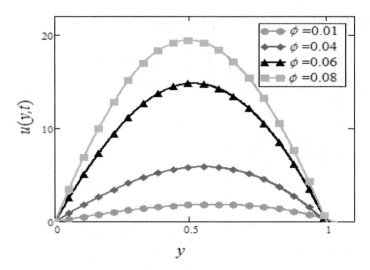

Figure 2. Velocity profiles for different values of ϕ of Ag in water based nanofluids when $Gr = 0.1$, $N = 0.1, r_c = 20$ nm, $Pe = 0.1, \lambda = 1, M = 1, K = 0.3, v_0 = 2, t = 5, \omega = 0.2$.

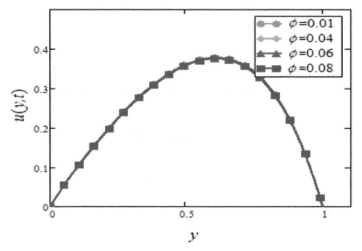

Figure 3. Velocity profiles for different values of ϕ of Ag in water based nanofluids when $Gr = 0.1$, $N = 0.1, r_c = 20$ nm, $Pe = 0.1, \lambda = 1, M = 1, K = 0.3, v_0 = -0.01, t = 5, \omega = 0.2$.

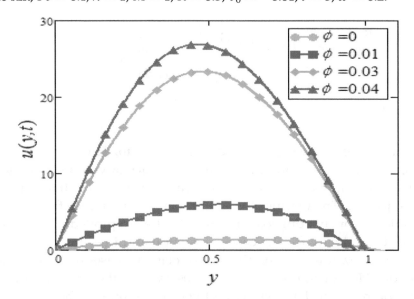

Figure 4. Velocity profiles for different values of ϕ of Ag in EG based nanofluids when $Gr = 0.1$, $N = 0.1, r_c = 20$ nm, $Pe = 0.1, \lambda = 1, M = 2, K = 3, v_0 = 4, t = 5, \omega = 0.2$.

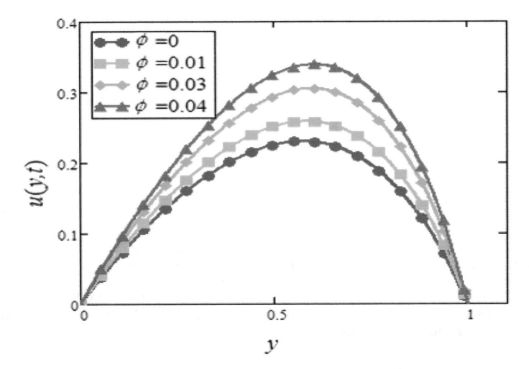

Figure 5. Velocity profiles for different values of ϕ of Ag in EG based nanofluids when $Gr = 0.1$, $N = 0.1$, $r_c = 20$ nm, $Pe = 0.1$, $\lambda = 1$, $M = 2$, $K = 3$, $t = 5$, $v_0 = -0.01$, $\omega = 0.2$.

Figures 6 and 7 are plotted for different values of Gr for both cases of suction and injection. It is noted from Figure 6 that the velocity of Ag in water-based nanofluids increases with the increase of Gr in the case of suction for Ag in water-based nanofluids while the velocity is decreased in the case of injection. The velocity in Figure 6, where $Gr = 0$, is not linear. However, the increasing values of Gr make it look like linear.

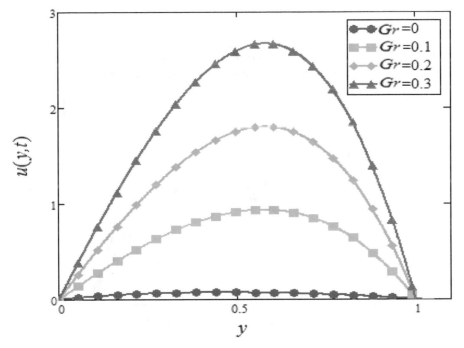

Figure 6. Velocity profiles for different values of Gr of Ag in water based nanofluids when $N = 0.1$, $Pe = 0.1$, $r_c = 20$ nm, $\phi = 0.04$, $\lambda = 1$, $M = 2$, $K = 3$, $v_0 = 10$, $t = 5$, $\omega = 0.2$.

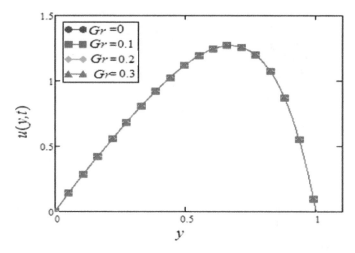

Figure 7. Velocity profiles for different values of Gr of Ag in water based nanofluids when $N = 0.1$, $Pe = 0.1$, $r_c = 20$ nm, $\phi = 0.04$, $Re = 0.1$, $\lambda = 1$, $M = 2$, $K = 3$, $v_0 = -1$, $t = 5$, $\omega = 0.2$.

Figures 8 and 9 were plotted to check the effect of K, the velocity of Ag in water-based nanofluids, for both cases of suction and injection. One can see from Figures 8 and 9 that the effect of suction, K, on the velocity of nanofluids is opposite to the case of injection.

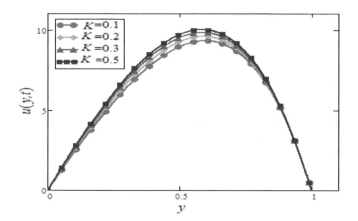

Figure 8. Velocity profiles for different values of K of Ag in water based nanofluids when $Gr = 0.1$, $N = 0.1$, $Pe = 0.1$, $r_c = 20$ nm, $\phi = 0.04$, $\lambda = 1$, $M = 2$, $v_0 = 6$, $t = 10$, $\omega = 0.2$.

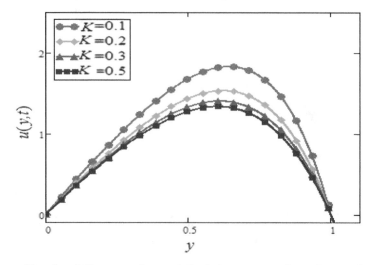

Figure 9. Velocity profiles for different values of K of Ag in water based nanofluids when $Gr = 0.1$, $N = 0.1$, $Pe = 0.1$, $r_c = 20$ nm, $\phi = 0.04$, $\lambda = 1$, $M = 2$, $v_0 = -0.01$, $t = 10$, $\omega = 0.2$.

The effect of the magnetic parameter, M, on the velocity profile is studied in Figures 10 and 11. For the case of suction, the $Ag - H_2O$ nanofluid's velocity profile decreases with increasing values of M. This effect is due to the Lorentz forces. Greater values of M correspond to stronger Lorentz forces, which reduces the nanofluid velocity. However, this trend reverses for the injection case.

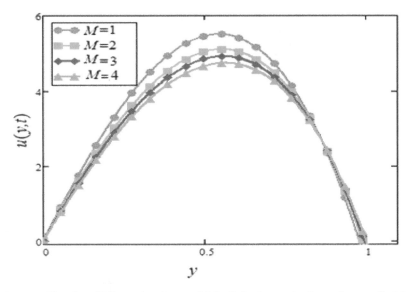

Figure 10. Velocity profiles for different values of M of Ag in water based nanofluids when $Gr = 0.1$, $N = 0.1$, $Pe = 0.1$, $r_c = 20$ nm, $\phi = 0.04$, $\lambda = 1$, $K = 0.3$ $v_0 = 5$, $t = 10$, $\omega = 0.2$.

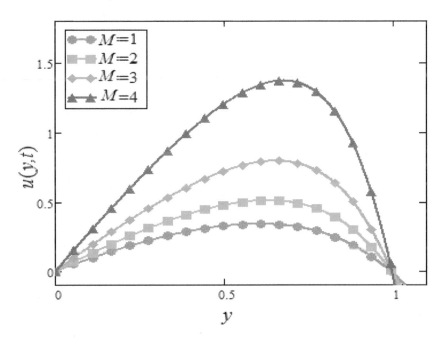

Figure 11. Velocity profiles for different values of M of Ag in water based nanofluids when $Gr = 0.1$, $N = 0.1$, $Pe = 0.1$, $r_c = 20$ nm, $\phi = 0.04$, $\lambda = 1$, $K = 0.3$ $v_0 = -0.01$ $t = 10$, $\omega = 0.2$.

Figures 12 and 13 shows that the velocity profiles of Ag in $Ag - H_2O$ nanofluids increase with the decrease of N in the case of suction. This physically means that an increase in N increases the conduction, which in turn decreases the viscosity of nanofluids. Decreasing the viscosity of nanofluids increases the velocity of nanofluids. However, the effect is the opposite due to the suction, whereas no variation is observed for injection. However, the velocity of zero radiation is greater than the velocity of nanofluids with radiation.

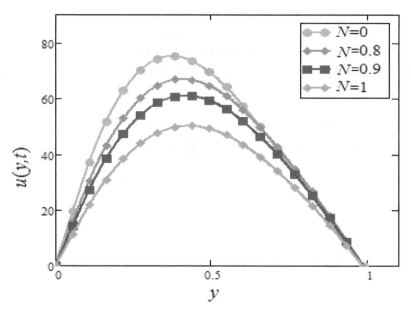

Figure 12. Velocity profiles for different values of N of Ag in water based nanofluids when $Gr = 0.1$, $Pe = 0.1$, $r_c = 20$ nm, $\phi = 0.04$, $\lambda = 1$, $M = 1$, $K = 1$, $v_0 = 7$, $t = 2$, $\omega = 0.2$.

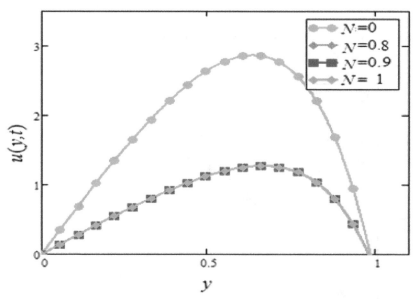

Figure 13. Velocity profiles for different values of N of Ag in water based nanofluids when $Gr = 0.1$, $Pe = 0.1$, $r_c = 20$ nm, $\phi = 0.04$, $\lambda = 1$, $M = 1$, $K = 1$, $v_0 = -1$, $t = 2$, $\omega = 0.2$.

The velocity profiles for different types of nanoparticles in water-based nanofluids are represented in Figure 14. It is clear that the velocity of Ag and Cu in water-based nanofluids is greater than TiO_2 and Al_2O_3 in water-based nanofluids. As mentioned in previous problems, different types of nanoparticles have different thermal conductivities and viscosities. It was concluded in our previous problems [31,32] that metallic nanoparticles, like Ag and Cu, had smaller velocities as compared to metallic oxide nanoparticles, like TiO_2 and Al_2O_3, due to high thermal conductivities and viscosities. However, the effect is the opposite to this problem because of the condition of permeable walls or suction. Due to these situations, different velocities have been observed.

The effect of ϕ in $Ag - H_2O$ nanofluids on the temperature profiles is shown in Figures 15 and 16 for the cases of suction and injection. It was found that the temperature of nanofluids increases with the increase of ϕ for the suction velocity whereas no significant variation is observed for the injection case.

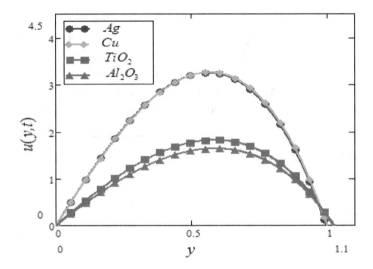

Figure 14. Velocity profiles for different types of nanoparticles in water based nanofluids when $Gr = 0.1$, $N = 0.1$, $Pe = 0.1$, $r_c = 20\,\mathrm{nm}$, $\lambda = 1$, $M = 2$, $K = 3$, $t = 5$, $\omega = 0.2$.

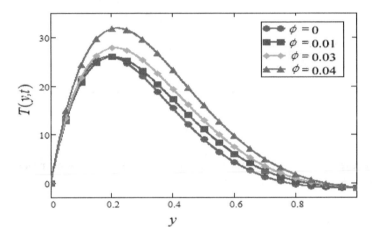

Figure 15. Temperature profiles for different values of ϕ of Ag in water based nanofluids when $r_c = 20\,\mathrm{nm}$, $N = 1$, $t = 1$, $v_0 = 10$, $\omega = 0.2$.

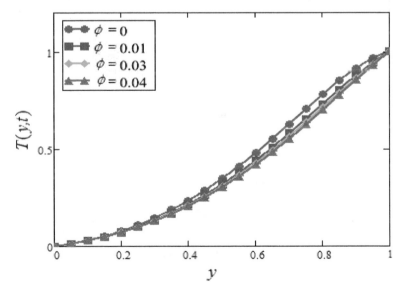

Figure 16. Temperature profiles for different values of ϕ of Ag in water based nanofluids when $r_c = 20\,\mathrm{nm}$, $N = 1$, $t = 1$, $v_0 = -1$, $\omega = 0.2$.

Figures 17 and 18 are sketched to show the effect of N on the temperature profiles of Ag in H_2O based nanofluids for both cases of suction and injection. The effects of different types of nanoparticles on the temperature of H_2O based nanofluids are plotted in Figure 19 for injection. It is observed that Cu in water-based nanofluid has the highest temperature followed by Ag, Al_2O_3, and TiO_2 in H_2O based nanofluids. This is due to the higher thermal conductivities of copper followed by Ag, Al_2O_3, and TiO_2 in water-based nanofluids. Due to these situations, different velocities were observed.

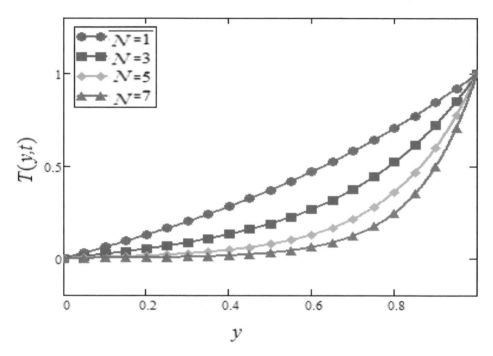

Figure 17. Temperature profiles for different values of N of Ag in water based nanofluids when $r_c = 20$ nm, $\phi = 0.04$, $t = 1$, $v_0 = -1$, $\omega = 0.2$.

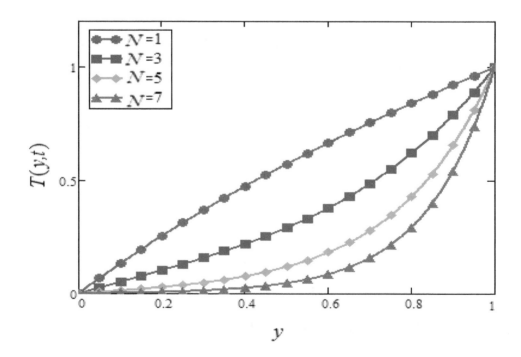

Figure 18. Temperature profiles for different values of N of Ag in water based nanofluids when $r_c = 20$ nm, $\phi = 0.04$, $t = 1$, $v_0 = 10$, $\omega = 0.2$.

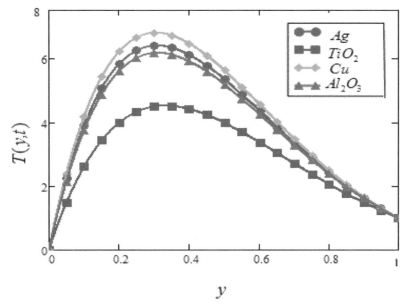

Figure 19. Temperature profiles for different types of nanoparticles in water based nanofluids when $r_c = 20$ nm, $N = 1$, $t = 1$, $v_0 = -1$, $\omega = 0.2$.

6. Conclusions

The channel flow of nanofluids in a porous medium with permeable walls was studied. The focal point of this research was to study the influence of permeable walls on momentum and heat transfer. The permeable parameter, which physically corresponds to suction and injection, was incorporated in both the momentum and energy equations. Expressions for the velocity and temperature were obtained. The effects of various parameters, such as thermal Grashof number, volume fraction, different types of nanoparticles, radiation, permeability, magnetic, suction, and injection, were studied in different plots. The concluding remarks are as follows:

1. It was found that the velocity of nanofluids increases with an increase of the volume fraction, radiation, and permeability parameter in the case of suction whereas an opposite behavior was noted in the case of injection.
2. The velocity of Ag nanofluids decreases with an increase of the magnetic parameter while the opposite behavior was noted in the case of injection.
3. The temperature of Ag nanofluids was found to decrease with an increase of ϕ for the extraction of fluid from the walls whereas a very small change was observed in the case of injection.
4. Finally, it was noticed that different types of nanoparticles have different effects on the velocity and temperature due to suction and injection.

Author Contributions: All the authors contributed equally to the conception of the idea, implementing and analyzing the experimental results, and writing the manuscript.

Nomenclature

H_2O	Water
$C_2H_6O_2$	Ethelyn glycol
$v(y,t)$	Velocity component in the x-direction
$T(y,t)$	Temperature
$v_\omega > 0$	Suction
$v_\omega < 0$	Injection
ρ_{nf}	Density of nanofluid
M	Magnetic parameter
Pe	Peclet number

μ_{nf} Dynamic viscosity of nanofluid
$(\rho\beta)_{nf}$ thermal expansion coefficient
g Acceleration due to gravity
$(\rho c_p)_{nf}$ Heat capacitance of nanofluids
k_{nf} The thermal conductivity of nanofluid
α Mean radiation absorption coefficient
Re Reynolds' number
Gr Grashof number
N Radiation parameter

References

1. Torda, T.P. Boundary Layer Control by Distributed Surface Suction or Injection. Bi-Parametric General Solution. *J. Math. Phys.* **1953**, *32*, 312–314. [CrossRef]
2. Berman, A.S. Laminar flow in channels with porous walls. *J. Appl. Phys.* **1953**, *24*, 1232–1235. [CrossRef]
3. Erickson, L.E.; Fan, L.T.; Fox, V.G. Heat and mass transfer on moving continuous flat plate with suction or injection. *Ind. Eng. Chem. Fundam.* **1966**, *5*, 19–25. [CrossRef]
4. Alamri, S.Z.; Ellahi, R.; Shehzad, N.; Zeeshan, A. Convective radiative plane Poiseuille flow of nanofluid through porous medium with slip: An application of Stefan blowing. *J. Mol. Liq.* **2019**, *273*, 292–304. [CrossRef]
5. Zeeshan, A.; Shehzad, N.; Ellahi, R.; Alamri, S.Z. Convective Poiseuille flow of Al 2 O 3-EG nanofluid in a porous wavy channel with thermal radiation. *Neural Comput. Appl.* **2018**, *30*, 3371–3382. [CrossRef]
6. Hassan, M.; Marinb, M.; Alsharifc, A.; Ellahide, R. Convective heat transfer flow of nanofluid in a porous medium over wavy surface. *Phys. Lett. A* **2018**, *382*, 2749–2753. [CrossRef]
7. Ellahi, R.; Zeeshan, A.; Shehzad, N.; Alamri, S.Z. Structural impact of kerosene-Al$_2$O$_3$ nanoliquid on MHD Poiseuille flow with variable thermal conductivity: Application of cooling process. *J. Mol. Liq.* **2018**, *264*, 607–615. [CrossRef]
8. Ijaz, N.; Zeeshan, A.; Bhatti, M.M.; Ellahi, R. Analytical study on liquid-solid particles interaction in the presence of heat and mass transfer through a wavy channel. *J. Mol. Liq.* **2018**, *250*, 80–87. [CrossRef]
9. Ali, F.; Aamina, A.; Khan, I.; Sheikh, N.A.; Saqib, M. Magnetohydrodynamic flow of brinkman-type engine oil based MoS 2-nanofluid in a rotating disk with Hall Effect. *Int. J. Heat Technol.* **2017**, *4*, 893–902.
10. Jan, S.A.A.; Ali, F.; Sheikh, N.A.; Khan, I.; Saqib, M.; Gohar, M. Engine oil based generalized brinkman-type nano-liquid with molybdenum disulphide nanoparticles of spherical shape: Atangana-Baleanu fractional model. *Numer. Methods Partial Differ. Equ.* **2018**, *34*, 1472–1488. [CrossRef]
11. Saqib, M.; Ali, F.; Khan, I.; Sheikh, N.A.; Khan, A. Entropy Generation in Different Types of Fractionalized Nanofluids. *Arab. J. Sci. Eng.* **2019**, *44*, 1–10. [CrossRef]
12. Saqib, M.; Ali, F.; Khan, I.; Sheikh, N.A.; Shafie, S.B. Convection in ethylene glycol-based molybdenum disulfide nanofluid. *J. Therm. Anal. Calorim.* **2019**, *135*, 523–532. [CrossRef]
13. Saqib, M.; Khan, I.; Shafie, S. Natural convection channel flow of CMC-based CNTs nanofluid. *Eur. Phys. J. Plus* **2018**, *133*, 549. [CrossRef]
14. Saqib, M.; Khan, I.; Shafie, S. Application of Atangana–Baleanu fractional derivative to MHD channel flow of CMC-based-CNT's nanofluid through a porous medium. *Chaos Solitons Fractals* **2018**, *116*, 79–85. [CrossRef]
15. Abbas, Z.; Sajid, M.; Hayat, T. MHD boundary-layer flow of an upper-convected Maxwell fluid in a porous channel. *Theor. Comput. Fluid Dyn.* **2006**, *20*, 229–238. [CrossRef]
16. Hayat, T.; Abbas, Z. Heat transfer analysis on the MHD flow of a second grade fluid in a channel with porous medium. *Chaos Solitons Fractals* **2008**, *38*, 556–567. [CrossRef]
17. Sheikholeslami, M.; Ashorynejad, H.R.; Domairry, D.; Hashim, I. Investigation of the laminar viscous flow in a semi-porous channel in the presence of uniform magnetic field using optimal homotopy asymptotic method. *Sains Malays.* **2012**, *41*, 1177–1229.
18. Ravikumar, V.; Raju, M.C.; Raju, G.S.S. MHD three dimensional Couette flow past a porous plate with heat transfer. *IOSR J. Math.* **2012**, *1*, 3–9. [CrossRef]
19. Bhatti, M.M.; Zeeshan, A.; Ellahi, R. Heat transfer with thermal radiation on MHD particle–fluid suspension induced by metachronal wave. *Pramana* **2017**, *89*, 48. [CrossRef]

20. Ma, Y.; Mohebbi, R.; Rashidi, M.M.; Yang, Z.; Sheremet, M.A. Numerical study of MHD nanofluid natural convection in a baffled U-shaped enclosure. *Int. J. Heat Mass Transf.* **2019**, *130*, 123–134. [CrossRef]

21. Upreti, H.; Pandey, A.K.; Kumar, M. MHD flow of Ag-water nanofluid over a flat porous plate with viscous-Ohmic dissipation, suction/injection and heat generation/absorption. *Alex. Eng. J.* **2018**, *57*, 1839–1847. [CrossRef]

22. Hosseinzadeh, K.; Alizadeh, M.; Ganji, D.D. Hydrothermal analysis on MHD squeezing nanofluid flow in parallel plates by analytical method. *Int. J. Mech. Mater. Eng.* **2018**, *13*, 4. [CrossRef]

23. Narayana, P.V.; Tarakaramu, N.; Makinde, O.D.; Venkateswarlu, B.; Sarojamma, G. MHD Stagnation Point Flow of Viscoelastic Nanofluid Past a Convectively Heated Stretching Surface. *Defect Diffus. Forum* **2018**, *387*, 106–120. [CrossRef]

24. Loganathan, P.; Chand, P.N.; Ganesan, P. Radiation effects on an unsteady natural convective flow of a nanofluid past an infinite vertical plate. *Nano* **2013**, *8*, 1350001. [CrossRef]

25. Qasim, M.; Khan, Z.H.; Khan, W.A.; Shah, I.A. MHD boundary layer slip flow and heat transfer of ferrofluid along a stretching cylinder with prescribed heat flux. *PLoS ONE* **2014**, *9*, e83930. [CrossRef] [PubMed]

26. Khalid, A.; Khan, I.; Shafie, S. Exact solutions for free convection flow of nanofluids with ramped wall temperature. *Eur. Phys. J. Plus* **2015**, *130*, 57. [CrossRef]

27. Das, S.; Jana, R.N. Natural convective magneto-nanofluid flow and radiative heat transfer past a moving vertical plate. *Alex. Eng. J.* **2015**, *54*, 55–64. [CrossRef]

28. Dhanai, R.; Rana, P.; Kumar, L. MHD mixed convection nanofluid flow and heat transfer over an inclined cylinder due to velocity and thermal slip effects: Buongiorno's model. *Powder Technol.* **2016**, *288*, 140–150. [CrossRef]

29. Reddy, J.V.R.; Sugunamma, V.; Sandeep, N.; Sulochana, C. Influence of chemical reaction, radiation and rotation on MHD nanofluid flow past a permeable flat plate in porous medium. *J. Niger. Math. Soc.* **2016**, *35*, 48–65. [CrossRef]

30. Xuan, Y.; Li, Q.; Hu, W. Aggregation structure and thermal conductivity of nanofluids. *AIChE J.* **2003**, *49*, 1038–1043. [CrossRef]

31. Maxwell, J.C.; Thompson, J.J. *A Treatise on Electricity and Magnetism*; Oxford University Press: Oxford, UK, 1904.

32. Brinkman, H.C. The viscosity of concentrated suspensions and solutions. *J. Chem. Phys.* **1952**, *20*, 571. [CrossRef]

33. Ali, F.; Saqib, M.; Khan, I.; Sheikh, N.A. Heat Transfer Analysis in Ethylene Glycol Based Molybdenum Disulfide Generalized Nanofluid via Atangana–Baleanu Fractional Derivative Approach. In *Fractional Derivatives with Mittag-Leffler Kernel*; Springer: Cham, Switzerland, 2019; pp. 217–233.

34. Hajmohammadi, M.R.; Maleki, H.; Lorenzini, G.; Nourazar, S.S. Effects of Cu and Ag nano-particles on flow and heat transfer from permeable surfaces. *Adv. Powder Technol.* **2015**, *26*, 193–199. [CrossRef]

35. Saqib, M.; Khan, I.; Shafie, S. New Direction of Atangana–Baleanu Fractional Derivative with Mittag-Leffler Kernel for Non-Newtonian Channel Flow. In *Fractional Derivatives with Mittag-Leffler Kernel*; Springer: Cham, Switzerland, 2019; pp. 253–268.

36. Ellahi, R.; Zeeshan, A.; Hussain, F.; Abbas, T. Study of shiny film coating on multi-fluid flows of a rotating disk suspended with nano-sized silver and gold particles: A comparative analysis. *Coatings* **2018**, *8*, 422. [CrossRef]

37. Ellahi, R.; Alamri, S.Z.; Basit, A.; Majeed, A. Effects of MHD and slip on heat transfer boundary layer flow over a moving plate based on specific entropy generation. *J. Taibah Univ. Sci.* **2018**, *12*, 476–482. [CrossRef]

38. Saqib, M.; Khan, I.; Shafie, S. Application of fractional differential equations to heat transfer in hybrid nanofluid: Modeling and solution via integral transforms. *Adv. Differ. Equ.* **2019**, *52*. [CrossRef]

39. Ellahi, R.; Tariq, M.H.; Hassan, M.; Vafai, K. On boundary layer magnetic flow of nano-Ferroliquid under the influence of low oscillating over stretchable rotating disk. *J. Mol. Liq.* **2017**, *229*, 339–345. [CrossRef]

40. Ellahi, R.; Raza, M.; Akbar, N.S. Study of peristaltic flow of nanofluid with entropy generation in a porous medium. *J. Porous Med.* **2017**, *20*, 461–478. [CrossRef]

Permissions

The contributors of this book come from diverse backgrounds, making this book a truly international effort. This book will bring forth new frontiers with its revolutionizing research information and detailed analysis of the nascent developments around the world.

We would like to thank all the contributing authors for lending their expertise to make the book truly unique. They have played a crucial role in the development of this book. Without their invaluable contributions this book wouldn't have been possible. They have made vital efforts to compile up to date information on the varied aspects of this subject to make this book a valuable addition to the collection of many professionals and students.

This book was conceptualized with the vision of imparting up-to-date information and advanced data in this field. To ensure the same, a matchless editorial board was set up. Every individual on the board went through rigorous rounds of assessment to prove their worth. After which they invested a large part of their time researching and compiling the most relevant data for our readers.

The editorial board has been involved in producing this book since its inception. They have spent rigorous hours researching and exploring the diverse topics which have resulted in the successful publishing of this book. They have passed on their knowledge of decades through this book. To expedite this challenging task, the publisher supported the team at every step. A small team of assistant editors was also appointed to further simplify the editing procedure and attain best results for the readers.

Apart from the editorial board, the designing team has also invested a significant amount of their time in understanding the subject and creating the most relevant covers. They scrutinized every image to scout for the most suitable representation of the subject and create an appropriate cover for the book.

The publishing team has been an ardent support to the editorial, designing and production team. Their endless efforts to recruit the best for this project, has resulted in the accomplishment of this book. They are a veteran in the field of academics and their pool of knowledge is as vast as their experience in printing. Their expertise and guidance has proved useful at every step. Their uncompromising quality standards have made this book an exceptional effort. Their encouragement from time to time has been an inspiration for everyone.

The publisher and the editorial board hope that this book will prove to be a valuable piece of knowledge for researchers, students, practitioners and scholars across the globe.

List of Contributors

Quentin J. Minaker and Jeffrey J. Defoe
Turbomachinery and Unsteady Flows Research Group, Department of Mechanical, Automotive and Materials Engineering, University of Windsor, 401 Sunset Ave., Windsor, ON N9B 3P4, Canada

Zahir Shah and Saeed Islam
Department of Mathematics, Abdul Wali Khan University, Mardan 23200, Pakistan

Asifa Tassaddiq
College of Computer and Information Sciences, Majmaah University, Al-Majmaah 11952, Saudi Arabia

A.M. Alklaibi
Department of Mechanical and Industrial Engineering, College of Engineering, Majmaah University, Majmaah 11952, Saudi Arabia

Ilyas Khan
Faculty of Mathematics and Statistics, Ton Duc Thang University, Ho Chi Minh City 72915, Vietnam
Department of Mathematics, College of Science Al-Zulfi, Majmaah University, Al-Majmaah 11952, Saudi Arabia

Iskander Tlili
Department of Mechanical and Industrial Engineering, College of Engineering, Majmaah University, Al-Majmaah 11952, Saudi Arabia

Muhammad Adil Sadiq
Department of Mathematics, DCC-KFUPM, KFUPM, Dhahran 31261, Saudi Arabia

Muhammad Mubashir Bhatti and Dong Qiang Lu
Shanghai Institute of Applied Mathematics and Mechanics, Shanghai University, Yanchang Road, Shanghai 200072, China
Shanghai Key Laboratory of Mechanics in Energy Engineering, Yanchang Road, Shanghai 200072, China

Muhammad Asif
Department of Mathematics, Abdul Wali Khan University, Mardan 23200, Pakistan

Sami Ul Haq
Department of Mathematics, Islamia College University, Peshawar 25000, Pakistan

Tawfeeq Abdullah Alkanhal
Department of Mechatronics and System Engineering, College of Engineering, Majmaah University, Majmaah 11952, Saudi Arabia

Zar Ali Khan
Department of Mathematics, University of Peshawar, Peshawar 25000, Pakistan

Kottakkaran Sooppy Nisar
Department of Mathematics, College of Arts and Science, Prince Sattam bin Abdulaziz University, Wadi Al-Dawaser 11991, Saudi Arabia

Haris Anwar and Muhammad Altaf Khan
Department of Mathematics, City University of Science and Information Technology, Peshawar 25000, Pakistan

Taza Gul
Department of Mathematics, City University of Science and Information Technology, Peshawar 25000, Pakistan
Department of Mathematics, Govt. Superior Science College Peshawar, Khyber Pakhtunkhwa, Peshawar 25000, Pakistan

Poom Kumam
KMUTT-Fixed Point Research Laboratory, Room SCL 802 Fixed Point Laboratory, Science Laboratory Building, Department of Mathematics, Faculty of Science, King Mongkut's University of Technology Thonburi (KMUTT), 126 Pracha-Uthit Road, Bang Mod, Thrung Khru, Bangkok 10140, Thailand
KMUTT-Fixed Point Theory and Applications Research Group, Theoretical and Computational Science Center (TaCS), Science Laboratory Building, Faculty of Science, King Mongkut's University of Technology Thonburi (KMUTT), 126 Pracha-Uthit Road, Bang Mod, Thrung Khru, Bangkok 10140, Thailand
Department of Medical Research, China Medical University Hospital, China Medical University, Taichung 40402, Taiwan

Umar Khan
Department of Mathematics and Statistics, Hazara University, Mansehra 21120, Pakistan

Adnan Abbasi
Department of Mathematics, Mohi-ud-Din Islamic University Nerian Sharif, Azad Jammu & Kashmir 12080, Pakistan

Naveed Ahmed and Syed Tauseef Mohyud-Din
Department of Mathematics, Faculty of Sciences, HITEC University Taxila Cantt, Punjab 47080, Pakistan

Sayer Obaid Alharbi
Department of Mathematics, College of Science Al-Zulfi, Majmaah University, Al-Majmaah 11952, Saudi Arabia

Saima Noor
Department of Mathematics, COMSATS University Islamabad, Abbottabad 22010, Pakistan

Waqar A. Khan
Department of Mechanical Engineering, College of Engineering, Prince Mohammad Bin Fahd University, Al Khobar 31952, Saudi Arabia

Rahmat Ellahi
Center for Modeling & Computer Simulation, Research Institute, King Fahd University of Petroleum & Minerals, Dhahran 31261, Saudi Arabia
Department of Mathematics & Statistics, FBAS, IIUI, Islamabad 44000, Pakistan

Ahmed Zeeshan
Department of Mathematics & Statistics, FBAS, IIUI, Islamabad 44000, Pakistan

Farooq Hussain
Department of Mathematics & Statistics, FBAS, IIUI, Islamabad 44000, Pakistan
Department of Mathematics, (FABS), BUITEMS, Quetta 87300, Pakistan

A. Asadollahi
Department of Mechanical Engineering & Energy Processes, Southern Illinois University, Carbondale, IL 62901, USA

Imran Ullah
College of Civil Engineering, National University of Sciences and Technology Islamabad, Islamabad 44000, Pakistan

Sharidan Shafie
Department of Mathematical Sciences, Faculty of Science, Universiti Teknologi Malaysia, UTM Johor Bahru 81310, Johor, Malaysia

Oluwole Daniel Makinde
Faculty of Military Science, Stellenbosch University, Private Bag X2, Saldanha 7395, South Africa

Khalil Ur Rehman
Department of Mathematics, Air University, PAF Complex E-9, Islamabad 44000, Pakistan

M. Y. Malik
Department of Mathematics, College of Sciences, King Khalid University, Abha 61413, Saudi Arabia

Waqar A Khan
Department of Mechanical Engineering, College of Engineering, Prince Mohammad Bin Fahd University, Al Khobar 31952, Kingdom of Saudi Arabia

S. O. Alharbi
Department of Mathematics, College of Science Al-Zulfi, Majmaah University, Al-Majmaah 11952, Saudi Arabia

Ibrahim M. Alarifi
Department of Mechanical and Industrial Engineering, College of Engineering, Majmaah University, Al-Majmaah 11952, Saudi Arabia

Ahmed G. Abokhalil
Department of Electrical Engineering, College of Engineering, Majmaah University, Al-Majmaah 11952, Saudi Arabia
Electrical Engineering Department, Assiut University, Assiut 71515, Egypt

M. Osman
Department of Mechanical and Industrial Engineering, College of Engineering, Majmaah University, Al-Majmaah 11952, Saudi Arabia
Mechanical Design Department, Faculty of Engineering Mataria, Helwan University, Cairo El-Mataria 11724, Egypt

Liaquat Ali Lund
Sindh Agriculture University, Tandojam Sindh 70060, Pakistan

Mossaad Ben Ayed
Computer Science Department, College of Science and Humanities at Alghat, Majmaah University, Al-Majmaah 11952, Saudi Arabia
Computer and Embedded System Laboratory, Sfax University, Sfax 3011, Tunisia

Hafedh Belmabrouk
Electronics and Microelectronics Laboratory, Faculty of Science of Monastir, University of Monastir, Monastir 5019, Tunisia
Department of Physics, College of Science at Zulfi, Majmaah University, Al Zulfi 11932, Saudi Arabia

Aisha M. Alqahtani
Department of Mathematics, Princess Nourah bint Abdulrahman University, Riyadh 11564, Saudi Arabia

Index

A

Acceleration, 33, 59-61, 80, 108-109, 155, 161, 207, 232

Activation Energy, 51, 148-149, 154-155, 160-161, 163-164

Assisting Flow, 58, 60, 211-215

Axial Velocity, 8, 10-11, 14-15, 17-19, 28, 31, 153-154, 160-161

B

Base Fluid, 42, 62, 120, 149, 160, 162

Boundary Conditions, 4, 27, 29, 53-55, 62-63, 68, 79, 109, 132, 166-167, 191, 202, 205, 207, 212-213, 215

Boundary Layer, 22, 30, 33, 41, 54-55, 58-59, 61-63, 65, 68, 77-78, 127, 130, 132, 154-155, 161, 164-167, 170-171, 175-176, 181-182, 190-191, 197, 202, 204, 206, 211-213, 215-218, 232-233

Brownian Diffusion, 61, 77, 153, 158, 160-161, 201

Buoyancy, 58-59, 61, 65, 78, 153, 163, 170, 203, 211-212

C

Casson Fluid, 51, 55, 63, 165-167, 170-171, 175-177, 181-182, 186-189, 191-192, 195, 197, 201-202, 205

Chemical Reaction, 52, 63-64, 148-149, 154-155, 160, 163-167, 170, 182, 186, 189-191, 202, 216, 233

Coaxial Tubes, 149, 160

Concentration Distribution, 46, 49, 182

Conservation Of Mass, 42, 46, 149

Constitutive Equations, 108-109

Couple Stress Fluid, 148-149, 158, 160, 162, 216

Couple Stress Parameter, 55, 153, 159-162

Curvature Parameter, 140, 144, 146, 165-166, 170-171, 176, 182, 188-189, 205

D

Dissimilar Values, 46, 49

E

Electromagnetic Force, 211, 213, 218

F

Fluid Concentration, 181-182

Fluid Flow, 43, 52-53, 59-61, 64, 146, 162-164, 166-167, 170, 175, 190-192, 197, 202, 205, 212, 216

Fluid Particles, 153, 170-171, 176, 182, 187

Fluid Temperature, 134, 154-155, 162, 166, 175-176, 197, 205

G

Grashof Number, 61, 153, 158, 160-161, 170, 173-174, 189, 231-232

Gravitational Field, 54-55

H

Heat Transfer, 42-43, 46, 51-55, 60-61, 63-64, 78, 118, 120, 124, 129-131, 133-134, 139, 142, 144-148, 162-163, 165-167, 176, 187, 189-191, 202-206, 211, 214-217, 231-233

I

Inclined Stretching Sheet, 54-55, 58-59, 61-63

Isotherms, 133, 140, 142, 158-159, 161

J

Jeffrey Fluid, 52, 54-55, 58, 60-63, 191

L

Lengthways, 204-206

M

Magnetic Field, 52, 54-55, 58-61, 64-65, 68, 76-77, 109, 118, 132, 135, 139, 147, 162, 166-167, 170, 175, 182, 190-191, 201-202, 204-206, 213, 215-216, 218, 232

Magnetic Parameter, 65, 68, 77, 126, 128, 130, 139-140, 144-145, 166, 170, 172-174, 177-178, 181, 183-185, 187-189, 218, 227, 231

Mass Transfer Rate, 165, 182, 187, 190

Mixed Convection, 52, 54, 58-59, 62-63, 134, 144-145, 163, 167, 191, 202-203, 205, 207, 211-218, 233

Molecular Diameter, 133, 145-146

Momentum Equation, 108-109, 207

Motion Parameter, 49, 65, 68, 77, 154-155, 158-161, 201

N

Nanoconcentration, 68, 77

Nanofluid, 42-43, 46, 49-55, 63-65, 68, 76, 78, 120, 124, 126, 129-133, 135, 139-140, 142-143, 145-147, 149, 153-154, 160-164, 166-167, 190-191, 197, 202-206, 215-218, 227, 230-233

Nanoparticles, 42, 46, 48, 53-54, 78, 120, 130-134, 139, 142-146, 148-149, 154-155, 161, 163, 191-192, 195, 201, 215-216, 218, 228-232

Nonlinear Hydroelastic Waves, 79-80, 94, 106

Numerical Solution, 65, 132, 192, 195

Nusselt Number, 42, 50, 55, 60-61, 68, 76-77, 128-130, 144, 146, 187, 189, 192, 200, 204, 206, 214-215

P

Peristaltic Motion, 148, 153, 158

Permeable Walls, 218, 228, 231

Polymer Extrusion, 65, 165

Porous Medium, 42, 52, 63-64, 78, 118, 162, 165-167, 170, 176, 190-191, 215-218, 231-233

R

Radiation Parameter, 55, 139-140, 144-146, 170, 180, 189, 232

Reaction Rate Constant, 155, 158, 160-161

Reynolds Number, 59, 68, 77, 201

Rotating Rigid Disk, 192, 201

S

Shooting Technique, 133, 204

Side Walls, 63-64, 108-109, 117-118

Skin Friction Coefficient, 54, 59, 68, 76-77, 170, 187-188, 204, 215

Slip Conditions, 78, 166, 191, 197

Solitary Waves, 79-81, 83, 85, 92-93, 95, 98, 106-107

Stagnation Point Flow, 65, 76-78, 163, 166, 191, 204-205, 215-216, 233

Steady Laminar Flow, 204, 206, 210, 215

Streamlines, 33, 133, 140-142, 158

Stretched Surface, 42, 63, 211

Stretching Parameter, 171-172, 174, 177, 179, 183, 188-189, 205

Stretching Surface, 42-43, 46, 50, 52-53, 55, 63, 132, 147, 165-166, 187, 190-191, 202, 217, 233

Suction Velocity, 228

T

Temperature Profiles, 130, 182, 203-205, 210, 218, 228-231

Thermal Boundary, 55, 59, 61, 65, 68, 77, 127, 130, 154, 167, 175-176, 212-213

Thermal Conductivity, 42-43, 51-53, 61, 78, 120, 131, 133-134, 145-149, 161, 163, 207, 216, 218, 232-233

Thermal Energy, 154

Thermal Expansion, 58-62, 207, 232

Thermal Radiation, 42, 52-53, 55, 63-64, 78, 131, 133-134, 145-147, 163-164, 166-167, 202-206, 215-216, 232

Thermophoresis, 42, 52, 61, 65, 68, 77, 153-155, 158-161, 192, 201-202, 205-206, 216

Thin Elastic Plate, 79, 81, 98, 106

U

Uniform Current, 79-81

V

Velocity Profiles, 113-114, 224-229

Velocity Ratio, 40, 204-206, 211-215

Viscosity, 62, 77-78, 120, 131, 133, 142-143, 145-147, 162, 170, 181, 190, 201-202, 206-207, 227, 232-233

Viscous Dissipation, 51-53, 58, 61-64, 134, 139, 147, 163, 165-167, 176, 191, 204-205, 210, 215-216

Viscous Fluid, 65, 108-109, 118, 165-166, 197, 204

Printed in the USA
CPSIA information can be obtained
at www.ICGtesting.com
JSHW051406091023
49903JS00006B/294